Lecture Notes in Control and Information Sciences 406

Editors: M. Thoma, F. Allgöwer, M. Morari

Alberto Bemporad, Maurice Heemels,
and Mikael Johansson

Networked Control Systems

Springer

Authors

Assoc.Prof. Alberto Bemporad
Università Siena
Fac. Ingegneria
Dipto. Ingegneria
dell'Informazione
Via Roma 56
53100 Siena
Italy

Assoc.Prof. Maurice Heemels
Eindhoven University of
Technology
Dept. Mechanical Engineering
Eindhoven
Netherlands

Prof. Mikael Johansson
Royal Institute of Technology (KTH)
School of Electrical Engineering
Automatic Control Lab.
Osquldas väg 10
100 44 Stockholm
Fl. 6
Sweden
E-mail: mikaelj@s3.kth.se

ISBN 978-0-85729-032-8 e-ISBN 978-0-85729-033-5

DOI 10.1007/978-0-85729-033-5

Lecture Notes in Control and Information Sciences ISSN 0170-8643

Library of Congress Control Number: 2010936459

Typeset & Cover Design: Scientific Publishing Services Pvt. Ltd., Chennai, India.

5 4 3 2 1 0

springer.com

Foreword

This book finds its origin in the WIDE PhD School on Networked Control Systems, which we organized in July 2009 in Siena, Italy. Having gathered experts on all the aspects of networked control systems, it was a small step to go from the summer school to the book, certainly given the enthusiasm of the lecturers at the school. We felt that a book collecting overviews on the important developments and open problems in the field of networked control systems could stimulate and support future research in this appealing area. Given the tremendous current interests in distributed control exploiting wired and wireless communication networks, the time seemed to be right for the book that lies now in front of you.

The goal of the book is to set out the core techniques and tools that are available for the modeling, analysis and design of networked control systems. Roughly speaking, the book consists of three parts. The first part presents architectures for distributed control systems and models of wired and wireless communication networks. In particular, in the first chapter important technological and architectural aspects on distributed control systems are discussed. The second chapter provides insight in the behavior of communication channels in terms of delays, packet loss and information constraints leading to suitable modeling paradigms for communication networks.

The second part focuses on decentralized and distributed control, estimation and optimization. The network aspect is here that not all information is available in one central controller, estimator or optimizer, but local agents have only limited information of the overall control, estimation or optimization problem and still aim at solving the central problem in an appropriate manner. Although the information might be limited for each individual agent, it is assumed that communication is ideal and imperfections such as communication delays, possible loss of information and quantization effects are ignored. Chapter 3 discusses distributed estimation and consensus problems and the fourth chapter surveys distributed optimization techniques. Chapter 5 and 6 provide overviews on decentralized and distributed control. The emphasis in chapter 5 is on decentralized and distributed model predictive control techniques.

While communication imperfections induced by the presence of a non-ideal and uncertain network channel are ignored in the second part of the book, they form the main topic of the third part. Methods for stability analysis and controller synthesis of control loops closed over communication channels are treated in chapter 7. Using appropriate models of networked control systems it is investigated how network-induced phenomena such as varying delays, varying sampling intervals, data loss and communication constraints influence the stability and performance of control loops. Chapter 8 studies the effects of the limited capacity of channels, *e.g.* limited bandwidth, on feedback control. Fundamental limitations in control using quantized information are discussed in detail in this chapter. Finally, chapter 9 treats control systems that do not use the conventional periodic sampling, but update their information based on specific discrete events. The resulting event-triggered feedback controllers have the potential to reduce the amount of communication required in networked systems while preserving the overall system's stability and performance. This last chapter provides in-depth analysis techniques for event-triggered control, estimation and optimization.

In summary, this book provides overviews on many facets of networked control systems. The writing of the book would have been impossible without the enormous efforts of the lecturers of the school and we are certainly indebted to them. We are also grateful to Davide Barcelli, who was responsible for the technical assembly of the separate contributions of the authors into one book. Finally, the support by the European Commission through the FP7-ICT-2007-2 thematic programme under project WIDE "Decentralized and wireless control of large-scale systems" (project no. 224168) for writing this book is greatly acknowledged.

We hope that you will enjoy reading this book and that it will be of assistance in your own research endeavors in the area of networked control systems.

April 2010

Alberto Bemporad, Trento, Italy
Maurice Heemels, Eindhoven, The Netherlands
Mikael Johansson, Stockholm, Sweden

Contents

List of Contributors

Davide Barcelli
Department of Information Engineering, University of Siena, Italy
barcelli@dii.unisi.it

Alberto Bemporad
Department of Mechanical and Structural Engineering,
University of Trento, Italy
bemporad@ing.unitn.it

Federica Garin
INRIA Grenoble Rhône-Alpes, France
federica.garin@inrialpes.fr

W.P.M.H. Heemels
Eindhoven University of Technology, Department of Mechanical Engineering, P.O. Box 513, 5600 MB Eindhoven, The Netherlands
M.Heemels@tue.nl

Hideaki Ishii
Department of Computational Intelligence and Systems Science, Tokyo Institute of Technology, 4259 Nagatsuta-cho, Midori-ku, Yokohama 226-8502, Japan
ishii@dis.titech.ac.jp

Riku Jäntti
Department of Communications and Networking Comnet Helsinki University of Technology TKK, Finland
riku.jantti@tkk.fi

Mikael Johansson
School of Electrical Engineering and ACCESS Linnaeus Center, Sweden
mikaelj@ee.kth.se

Kyoung-Dae Kim
Department of Electrical and Computer Engineering, University of Illinois at Urbana-Champaign, USA
kkim50@illinois.edu

P. R. Kumar
Department of Electrical and Computer Engineering, University of Illinois at Urbana-Champaign, USA
prkumar@illinois.edu

Sanjay Lall
Department of Electrical Engineering and Department of Aeronautics and Astronautics, Stanford University, Stanford, CA 94305, USA
lall@stanford.edu

Michael Lemmon
University of Notre Dame,
Notre Dame, Indiana, USA
lemmon@nd.edu

Luca Schenato
Department of Information
Engineering, University of Padova,
Italy
schenato@dei.unipd.it

John Swigart
Department of Aeronautics and Astronautics, Stanford University, Stanford,
CA 94305, USA
jswigart@stanford.edu

Bo Yang
Department of Automation,
Shanghai Jiao Tong University,
China
bo.yang@sjtu.edu.cn

N. van de Wouw
Eindhoven University of Technology,
Department of Mechanical
Engineering,
P.O. Box 513,
5600 MB Eindhoven,
The Netherlands
N.v.d.Wouw@tue.nl

Chapter 1
The Importance, Design and Implementation of a Middleware for Networked Control Systems

Kyoung-Dae Kim and P.R. Kumar

Abstract. Due to the advancement of computing and communication technology, networked control systems may soon become prevalent in many control applications. While the capability of employing the communication network in the control loop certainly provides many benefits, it also raises several challenges which need to be overcome to utilize the benefits.

In this chapter, we focus on one major challenge: a middleware framework that enables a networked control system to be implemented. Indeed our thesis is that a middleware for networked control system is important for the future of networked control systems.

We discuss the fundamental issues which need to be considered in the design and development of an appropriate middleware for networked control systems. We describe *Etherware*, a middleware for networked control system which has been developed at the University of Illinois, as an example of such a middleware framework, to illustrate how these issues can be addressed in the design of a middleware. Using a networked inverted pendulum control system as an example, we demonstrate the powerful capabilities provided by Etherware for a networked control system.

1.1 Introduction

Over the past several decades, communication and computing technologies have advanced tremendously. Consequently, the platform for the control system itself also has changed with the emergence of networked control. In general, a networked control system is a system whose constituents such as sensors, actuators, and controllers are distributed over a network, and their corresponding control-loops are formed through a network layer. Thus, the scale of the networked control system

Kyoung-Dae Kim · P.R. Kumar
Department of Electrical and Computer Engineering,
University of Illinois at Urbana-Champaign, USA
e-mail: kkim50@illinois.edu, prkumar@illinois.edu

A. Bemporad, M. Heemels, and M. Johansson: Networked Control Systems, LNCIS 406, pp. 1–29.
springerlink.com

is typically much larger than that of classical control systems. An example of such a system, an automatic traffic control system, established in the IT Convergence Laboratory at the University of Illinois, is shown in Figure 1.1 (see [12]).

In addition to the scale of the system, the complexity of a networked control system is also greater. Due to the existence of the networked communication and computation system below the control application layer, several challenging issues such as communication delay, the interface between a control application and the network layer, platform heterogeneity, clock differences between the computers, and others, arise. Clearly, all of these constitute an extraordinarily big burden on control engineers if they have to address these issues too, while designing of the control loops.

One solution to release these burdens from the control engineer is to interpose an abstraction layer between the application layer and the underlying networked communication and computation layer. Such an abstraction layer can encapsulate the complexity of the underlying system so that the application layer can have a much simpler view of the system. This can significantly simplify and shorten the development of a networked control application. Typically, such an abstraction can be realized as a software framework, called a *middleware*. When such a middleware is designed and developed, it is important to consider the domain requirements which capture all the characteristics of the application domain. Thus, as a first step toward the development of the middleware for networked control systems, it is necessary to understand the fundamental characteristics of networked control systems and then establish corresponding requirements for the middleware framework.

Etherware is such a middleware for networked control systems which has been developed at the University of Illinois [4, 13]. In the sequel, we extensively discuss how Etherware is designed and how it works to support the domain requirements established from the domain characteristics, since it can serve as an exemplar of middleware for networked control systems. We also present a particularly demanding application that we have implemented, a networked inverted pendulum control system, as a case study of a particular networked control system which is implemented on top of Etherware, in order to demonstrate the usefulness of a middleware framework.

1.2 Networked Control Systems

1.2.1 Domain Characteristics

There are many potential examples of networked control systems in various application areas, such as smart power grids, intelligent traffic control systems, automatic warehouse management systems, and so forth. In this section, we investigate the common characteristics which are shared by many networked control systems in many application domains.

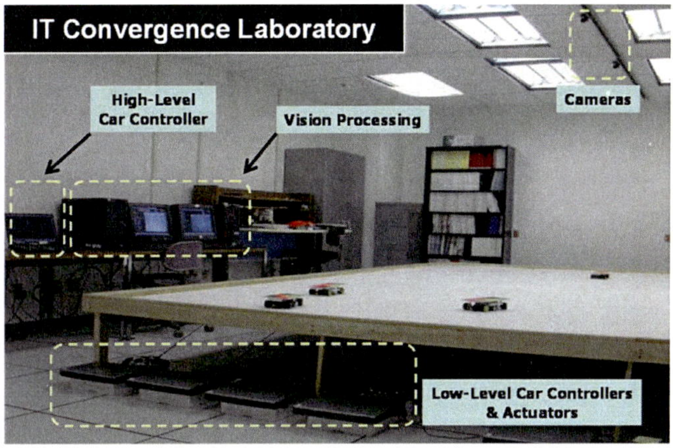

Fig. 1.1 Traffic control Testbed in IT Convergence Laboratory at the University of Illinois

1.2.1.1 Large-Scale

In a networked control system, the control loop is typically formed through the underlying communication network. Thus, the physical distance between the entities in the loop is not an issue anymore. Also, the communication network allows us to form multiple control loops through it so that multiple control objectives can be achieved at the same time. The Testbed example shown in Figure 1.1 is a good example which has such characteristics. In the testbed, a vision system is used to detect the state of the moving vehicle, and a low-level controller controls the vehicle to follow a given trajectory generated by a high-level controller. The inner control loop for tracking the given reference trajectory is formed through a communication network since all these elements are running at different computing nodes. Besides this inner control loop, there is another control loop formed through the same communication network to achieve a different slightly higher level control objective, which is collision avoidance between vehicles. In addition to all these, we also have another control loop in the testbed for runtime system management, such as upgrading or migrating some software modules to optimize the overall system performance.

1.2.1.2 Openness

In a classical control system, it is typically not allowed to change the running system. However, networked control systems are open to runtime system reconfiguration. While the system is running, new entities can join or leave the system, and also an existing entity can be replaced or even migrated to other location in a networked control system. Depending on the situation, the information flow which forms a control loop can be dynamically changed at runtime as well. The testbed system in Figure 1.1 can exhibit all of these dynamic reconfiguration features. Clearly, vehicles can join or leave the traffic system dynamically. If a better traffic control algorithm is developed, then the existing traffic controller can be replaced by the new one with a better algorithm, so that the overall traffic control performance can be improved. Also, depending on the network traffic, a traffic controller may need to be migrated to another computing node to provide better traffic control performance. Once a controller is migrated, then the corresponding control loop also has to be reconfigured accordingly.

1.2.1.3 Time-Criticality

A delay in a control loop typically affects the performance and the stability of the control system. Therefore, a control system is in fact a time-critical system in most cases. In fact, time-criticality is one of the fundamental characteristics of any control system, not just for a networked control system. Considering the fact that the computation and communication network are in the middle of control loop, the time-criticality might an even be more challenging issue for a networked control system.

1.2.1.4 Safety-Criticality

In many cases, a control system is indeed a safety-critical system which can cause severe consequences once the system fails. As shown in the testbed, a vehicle control system can be an example of such a safety-critical control system. In a networked control system, it becomes is harder to achieve the safety-guarantee due to the existence of the computation and communication network.

1.2.2 Domain Requirements

The following are some of the requirements for a middleware framework for networked control systems.

1.2.2.1 Operational Requirements

As we discussed in Section 1.2.1, entities which constitute a networked control system typically run on different computing nodes over the network. This distributed nature of a control system in fact causes several issues which have to be resolved for correct operation of a control system. The existence of the network itself can cause several difficulties in the development of a networked control system. Among them, the *location difference* and *clock difference* between entities in the control loop are two essential issues caused by the distributed nature. Thus, as an underlying platform of a networked control system, a middleware framework is necessary to provide an abstraction about the networked system which hides all such issues, so that a networked control system can be easily developed by the control designer. Besides these two requirements, it is also required to provide a mechanism which supports semantic addressing (or context-aware addressing) so that the portability and reusability of the application code can be enhanced.

1.2.2.2 Management Requirements

As explained in Section 1.2.1, a networked control system is typically subject to runtime system reconfiguration since it is an open system. Even though it is still possible to implement all such functionalities in an application layer, it becomes much easier and more efficient to develop and manage a networked control system if the underlying platform is equipped with some functionalities which can be used for runtime reconfiguration. Thus, as a platform for a networked control system, a middleware framework is required to provide some mechanisms for *runtime system management* which enables continuous system evolution.

1.2.2.3 Non-functional Requirements

The non-functional requirements [1] for a middleware framework are induced from both the time-critical and safety-critical characteristics of a networked control system. The time-criticality requires a control system to behave in a predictably timely manner so as to minimize the effect of delay. Thus, a middleware framework is required to provide some mechanisms which support the *timeliness* behavior of the control system. Also, the safety-criticality of a control system requires that the middleware framework itself be error-free, and also provide some mechanisms to tolerate faults which can occur in the application layer, to achieve overall *reliability*.

1.3 Middleware for Networked Control Systems

1.3.1 Middleware Fundamentals

A middleware is a software framework running in between an application and the underlying platform such as an operating system. Even a control system application running on a single computer can benefit from a middleware framework. However, the true value of a middleware is for a system which involves the features of both *heterogeneity* and *distributed operation*. Since these two fundamental features of distributed systems add a lot more complexity to the application, it is much harder to develop an application in general. Therefore, it is important to have a simpler abstract model of the system which hides all the complex details of the underlying system from an application developer. A middleware framework can provide such an abstraction of the system to the application developer so that she can develop an application easily on top of the abstraction. In this way, a distributed application can be developed more rapidly and reliably. In addition to rapid application development, an application can also be made more reliable since many of the commonly

[1] It should be noted that the phrase "non-functional requirements" can be used in dramatically opposite ways in different communities: with respect to the middleware designer, both a naming service or communication mechanism are both functional requirements, but achieving control loop stability is a non-functional requirement. From the viewpoint of the control designer, the reverse is true. In this paper, the viewpoint is that of the middleware designer.

used functionalities which usually require expertise to handle low level complexities can be developed and provided to the application developer by a middleware as a form of middleware service. An example is *component reuse* which can lead to a component economy.

1.3.1.1 Communication Mechanisms

In a networked control application, the distribution of the control system application over different nodes requires the interaction among components to occur over a network. For network programming, application programming interfaces such as *sockets* are provided by an operating system. But these are usually low level and require some expertise to use. Also, they are tightly coupled to the underlying computer platform. Thus, they are not appropriate to be used in a platform independent way for developing a distributed application in general.

In contrast, a middleware framework can provide simpler *inter-application communication mechanisms* to the application layer by encapsulating these low level network programming interfaces. A distributed application can then easily be developed using the inter-application communication mechanisms provided by the middleware, which allows components to interact with each other over a network without worrying about the low level network programming which is typically tedious and error prone.

In provisioning such communication mechanisms, there are roughly two forms of mechanisms which can be provided by a middleware, *message-oriented communication* and *request-oriented communication* [18]. In message-oriented communication, the message sender transmits a message to the receiver but the receiver does not respond to the sender. In contrast, the receiver replies with a response message when it receives a message from the sender in the request-oriented communication. Thus, the message-oriented communication can be considered as a one-directional communication model while the request-oriented communication can be considered as bi-directional communication. Each of these communication mechanisms can be further classified as *synchronous communication* or *asynchronous communication*. The sender is passive (*i.e.*, its execution is blocked) in synchronous communication, while it is active during the communication process in asynchronous communication. Considering the characteristics of a distributed system, the communication mechanism provided by a middleware is most fundamental to providing an abstraction of the original distributed system that eliminates several issues related to its distributed nature.

1.3.1.2 Naming Service

The other useful functionality which can be provided by a middleware framework is a *naming service* which allows an application to easily find or communicate with other applications in a distributed system. It is still possible to develop a distributed application without having such a naming service; however it would require the explicit specification of the physical network address in the source code of the applications. A naming service provided by a middleware can eliminate this otherwise undesirable necessity. In fact, from the software engineering point of view, a middleware's naming service improves significantly the portability and reusability of the source code by breaking the connection between the application code and the underlying platform.

1.3.1.3 Other Domain Specific Services

Besides the above inter-application communication mechanisms and naming service, there are many other functionalities which can be provided as middleware services, such as security service, transaction service, event service, and so on [27].

1.3.2 Etherware

In this section, we continue our discussion about the middleware framework for networked control systems in the context of a specific middleware framework, called *Etherware* [4], which has been developed at the University of Illinois.

1.3.2.1 Domainware for Networked Control Systems

Etherware is a middleware framework developed specifically for the networked control application domain. The main objective of Etherware is to provide a software framework which enables a rapid, reliable, and evolvable networked control application development. A networked control application can be easily developed in Etherware since it supports *component-based application development*. A software component is a software module which provides a set of functionalities through a set of pre-defined interfaces. In Etherware-based applications, the pre-defined interface is used for interaction between a component and Etherware, and components interact with each other through Etherware. Thus, Etherware itself provides a virtual communication layer to the application layer. Etherware uses the message exchange mechanism for component interaction. More specifically, a component needs to create a message and sends it to Etherware. Then Etherware delivers the sent message to the receiver though its message delivering mechanisms. In Etherware, every message is an instance of *Message* class which is a well-defined XML document object [29]. Listing 1.1 shows the XML structure of the Message class.

```
<EtherMsg type=... rel=... >
  <profile name=... ></profile>
  <content> ... </content>
  <ts value=... ></ts>
</EtherMsg>
```

Listing 1.1 XML structure of an Etherware Message

The `type` attribute of the `EtherMsg` element is used to specify the name of the message. The name of the receiver component can be specified in the `name` attribute of `profile` element. In the `content` element, any information concerning the interaction semantics can be specified. The clock time when the message is created is specified in the `value` attribute of `ts` element.

1.3.2.2 Architecture

The concept of microkernel architecture in an operating system [25] is used in the design of Etherware architecture. Therefore, only the minimal invariant functionalities are implemented in the core module of Etherware, called *Etherware Kernel*, while all the other functionalities are implemented as *Etherware Components*. As shown in Figure 1.2, Etherware components can be classified further into two different layers. The top layer contains components which implement the application logic, called *application components*, while the bottom layer is for components which provide functionalities to support several fundamental domain requirements, called *service components*. The details of these components are explained in the following sections.

In this section, we discuss the Etherware Kernel which is a runtime platform for Etherware components. As a platform for component execution, Etherware Kernel is responsible for both component life-cycle management and message delivery between components. To deliver a message from one component to another, Kernel uses an object, called *job*, as its scheduling entity. When a new message arrives in the Kernel, Kernel creates a new job which contains the sent message itself and the address of the recipient component. Then the Etherware Scheduler enqueues the job into a job queue. The enqueued jobs are processed one by one by a job processing software module, called *Dispatcher*. When a Dispatcher processes a job, it first extracts the address information from the job and then it delivers the encapsulated message to the receiver component. During this delivery process, some of the service components can be called by Dispatcher depending on the delivery information specified in the message.

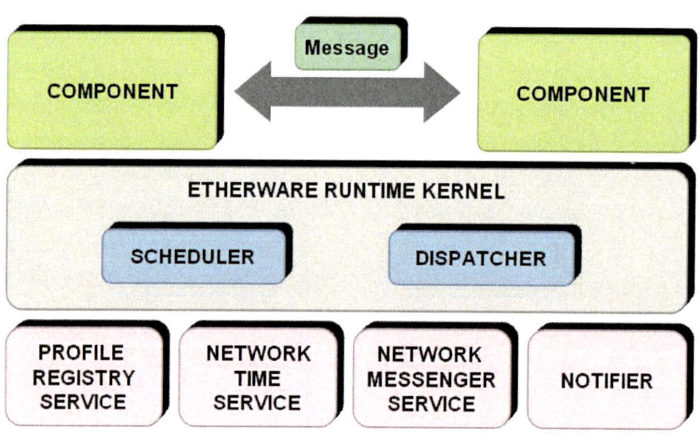

Fig. 1.2 Etherware architecture

1.3.2.3 Component Model

Etherware's component model shown in Figure 1.3 provides the framework in which an application component can easily be developed. In designing the component model, several *software design patterns* [7] were used. *Shell* plays a central role in the component model. First, it manages the life cycle of a component which is encapsulated by it. Shell creates or destroys an instance of component as needed. Second, Shell in a component model provides a channel which allows a component to interact with other components.

Basic design to implement Shell is based on the *Facade* design pattern. The *Strategy* design pattern was used to design the component interface which defines a uniform interface between Shell and all components. Due to this Strategy design pattern, Shell can do runtime component upgrade since every component implements the same interface called *Component Interface*. To support runtime component migration, another design pattern called *Memento* was used in the component model. For *component migration*, it is not enough to move a component by stopping at one place and by restarting at other place. To reduce the performance degradation due to component migration, the execution state of a component should be continued smoothly before and after the migration. The Memento design pattern was adopted to support this feature.

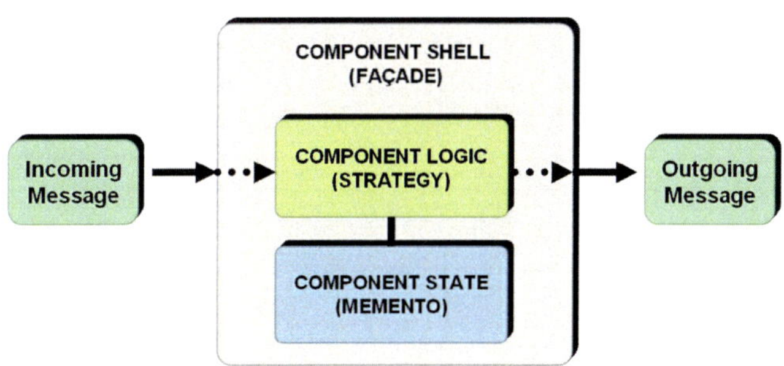

Fig. 1.3 Etherware component model

1.3.2.4 Services

Etherware supports several fundamental functionalities which are commonly required for networked control system applications as Etherware services. *ProfileRegistry* is a naming service which is implemented in Etherware to support the semantic addressing requirement. It maintains information about the profile of a component and its network address. *NetworkMessenger* provides the service of message delivery over the network. It encapsulates all the details about network information such as protocol and network address. So, NetworkMessenger is the Etherware service which supports the domain requirement of hiding location discrepancy. NetworkMessenger is called only when a message is destined for a remote component since Etherware Kernel delivers the message directly if it is for local component.

Etherware resolves the time discrepancy issue of distributed systems by implementing the *NetworkTime* service. NetworkTime service translates a time stamp from the clock of a remote computing node to that of a local machine for every message which is received by NetworkMessenger. For this purpose, NetworkTime service maintains the clock offset and skew for every other computing node where an other NetworkTime service is running, by periodically exchanging *Ping* and *Response* messages. The *Notifier* service provides a time-triggered message service to Etherware. Basically, Etherware is an *event-driven system* such that a component gets executed only when it receives a message. However, in many cases, control actions need to be performed based on the *time*. In such situations, the Notifier service enables a component to execute at the time that it has to, by sending a notification message to that component.

1.4 Real-Time Operation of Networked Control Systems

Now we turn to the issues raised by the real-time operation that is necessary for control systems.

1.4.1 Real-Time System Fundamentals

A real-time system is a system whose correctness depends not only on the logical result of the computation but also on the *time* at which the results are produced [23]. Thus, timeliness is a critical attribute of any real-time task. More specifically, the main objective of a real-time system is to meet the timing requirements of each real-time task. However, this does not necessarily mean that a real-time system is a fast system. Fast computing may *help* in meeting timing requirements. However, fast computing alone does not *guarantee* meeting timing requirements. In fact, a real-time system is more precisely a system which is predictable, whether fast or slow [23, 6]. In general, *predictability* is considered to be one of the most fundamental attributes of any real-time system. What one basically requires is that the system should behave in such a way that the execution behavior of the running task set can be precisely described from the information about both the system itself and the task set.

Once a system becomes predictable, then it is possible to address the issue of schedulability of a given real-time *task set*. Each *task* in the task set can be regarded as a collection of *jobs*. Each job may have a *deadline* (also called absolute deadline), which is the time by which the job must be completed. A task is said to have met its timing requirements if all the jobs in the task are completed by their deadlines. In a real-time system, a set of real-time tasks is said to be *schedulable* if all tasks in the set can complete their execution, while satisfying all their timing requirements, under some task scheduling algorithm (or scheduling policy). Otherwise, the task set is said to be *unschedulable*. A scheduling policy is a set of rules which determine the execution order, called *schedule*, among the tasks in a given task set. However, determining schedules in a scheduling problem consisting of a set of n jobs $J = \{J_1, J_2, ..., J_n\}$, a set of m processors $P = \{P_1, P_2, ..., P_m\}$, and a set of r types of resources $R = \{R_1, R_2, ..., R_r\}$, is known to be NP-complete [6]. Therefore, it is important to consider a scheduling problem under additional assumption so that the problem becomes simple enough with respect to computational tractability while still preserving its practicality. Later, in this section, we introduce some of these fundamental real-time scheduling theories[2]. Before discussing real-time scheduling, we first highlight the difference between a *hard* real-time system and a *soft* real-time system.

1.4.1.1 Hard Real-Time vs. Soft Real-Time

A real-time task has timing constraints which have to be met for its correctness. As noted earlier, one important time constraint is the *deadline* by which a job in a real-time task has to complete its execution. Depending on the severity of the consequences which could occur by failure to meet the deadline constraints, real-time tasks are usually classified into two groups, *hard real-time* tasks and *soft real-time* tasks. A hard real-time task is a real-time task whose deadline constraints are very strict. The consequence of a deadline miss of a job in a hard real-time task could be catastrophic. In fact, many control tasks have the characteristics of a hard real-time task. One can easily imagine that an aircraft flight control task which has the highest priority among the tasks in any avionics system is indeed a hard real-time task. In contrast, the deadline constraints of a soft real-time task are not as strict as that of a hard real-time task. In a soft real-time system, an occasional miss of a deadline does not cause a catastrophic situation. However, the performance of the task might be affected by the rate or frequency of deadline misses. On-line multimedia streaming is a typical example of a soft real-time task.

1.4.1.2 Processor Utilization Bound

In many control systems, the sensing and control actions are typically periodic actions. Thus, a periodic task model in real-time scheduling theory can be used to capture the fundamental behavioral characteristics of control systems. In real-time scheduling theory, a periodic task can be described by several parameters such as the *execution time* (C), the *period* (T), and the *relative deadline* (D). The *period* is that length of the time interval between successive arrivals of jobs in the task. The *relative deadline* is the time interval from the arrival of a job to its deadline, which is the latest time by which jobs must be completed. The *execution time* is the amount of the processor's time that the job needs in order to be completed. For simplicity in schedulability analysis, it is usually assumed that the relative deadline of a job is the same as the period. With this assumption, the demand for the processor's time of a task set consisting of n periodic tasks can be characterized by a parameter U, called *processor utilization factor*, which is defined as follows:

[2] For more details and comprehensive coverage of real-time scheduling, we refer the reader to [6, 22].

$$U = \sum_{i=1}^{n} \frac{C_i}{T_i}. \tag{1.1}$$

For a given scheduling policy and a given periodic task set, there exists a number U_{ub}, such that the schedulability of the task set is guaranteed if the process utilization factor of the task set is below U_{ub}. Above this upper bound, the given task set maybe unschedulable. Note that the value of U_{ub} depends not only on the scheduling policy but also on the task set. A critical quantity is the least upper bound U_{lub}, defined as the minimum among all U_{ub} over all task sets. U_{lub} can be used the threshold so that it provides a sufficiency condition to test the schedulability of any task set under a specific scheduling policy.

1.4.1.3 Rate Monotonic Scheduling

Rate monotonic (RM) scheduling is central to several fundamental results in real-time scheduling. The rate of a task is defined as the reciprocal of its period. In RM scheduling, the execution priorities are statically assigned to tasks based on the rate of each task in a given set of periodic tasks. The higher the rate (*i.e.*, the shorter the period), the higher the priority. RM scheduling is known to be optimal among all fixed-priority (or static) scheduling policies for periodic task sets [15]. It is optimal in the sense that if any other fixed-priority scheduling policy can schedule a given periodic task set, then the RM scheduling can also schedule the task set. In other words, there is no other fixed-priority scheduling policy which can schedule a given task set that is not schedulable under RM scheduling. In [15], the *least* upper bound of the processor utilization which guarantees schedulability of a periodic task set consisting of n tasks is shown to be.

$$U_{lub} = n(2^{1/n} - 1). \tag{1.2}$$

If we take the limit as $n \to +\infty$ in (1.3), we obtain U_{lub} of RM scheduling for any periodic task set with any number of tasks in it:

$$\lim_{n \to \infty} U_{lub} = \ln 2 \simeq 0.69. \tag{1.3}$$

By using this condition, it is very easy to check the schedulability of a given task set under RM scheduling. If the processor utilization demand is below U_{lub}, then the task set is schedulable. Otherwise, there is no guarantee for a given task set to be schedulable. In this case, the task set can be either schedulable or unschedulable. To determine the schedulability of such task sets, it is necessary to employ iterative response time analysis which is a task set dependent, schedulability analysis technique [6].

1.4.1.4 Earliest Deadline First Scheduling

While RM scheduling is an optimal *static* real-time scheduling policy, the *earliest deadline first (EDF)* is an optimal *dynamic* real-time scheduling policy. In RM scheduling, a priority is assigned to a task and the assigned priority is never changed during the task's periodic execution. However, in EDF scheduling, a priority is assigned to each instance of a task, *i.e.*, to each *job*, based on the current job's absolute deadline. Therefore, a priority assigned to a task keeps changing depending on its current execution state. In [15], the necessary and sufficient conditions for the schedulability of a set of periodic tasks under the EDF scheduling is shown to be:

$$U = \sum_{i=1}^{n} \frac{C_i}{T_i} \leq 1 = U_{lub}. \tag{1.4}$$

where C_i is the processor time necessary for executing the task i without interruption and T_i is the period of task i. Note that the optimality of EDF is immediate from (1.4) since the least utilization upper bound it provides for EDF scheduling is 100 percent of the processor time.

1.4.1.5 Resource Sharing Protocol

During its execution, a real-time task may access many resources such as data structures, files, memory areas, peripheral devices, a set of variables and so on. Also, in any real-time computing system, multiple tasks running concurrently may access the same resource. In some cases, accesses of the resource have to be mutually exclusive for the integrity of the sate of the shared resource at the same time. For this purpose, operating systems which manage the resources provide synchronization mechanisms to allow mutually exclusive access to the shared resource. To synchronize concurrent access from multiple tasks, the synchronization mechanism blocks a task when it tries to access a resource which is already occupied by another task, until it is released by the latter task. However, synchronized resource sharing causes an undesirable phenomenon in real-time systems. In real-time systems, a task with the highest priority should be able to continue its execution under any circumstances. However, this may not be true any more under synchronized resource sharing. A higher priority task can be blocked by a lower priority task. This is called a *priority inversion*. Figure 1.4 illustrates an example of priority inversion [6]. In this example, J_1 has the highest priority, while J_2 and J_3 have intermediate and lowest priority, respectively.

In general, the duration of priority inversion can be potentially unbounded since any intermediate priority task such as J_2 in Figure 1.4 could indirectly block the highest priority. This in turn means that a real-time task can fail to meet its timing constraints due to priority inversion. To overcome this issue, a real-time resource sharing protocol, called *priority inheritance protocol* (*PIP*), was proposed in [21]. The basic idea of the PIP is that a task which currently holds the shared resource inherits temporarily the highest priority among the blocked tasks, until it releases the resource. After releasing the resource, it recovers its original priority. In this manner, the blocking task will never be preempted by any intermediate priority task while it is accessing a resource. By adding PIP for resource sharing among real-time

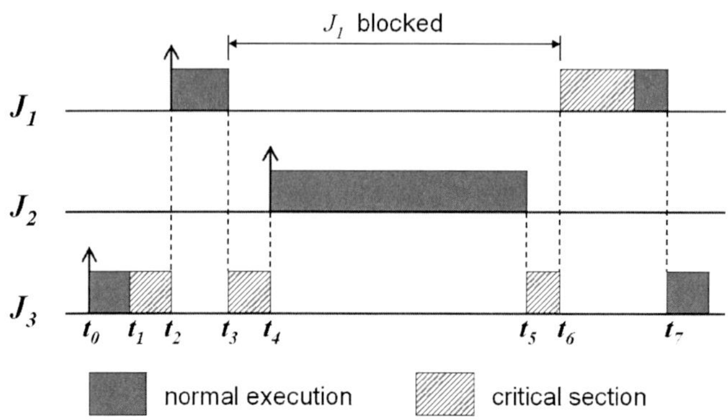

Fig. 1.4 An example of priority inversion

tasks in a given task set, the original RM schedulability test was extended in [21]. Any set of n periodic tasks using the priority inheritance Protocol can be scheduled by the rate monotonic algorithm if the following conditions are satisfied:

$$\forall i,\ 1 \leq i \leq n, \quad \sum_{k=1}^{i} \frac{C_k}{T_k} + \frac{B_i}{T_i} \leq i(2^{1/i} - 1). \tag{1.5}$$

where B_i is the maximum blocking time, due to lower-priority tasks, that a task J_i may experience. C and T are the execution time and period of a task, as explained above.

1.4.2 *Real-Time Support in Etherware*

In this section, we describe how Etherware supports the real-time requirements of a middleware for networked control systems.

1.4.2.1 Quality of Service (QoS) of Message Delivery

In Etherware, each Etherware *component* can be considered as a task in a real-time scheduling model. Also, each message sent by a component can be thought of as an instance of the task, a *job* in the real-time scheduling model. As mentioned in Section 1.4.1, each real-time task has a set of timing constraints such as period and relative deadline which are required to be met. In Etherware, these timing constraints can be specified as a QoS specification in each message sent by a component [13]. Such information is encoded as a QoS element in the Message class object which itself is an well-defined XML document, as explained in Section 1.3.2.1. Basically, Etherware defines QoS as a collection of attributes of an application which are used in scheduling for execution. Hence, any information which affects the scheduling can be specified as a constraint in QoS specification. In our current implementation of Etherware, the period (`period`), the relative deadline (simply called `deadline`), the worst case execution time (`wcet`), and the importance of a message (`crit`), are defined for QoS specification of a message, as illustrated in Listing 1.2.

```
<EtherMsg type=... rel=... >
  <profile name=... ></profile>
  <content> ... </content>
  <ts value=... ></ts>
  <QoS crit=...  period=...  deadline=... wcet=...>
  </QoS>
</EtherMsg>
```

Listing 1.2 QoS specification in Etherware Message

1.4.2.2 Priority and Concurrent Processing

Concurrency is a fundamental feature of any real system. Multiple tasks may be released concurrently, and a real-time scheduling policy prioritizes the execution order based on some rules, while satisfying their timing constraints. Thus, concurrency and priority are two key aspects of any real-time system.

As explained in Section 1.3.2.2, Dispatcher is a software module inside the Etherware Kernel, for processing a *job* (a scheduling entity in Etherware Kernel). For concurrent message processing, multiple Dispatchers can be used to form a *dispatching module* as shown in Figure 1.5. For real-time message processing, Etherware uses a hierarchical prioritization mechanism [8]. First, each Dispatcher in a dispatching module is assigned a fixed priority[3] so that each message enqueued in a Dispatcher is processed at the fixed priority. The specific number of Dispatchers in a dispatching module and their corresponding priority levels are determined by a user-provided specification, called a *Thread Scheduling Rule* (*TSR*). Second, the job queue of a Dispatcher is a prioritized queue which orders jobs in the queue based on some information specified for a job. Owing to this hierarchical mechanism, Etherware can support various types of real-time scheduling policies, such as RM scheduling and non-preemptive EDF. An example of the implementation of a RM scheduling policy implementation is shown in Listing 1.3.

For predictable behavior, Etherware utilizes the priority-based scheduling mechanism of the underlying platform upon which Etherware is executed. This means that the execution order among Dispatchers is determined by the underlying operating system platform based on the fixed priorities assigned to each of them. Thus, Etherware Scheduler does not need to handle such priority-based scheduling. It is automatically taken care of by the platform. Instead, Etherware Scheduler performs the scheduling action at a higher level. Specifically, it determines the right place where a job should be put in the dispatching module. Due to the hierarchical nature

[3] The specific priority set is given by the underlying software platform.

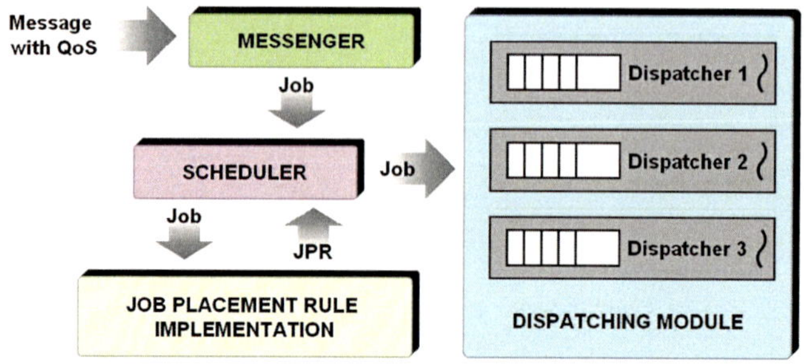

Fig. 1.5 Real-time scheduling mechanism with three Dispatchers

of the prioritization in message processing, Etherware Scheduler uses two pieces of information, one to select a Dispatcher in dispatching module, and the other to find the right position in the job queue of the selected Dispatcher. This information is provided to Etherware Scheduler at runtime by a software module which implements a user defined scheduling rule, called the *Job Placement Rule* (*JPR*), which maps the QoS information specified in a message contained in a job to the position in the Etherware's dispatching module. Listing 1.3 is an example of the JPR implementation which utilizes the period information in a message to implement the RM scheduling policy. The corresponding experimental results are shown in Table 1.1.

```
/* QoS : period(P) in millisecond */
JPR queryJPR(Job job) {
 JPR jpr;
 if(job.P=80) jpr=(Disp#1, NULL);
 else if(job.P=200) jpr=(Disp#2, NULL);
 else if(job.P=350) jpr=(Disp#3, NULL);
 return jpr;
}
```

Listing 1.3 A pseudo-code example of JPR which implements the RM scheduling policy

To test the real-time performance of Etherware, experiments were performed under several different conditions, depending on the execution time of a task with 200 (ms) period. In each of these different conditions, the activation periods and the execution times per activation of each task are measured to see how much these changes affect the execution behaviors of two other periodic tasks. As shown in Table 1.1, the task with the shortest period is not affected at all by the execution time changes of the task with the second shortest period. In contrast, the third task, which has the longest period, is affected a lot in its execution behavior by these changes.

Thus, this result shows that the periodic tasks are in fact correctly scheduled under the RM scheduling policy.

Table 1.1 Experimental results of task execution under RM scheduling policy

Test	(mean, jitter) of Execution Time (ms)			(mean, jitter) of Period (ms)		
Case	Task(80ms)	Task(200ms)	Task(350ms)	Task(80ms)	Task(200ms)	Task(350ms)
1	(14.5, 1.1)	(42.4, 15.8)	(49.0, 51.2)	(80.9, 1.1)	(200.9, 29.5)	(350.9, 96.9)
2	(14.5, 0.4)	(87.6, 16.0)	(56.3, 102.0)	(80.9, 1.2)	(200.9, 29.6)	(350.9, 156.1)
3	(14.5, 0.3)	(129.9, 16.5)	(64.7, 153.0)	(80.9, 1.2)	(200.9, 29.5)	(350.8, 192.0)
4	(14.5, 0.3)	(175.6, 17.6)	(350.8, 191.9)	(80.9, 2.0)	(200.9, 29.3)	(350.4, 191.9)

1.5 Reliability for Networked Control Systems

1.5.1 Fundamentals of Reliable System

The reliability of a system is usually defined as the probability that a system will perform its functionality correctly throughout a duration of time. A technical measure of the reliability is the mean time between failures (MTBF), which is the sum of the mean time to failure (MTTF) and the mean time to repair (MTTR). The MTTF is a measure of how long a system is expected to operate correctly before a failure occurs, while the MTTR measures the difficulty of recovering a system after its failure. As the definition indicates, the failure of a system is at the heart of the discussion of the reliability. Therefore, we first introduce the definition of failure and two other fundamental concepts, *error* and *fault*.

1.5.1.1 Fault, Error, and Failure

Typically, all details about what are the acceptable behaviors of a system are described in a system specification. If the behavior of the system deviates from the specified acceptable behaviors, then this is called *failure* of the system. The immediate cause of a failure is called an error. An *error* is a part of an erroneous state, a system state which could lead to a failure by a sequence of valid state transitions. Finally, a *fault* is defined to be the cause of an error. Thus, a fault is the root cause of a failure.

Depending on the view point, faults can be classified in many different ways [20]. A fault can be either *transient* or *permanent* based on the time duration that a fault can exist. Typically, a transient fault is much more problematic than a permanent fault since it is usually harder to diagnose. A fault can also be classified as a *design fault* or *operational fault* based on the underlying cause of the fault. The other classifications of faults are based on the symptoms caused by the fault, *e.g.*, *crash faults*, *timing faults*, *omission faults*, and *Byzantine faults*. A crash fault is a fault which causes a system to crash so that it can never return to a valid operational state. When a system experiences an omission fault, it fails to perform its designated service even though it is still operating. Under a timing fault, a service provided by a system can be delayed. Lastly, a system behaves unpredictably with Byzantine faults. There is no specific patterns of symptoms caused by Byzantine faults.

1.5.1.2 Fault Prevention

One approach for achieving reliability of a system is to prevent system failures by ensuring that all possible causes of unreliability have been removed from the system before the system is deployed for its operation [1]. This is called *fault prevention*. As a first step toward fault prevention, a system needs to be carefully developed so that all faults which can be anticipated are removed during the development process. Various techniques or methodologies from software engineering, or formal methods, can be used during this stage to attempt to avoid introducing any faults into the system design. However, it is not usually possible to guarantee that a developed system is completely free from faults. No matter how thorough the development process, there may always exist faults. The faults in the constructed system are therefore attempted to be removed through some experimental validation process. The implemented system is tested under various operating conditions to expose any existing faults, so that they can be removed. In some cases, artificial faults are introduced into the implemented system. This technique is called *software fault injection (SFI)* [28]. SFI tries to determine what could happen when faults are activated. The information collected through the SFI process can be used to both improve reliability of the system and also to estimate the resilience of the implemented system to faults.

1.5.1.3 Fault Tolerance

In general, the application of fault prevention techniques to a system is not sufficient to achieve high reliability. Given this fact, it is usually required that a system be fault tolerant in order to provide reliability despite the presence of faults. *Fault tolerance* is the ability of a system to perform its function correctly even in the presence of internal faults [20]. There are four distinct activities which provide the general means by which fault tolerance can be implemented [1]. These four activities constitute the basic principles which underly all fault tolerant systems. Toward fault tolerance, errors caused by faults have to be detected first. Thus, *error detection* is the first step for fault tolerance. The common techniques for error detection are replication checks, timing checks, coding checks, and so on (see [1] for details). Once an error is detected, it may be necessary to asses the extent to which the system state has been damaged by the fault which manifested the detected error. This is known as *damage confinement*. The next step is *error recovery* which recovers the system from the erroneous system state to a valid error-free state. In the error recovery phase, there are basically two different approaches to recover the system state, backward error recovery and forward error recovery. In backward error recovery, the system state is restored to a past valid state which was checkpointed during normal operations. On the other hand, forward error recovery techniques drive the system to a new valid state which is produced by manipulating some portion of the erroneous current state. Finally, the forth step of fault tolerance is *fault treatment*. To prevent reoccurrence of the error, it is necessary to identify and treat the fault. One issue in this procedure is that it may be hard or take a long time to identify faults which caused the errors, since the relationship between a fault and the corresponding errors is typically very complex.

1.5.1.4 Software Fault Tolerance

In this section, we introduce two main techniques for software fault tolerance, *recovery block scheme* and *N-version programming*. Basically, both the techniques depend on the effective utilization of redundancy with the expectation that components built differently should fail differently [26]. The basic configuration of the recovery block scheme is illustrated in Figure 1.6. The first step in the recovery block scheme is checkpointing the current valid system state before executing any modules. Then the primary module is entered. Once the primary module completes its execution, the execution result of the primary module is tested by the acceptance test module to detect any error from the primary module if there is any. If the result is valid, then it is propagated as the output of the block. If any error is detected, then the error recovery process restores the primary module with the checkpointed state. Following the recovery, the same sequence of executions is repeated, except that the next module is used in place of the module that failed. If all of the modules fail, then it is considered as the failure of the recover block. One issue with the recovery block scheme is that the acceptance test module is highly application dependent [26]. Hence, its error detection logic is usually required to be implemented by the module developer.

While the recovery block scheme requires an application dependent acceptance test module, the N-version programming model can use a generic decision algorithm to select the correct output [26]. Usually, voting algorithms such as Formalized Majority Voter, Generalized Media Voter, Formalized Plurality Voter, and Weighted Averaging Techniques, can be used as the generic decision algorithms in the design

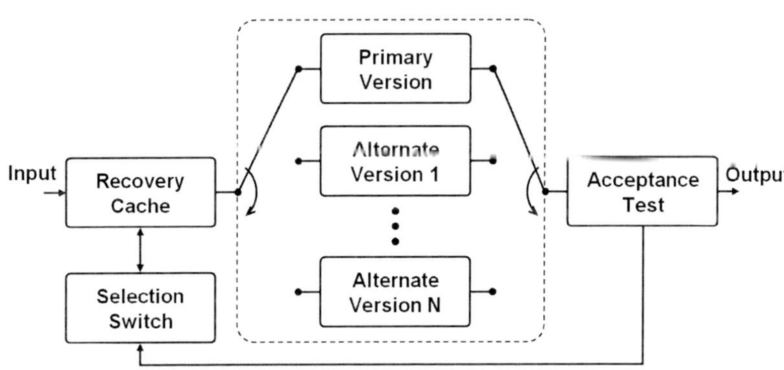

Fig. 1.6 Recovery block scheme

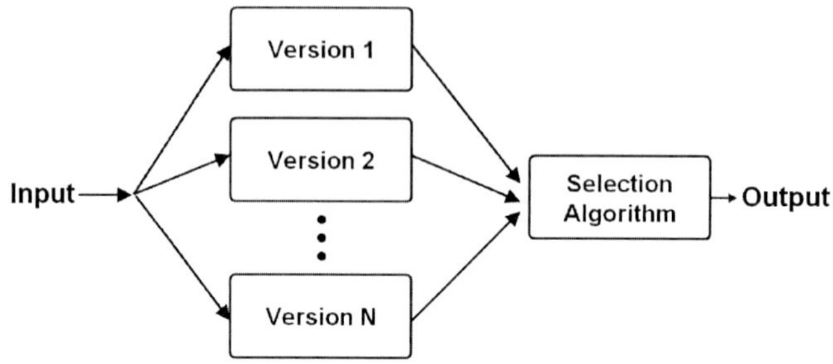

Fig. 1.7 N-version programming model

of the selection module (see [16] for details). Another significant difference between N-version programming and the recovery block is that the output is determined through a voting process. In a voting process, each of the modules processes the input first, and then their execution results are collected by the selection module. The selection algorithm determines the output based on the outputs from all modules and some decision rules. This procedure is illustrated in Figure 1.7.

1.5.2 Reliability Support in Etherware

In this section, we describe how Etherware supports the reliability requirement of middleware for networked control systems.

1.5.2.1 Local Temporal Autonomy

Local temporal autonomy (*LTA*) is defined as the ability to operate correctly for a while in the presence of a failure of the other system components [9]. To accommodate the failure of a network connection, nodes or components in a networked control system, several design principles based on the LTA were proposed in [9, 19]. The basic idea of these design principles is to reduce the inter-dependency between the interacting components. These principles can be easily understood by considering specific examples. As shown in the top figure in Figure 1.8, in a networked control system, a controller relies on both the information from sensors and the communication network through which the sensor information is delivered. If one of these fails, a controller also fails. The proposed design principle suggests that an estimator be used, which is collocated, with the controller to estimate the sensor data. Then, any transient failure of either the sensor or communication network becomes transparent to the controller.

Another example is the case when a remote controller sends its computed control value to an actuator component which is collocated on the target plant and delivers the control to the plant. In this situation illustrated in the bottom figure in Figure 1.8, the proposed design principle first employs a controller to compute multiple steps of future control values, and then sends them en bloc to the actuator. Secondly, an

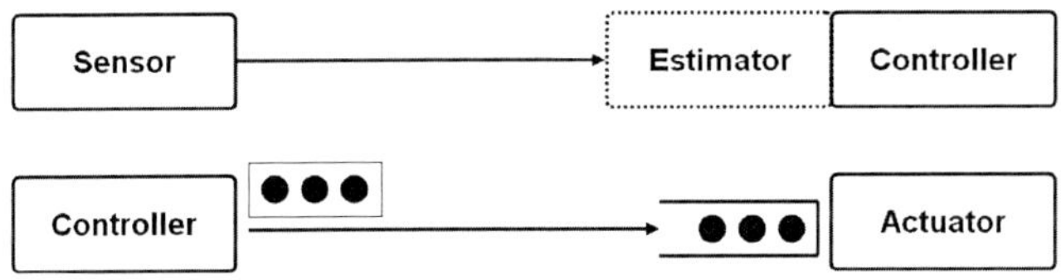

Fig. 1.8 The LTA-based design principles

actuator is equipped with a buffer to hold the block of control values delivered from a controller. Clearly, this block computation and buffer mechanism can reduce dependence between a controller and an actuator so that any transient failure of either a controller or the communication network is transparent to the actuator. Etherware provides support for both these strategies [19].

1.5.2.2 Component Model for Fault-Tolerance

In addition to the LTA-based design principles, a *fault-tolerant component model* (*FT component model*) was proposed to support the reliability requirements in [13]. Basically, the FT component model is an extension of the original component model of Etherware. As shown in Figure 1.9, it is designed to achieve both redundancy based fault-tolerance, and systematic fault detection and management. The redundancy based fault-tolerance is achieved by allowing multiple components, the primary and other replica components, to be executed within a Shell. The systematic fault detection and management is possible through the fault detector and fault handler elements in a Shell. Among the elements encapsulated in a Shell, the fault manager plays the central role in the fault tolerant operation. It coordinates all the interactions among elements within the FT component model. The *fault management policy* (*FM policy*) provides decisions about how to coordinate them. Depending on the FM policy, the FT component model itself can behave similar to either the recovery block mechanism or the N-version programming mechanism as explained in Section 1.5.1 (see [13] for details about the FM policy).

The fault detector can be used to detect a design fault related to a component's operational semantics. As an example, if the value computed by a component is typically expected to be both within some range and smooth, then the fault detector checks if the result from a component satisfies these conditions. If not, it reports this as a design fault to the fault manager, which in turn calls the fault handler to handle this fault. If there is a crash fault or a timing fault from other Etherware components, then these faults can also be managed by the fault handler. Unlike the design fault, detecting these faults cannot be done within a Shell since it usually requires a time-based delay detection mechanism such as a watchdog timer. Therefore, an additional Etherware service, called *interaction fault detection service* (*IFDS*), was proposed, and is being developed to detect interaction delays caused by another component's crash or timing fault.

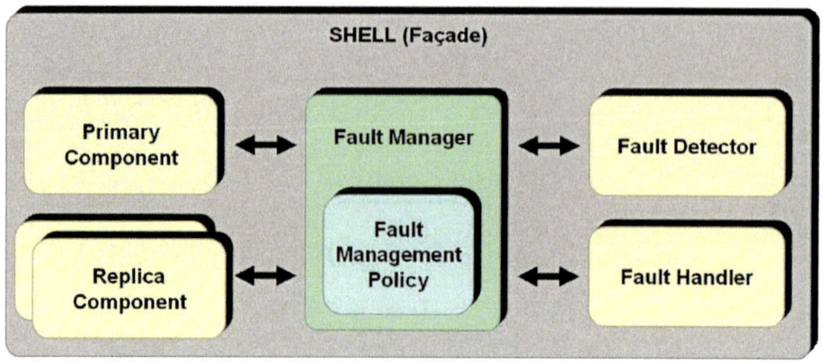

Fig. 1.9 Fault-tolerant component model

1.6 Case Study: Networked Inverted Pendulum Control System

In this section, to exemplify the usefulness of a middleware framework, specifically Etherware in this case study, in implementing a networked control system, we present the experimental results of using Etherware to control an inverted pendulum control system. Basically, two conclusions can be made which are supported by this case study. First, it can be easy to both develop and manage a networked control system with proper support from a middleware. Second, it is important to consider the non-functional requirements such as real-time and reliability, for correct operation of a networked control system.

Next, we briefly introduce the inverted pendulum control system which is used in the experiment.

1.6.1 Inverted Pendulum Control System

Figure 1.10 shows the inverted-pendulum system that is used in our experiment. As shown in the figure, it has two links. The link in the base is an active one which is actuated by a DC motor attached to it, while the other one is a passive link. The inverted pendulum system itself is equipped with a DSP board to measure the joint angles of both links and to apply the PWM signal to a DC motor. In our experiment, a controller, which implements a simple state feedback control law, is developed as an Etherware component, and it runs in a laptop which is attached to the inverted pendulum system through an RS-232C serial communication channel. For real-time operation of Etherware, we use the Sun JavaRTS[4] as the underlying platform of Etherware[5].

[4] Sun JavaRTS is a real-time Java virtual machine which implements the real-time Java specification (RTSJ). For details about Sun JavaRTS and RTSJ, we refer the reader to [24, 5].

[5] Note that current Etherware is developed with Java programming language. Therefore, Etherware requires a Java virtual machine which provides the fixed-priority based scheduling mechanism for correct operation of Etherware's real-time Scheduler.

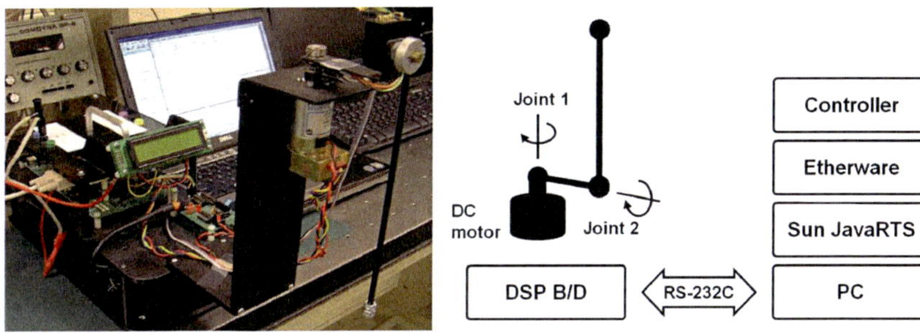

Fig. 1.10 Inverted pendulum control system

With support from Etherware's notification mechanism, explained in Section 1.3.2.4, the controller component is periodically activated to output its control action. In our experiment, the period is set to 15ms. In each period, the controller begins its execution by requesting the angle data from the DSP. Once it receives the measured angle data, it then computes a control output value and sends it back to the DSP so that the control command can be applied to control the inverted pendulum.

Thus, a periodic control action requires multiple interactions between a controller and the inverted pendulum system through the RS-232C communication network. Therefore, in addition to the predictable activation of a controller component in Etherware, the predictability and the reliability of the RS-232C communication network are also essential factors affecting the success of the inverted pendulum system. As explained in Section 1.4.2, the predictability of the periodic activation of a controller component is guaranteed by the real-time scheduling mechanisms of Etherware. However, the predictability and the reliability of the RS-232C communication network is not guaranteed by the underlying platform in our implementation. Therefore, we adopted the state estimator LTA design principle to tolerate occasional communication errors, so as to achieve better periodic performance over the unreliable communication layer.

Figure 1.11 is a still oscilloscope image which captures the serial communication between the DSP and PC on which the Etherware-based controller is running. From the figure, we can observe the periodic interaction between the Etherware-based controller and the DSP. The upper signal in the scope image is the signal for feedback of angle data from DSP to PC, and the lower one is from PC to DSP for requesting angle data and sending a control command.

1.6.2 *Periodic Control under Stress*

The real-time performance of Etherware is verified through a stress condition that is imposed alongside the periodic control of inverted pendulum. To stress the computer on which a periodic control task is running, an extra computational task is

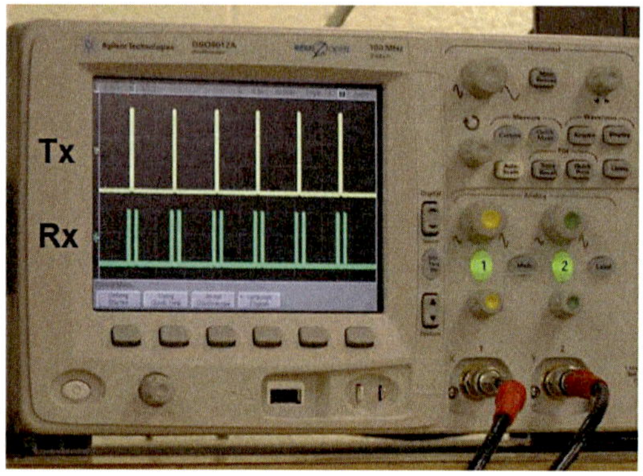

Fig. 1.11 Periodic sensing and control action over RS-232C communication

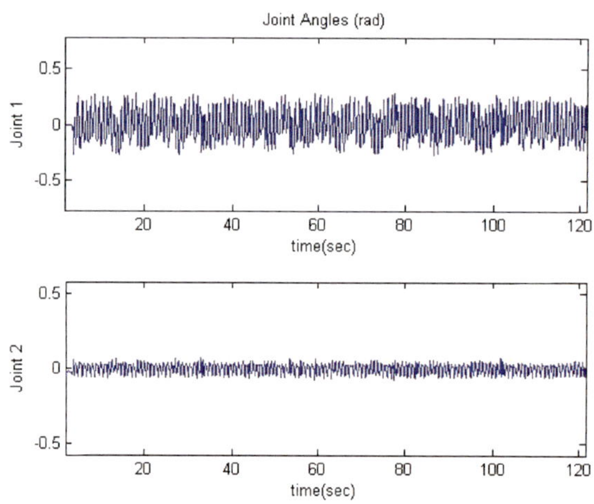

Fig. 1.12 Periodic control of an inverted pendulum under stress

made to run concurrently with the control task. In this experiment, the stress task is also executed periodically with a period of 5 seconds. Once it begins its execution, the stress task requires about 1 second to finish its computation. Considering that the period of control task is only 15ms, as explained in the above section, 1 second is long enough to disturb the stability of the inverted pendulum, if the real-time performance had not been supported by Etherware. To achieve timely execution behavior of the periodic control task, the periodic notification message from Notifier to the periodic control task is specified to have a higher priority than that of the stress task. Figure 1.12 shows the experimental result of this experiment. In the experiment, the stress task starts its execution around 20 seconds after the system starts. As shown in the result, there is no apparent adverse affect on control performance even under the stress condition. This demonstrate Etherware's real-time performance.

1.6.3 Runtime System Management

As explained in Section 1.3.2, Etherware provides mechanisms which support the continuous evolution of a system after its deployment. The component model is at the heart of such mechanisms. *The combination of these mechanisms for flexibility, and the real-time mechanisms for temporal predictability, generate capabilities which enable us to do runtime management of a system, especially a time-critical control system, without sacrificing the system's stability*. The specific capabilities that we aim to provide are *controller upgrade* and *controller migration*. In this section, we demonstrate these capabilities of Etherware.

1.6.3.1 Controller Upgrade

In this experiment, we show Etherware's capability for *runtime component upgrade*. More specifically, a running inverted pendulum controller is replaced with a new controller which has better control performance, while still maintaining the stability of the inverted pendulum. To perform this upgrading process correctly without violating the stability of running system, it is important to externalize the state of the running controller and recover the state with a new controller timely. Etherware supports this operation though its component model which enables component upgrade and its real-time scheduling mechanism.

Figure 1.13 shows the configuration of an application for this runtime controller upgrade experiment. The periodic controller is running on a computer which is directly connected to the inverted pendulum system through a serial port. On the other computer, a component which requests the controller upgrade is running. In the experiment, the requester component sends a request message for better control performance around 30 seconds after the system starts. As shown in the figure, the control performance is improved around 30 seconds, at the time when the controller upgrade is requested. Thus, this result demonstrates Etherware's capability of real-time component upgrade.

1.6.3.2 Controller Migration

In this experiment, we demonstrate Etherware's capability for *runtime component migration*. More specifically, a controller which controls the inverted pendulum is migrated from one computer to another at runtime while preserving the stability of the inverted pendulum. This type of capability can be very useful in a wide range of applications in optimizing the behavior of control systems. For example, if the network causes long delay from a specific computer in the network, then one may want to relocate the controller logic, *i.e.*, component, to another computer which has less delay. For such runtime migration, several more steps of actions have to be performed, in addition to the runtime state externalization, as explained in previous section. Specifically, the externalized state itself has to be migrated to the destination where the controller will be migrated. Also, once a new instance of the controller is created at the destination node, the migrated state has to be recovered with the

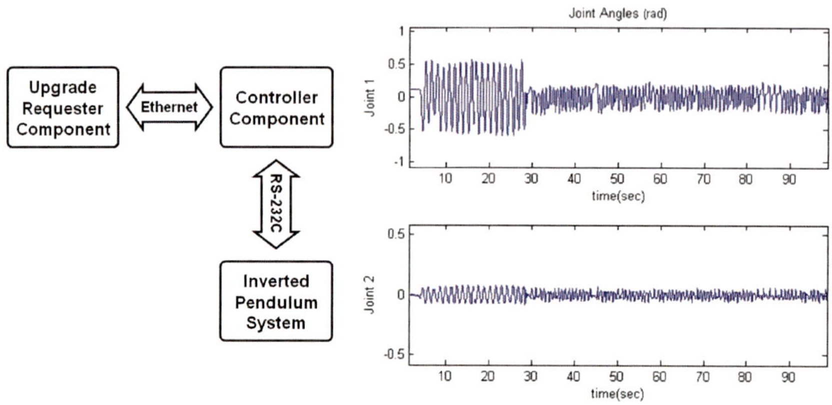

Fig. 1.13 Joint angles of the inverted pendulum under runtime controller upgrade

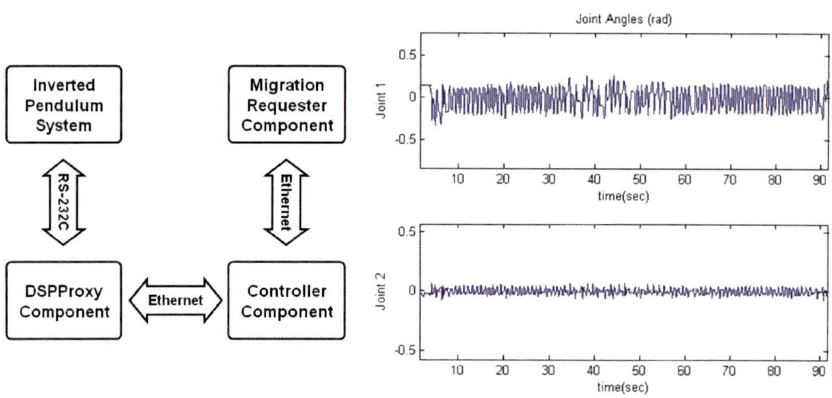

Fig. 1.14 Joint angles of the inverted pendulum under runtime controller migration

controller. Thus, component migration is a much more complex task than upgrade, in general. Furthermore, all these actions have to be performed in a timely manner so as to preserve the stability of the inverted pendulum. Therefore, Etherware's mechanisms for flexibility and predictability are essential to perform this runtime management operation. Moreover, Etherware's facilities make quite simple the development and deployment of such advanced capabilities.

Figure 1.14 illustrates the application configuration for the runtime controller migration experiment. Initially, the inverted pendulum is controlled over the network by a controller which runs at remote computer. At the computer which is directly connected to the inverted pendulum, a component, called DSPProxy, is running to mediate the interaction between the controller and the DSP in the inverted pendulum system. At a third computer, another component which requests the controller migration process is running. In this experiment, the requester component sends out a request message, which requests the migration of the controller from its current location to the computer which has the direct connection to the inverted pendulum. This request is made around 40 seconds after the system starts. Then Etherware performs the controller migration process. As shown in Figure 1.14, the stability is maintained even while the controller is migrating to a new computer node. Thus, this result demonstrates Etherware's capability of real-time component migration.

1.7 Conclusion

In this paper, we have discussed fundamental characteristics which are common to any networked control systems. The four characteristics, which are large-scaleness, openness, time-criticality, and safety-criticality, are identified in Section 1.2.1. Due to these fundamental characteristics, there is a need for a middleware, which enables us to realize such complex control systems. Indeed our thesis is that such a middleware is important for the future of networked control systems.

As an underlying platform for networked control systems, a middleware is required to satisfy domain requirements which are necessitated by the domain characteristics. There are several fundamental functionalities that have to be provided by any middleware to satisfy both the operational and management domain requirements which are induced by the characteristics of a distributed system. Etherware, a middleware developed at the University of Illinois, is an example of such a middleware for networked control system.

In addition to the functional domain requirements such as operational and management requirements, a middleware framework is typically expected to also support the non-functional domain requirements such as timeliness and reliability. We have presented Etherware's approach to support the timeliness requirements. We have also addressed the issue of faults and approaches toward fault prevention and tolerance, Etherware's LTA-based design principles, and the fault-tolerant component model.

The performance of a networked inverted pendulum control system is provided as a case study of a middleware based networked control system. In the presented system, Etherware is used as a middleware framework which for rapid and evolvable control application development. We have exhibited complex and important run time capabilities such as controller migration and controller update, to highlight the sophisticated capabilities that a middleware such as Etherware can provide. We have thus experimentally demonstrated Etherware's flexibility and temporal predictability properties.

Acknowledgements. This material is based upon work partially supported by AFOSR under Contract No. FA9550-09-0121, ARO under Contract No. W911NF-08-1-0238, and NSF under Contract Nos. NSF ECCS-0701604, CNS-07-21992, and CCR-0325716.

References

1. Anderson, T., Lee: Fault Tolerance: Principles and Practice. Prentice-Hall, Englewood Cliffs (1981)
2. Arnold, K., Gosling, J., Holmes, D.: Java(TM) Programming Language, 4th edn. Prentice Hall PTR, Englewood Cliffs (2005)
3. Bacon, D.F., Cheng, P., Rajan, V.T.: The metronome: A simpler approach to garbage collection in real-time systems. In: Workshop on Java Technologies for Real-Time and Embedded Systems (JTRES), OTM Workshops, pp. 466–478 (2003)
4. Baliga, G.: A Middleware Framework for Networked Control Systems. PhD thesis, Department of Computer Science, University of Illinois at Urbana-Champaign (2005)
5. Bollella, G., Brosgol, B., Gosling, J., Dibble, P., Furr, S., Turnbull, M.: The Real-Time Specification for Java, 1st edn. Addison Wesley Longman, Amsterdam (2000)
6. Buttazzo, G.C.: Hard Real-Time Computing Systems: Predictable Scheduling Algorithms and Applications, 2nd edn. Springer, Heidelberg (2004)
7. Gamma, E., Helm, R., Johnson, R., Vlissides, J.M.: Design Patterns: Elements of Reusable Object-Oriented Software. Addison-Wesley, Reading (1994)
8. Gill, C.D., Levine, D.L., Schmidt, D.C.: The design and performance of a real-time CORBA scheduling service. Real-time Systems, The International Journal of Time-Critical Computing Systems, special issue on Real-time Middleware 20 (1999)
9. Graham, S., Baliga, G., Kumar, P.R.: Issues in the convergence of control with communication and computing: proliferation, architecture, design, services, and middleware. In: 43rd IEEE Conference on Decision and Control, CDC 2004, vol. 2, pp. 1466–1471 (2004)
10. Graham, S., Baliga, G., Kumar, P.R.: Abstractions, architecture, mechanisms, and a middleware for networked control. IEEE Transactions on Automatic Control 54(7), 1490–1503 (2009)
11. Henriksson, R.: Scheduling Garbage Collection in Embedded Systems. PhD thesis, Lund University (1998)
12. IT Convergence Laboratory, University of Illinois. Testbed of a traffic control system, `http://decision.csl.illinois.edu/~prkumar/testbed/videoclips.html`
13. Kim, K.-D., Kumar, P.R.: Architecture and mechanism design for real-time and fault-tolerant Etherware for networked control. In: Proceedings of the 17th IFAC World Congress (July 2008)
14. Lindholm, T., Yellin, F.: Java(TM) Virtual Machine Specification. Prentice Hall PTR, Englewood Cliffs (1999)
15. Liu, C.L., Layland, J.W.: Scheduling algorithms for multiprogramming in a hard-real-time environment. Journal of the ACM 20(1), 46–61 (1973)
16. Lorczak, P.R., Caglayan, A.K., Eckhardt, D.E.: A theoretical investigation of generalized voters for redundant systems. In: IEEE Symposium on Fault-Tolerant Computing (1989)
17. OMG, CORBA, `http://www.corba.org/`
18. Puder, A., Römer, K., Pilhofer, F.: Distributed Systems Architecture: A Middleware Approach. Morgan Kaufmann, San Francisco (2005)
19. Robinson, C.L., Baliga, G., Kumar, R.R.: Design patterns for robust and evolvable networked control. In: Proceedings of the 3rd Annual Conference on Systems Engineering Research, New Jersey, USA (March 2005)
20. Selic, B.: Fault tolerance techniques for distributed systems, `http://www.ibm.com/developerworks/rational/library/114.html`

21. Sha, L., Rajkumar, R., Lehoczky, J.P.: Priority inheritance protocols: an approach to real-time synchronization. IEEE Transactions on Computers 39(9), 1175–1185 (1990)
22. Sha, L., Abdelzaher, T., Arzen, K.-E., Cervin, A., Baker, T., Burns, A., Buttazzo, G., Caccamo, M., Lehoczky, J., Mok, A.K.: Real time scheduling theory: A historical perspective. Real-Time Systems 28(2-3), 101–155 (2004)
23. Stankovic, J.A.: Misconceptions about real-time computing: A serious problem for next-generation systems. Computer 21(10), 10–19 (1988)
24. Sun Microsystems, Sun Java Real-Tim Systems, `http://www.sun.com/`
25. Tanenbaum, A.S.: Modern Operating Systems, 3rd edn. Prentice Hall, Englewood Cliffs (2007)
26. Torres-pomales, W.: Software fault tolerance: A tutorial. Technical report, National Aeronautics and Space Administration (2000)
27. Vinoski, S.: An overview of middleware. In: Llamosí, A., Strohmeier, A. (eds.) Ada-Europe 2004. LNCS, vol. 3063, pp. 35–51. Springer, Heidelberg (2004)
28. Voas, J.M., McGraw, G.: Software Fault Injection: Inoculating Programs Against Errors. John Wiley & Sons, Chichester (1998)
29. World Wide Web Consortium. XML, `http://www.w3.org/XML/`

Chapter 2
Wireless Networking for Control: Technologies and Models

Mikael Johansson and Riku Jäntti

Abstract. This chapter discusses technologies and models for low power wireless industrial communication. The aim of the text is to narrow the gap between the models used in the theoretical control literature with models that arise when tools from communication theory are used to model emerging standards for industrial wireless. The chapter provides a tutorial overview covering basic concepts and models for wireless propagation, medium access control, multi-hop networking, routing and transport protocols. Throughout, an effort is made to describe both key technologies and associated models of control-relevant characteristics such as latency and loss. Some existing and emerging specifications and standards, including Zigbee, WirelessHART and ISA100, are described in some detail, and links are made between the developed models and useful network abstractions for control design.

2.1 Introduction

Since the first application of wireless in industrial control almost a 100 years ago [41], the number of actual deployments has remained small. For a long time, the market has been limited to a specific target applications (*e.g.* wireless remote controls) engineered using customized technologies and sometimes even operating on licensed spectrum. There is a growing consensus that this trend is now about to change: the enormous success of short-range wireless in home and office applications has raised consumer confidence in wireless technologies; the emergence of standardized, low cost, low-power radios has made industrial wireless economically

Mikael Johansson
School of Electrical Engineering, KTH, SE-10044 Stockholm, Sweden
e-mail: `mikaelj@ee.kth.se`

Riku Jäntti
Department of Communications and Networking
Comnet Helsinki University of Technology TKK
e-mail: `riku.jantti@tkk.fi`

A. Bemporad, M. Heemels, and M. Johansson: Networked Control Systems, LNCIS 406, pp. 31–74.
springerlink.com

attractive compared with cabled sensors [28]. Intense efforts on wireless sensor networks [33] and networked control [4] indicate that a large class of industrial processes could be reliably controlled despite deficiencies of the wireless medium. New standards for industrial wireless communication, such as WirelessHART and ISA100 have recently been approved, and stardards-compliant technology is hitting the market. All together, this raises expectations of a wide deployment of industrial wireless [38].

The development of controllers operating over wireless networks is inherently a problem of co-design, and good system designs require insight and understanding of the traditionally separate disciplines of control and communications. This chapter reviews technologies and models for industrial wireless networking in an attempt to narrow the gap between models used in the theoretical control literature and the models that arise when emerging standards for industrial wireless networks are modelled using tools from communication theory. Covering a wide area of topics, from wireless propagation and medium access control to routing and transport protocols for multi-hop wireless networking, the chapter is admittedly broad rather than deep. Nevertheless, a range of useful models for control-relevant quantities such as latency and loss in industrial wireless networks are given, and several existing and emerging specifications and standards, including Zigbee, WirelessHART and ISA100, are described in some detail. Pointers to books and selected key publications are given throughout to guide the reader to good entry-points for more in-depth studies.

The chapter is organized as follows. We begin by a review of wireless propagation and models for the behavior of a single wireless link. We then describe various medium access control techniques for sharing the wireless spectrum among multiple wireless links. Next, we discuss multi-hop wireless networking, routing and protocols for end-to-end communication. With this basic understanding of wireless communication at hand, we describe several specifications and standards for low-power wireless networking. Finally, we establish links between the developed models and useful network abstractions for control design.

2.2 Understanding the Single Link

Aim: to describe the concepts of wireless propagation and outage, leading to reasonable models of the raw packet-error rates in the 2.4 GHz ISM-band.

2.2.1 Wireless Propagation and Outage

Radio propagation refers to the way that radio waves behave when they are transmitted and propagate from one point to another. Radio waves are affected by the same phenomena as light waves, including reflection, diffraction, absorption and scattering. The impact of the wireless channel on the transmitted signal is customarily divided into *large-scale effects* and *small-scale effects*. Large scale effects include signal attenuation due to the propagation distance and shadowing caused by large stationary obstacles in the radio-path. Small-scale effects are caused by signal reflections and movement of the receiver, transmitter or small objects in the radio path. Reflections cause multiple copies of the signal to arrive at the receiver with different attenuation and propagation delay. The superposition of these signals at the receiver causes time dispersion of the original signal. In the frequency domain, this gives raise to frequency-selective fading where different frequency components of the signal experience different attenuation and phase shift. Movement of the transmitter, receiver or small objects in the environment cause dispersion of power in frequency. This is effect is called Doppler spread and gives rise to time-selective fading, *i.e.* it makes the channel response time-varying. In certain conditions, the signal may be absorbed or multi-path components may interfere each other destructively such that the signal vanishes completely at the receiver. This phenomenon is called *erasure fading*.

Small-Scale Effects

Because of multipath reflections, the impulse response of a wireless channel looks likes a series of pulses. It is customary to characterize the delay profile of the channel by observing the mean energy $p(\tau)$ for each delay τ. The maximum delay spread is the total time interval during which reflections with significant energy arrive. A common parameter for quantifying the multipath behavior of the channel is the *root-mean-square (RMS) delay spread* σ_τ. The corresponding quantity in the frequency domain is the *coherence bandwidth* of the channel, the frequency range over which two transmitted sinusoidal signals are likely to see the phase shift and attenuation. The coherence bandwidth $B_{C.50}$ where the correlation of the channel exceeds 0.5 is approximately

$$B_{C.50} \approx \frac{1}{5\sigma_\tau}$$

If the bandwidth of the transmitted signal B is less than $B_{C.50}$, the fading is called frequency-non-selective or *flat fading* and can be modelled with impulse response having only one tap (one radio path); otherwise the channel fading is called *frequency-selective*. In frequency selective channel an equalizer (adaptive tapped delay line filter) is needed at the receivers in order to keep the bit error probability at tolerable levels. Such a filter is not needed in flat faded channel.

Measurements on the 1.3 GHz frequency band suggests that typical values for RMS delay spread in factory environment lie between 100 ns and 200 ns, but values as high as 300 ns are possible [48]. These values are one order of magnitude higher than what has been reported for office environments. Measurements on the 2.4 GHz band suggest that the mean RMS delay between 16 ns (mine in granite) to 85 ns (transformer station) translating to coherence bandwidth between 1250 kHz to 227 kHz [59]. The results also indicate that the RMS delay varies a lot between links in different positions within the same plant. The RMS delay spread observed in a site tends to follow a (truncated) normal distribution. In the two examples mentioned above, the standard deviation was 8 ns and 89 ns, respectively. The mean values tend to be smaller than in office environment where measurement

Table 2.1 Comparative table of sample mean and standard deviation of the RMS delay spread and the coherence bandwidth related to the mean RMS delay spread

Site	Mean RMS delay spread (ns)	Standard deviation RMS delay spread (ns)	Coherence bandwidth $B_{C,50}$ (MHz)
Petrochemical plant	38	9	5.26
Transformer station	85	89	2.27
Manufacturing plant	44	24	4.55
Carpark amongst multi-story buildings	74	107	2.63
Mine in granite	16	8	12.50
Coal mine	23	9	8.70

reports indicate RMS delay spreads between 45 ns and 420 ns [32]. Table 2.1 summarizes the measurement results reported in [26].

To understand how small-scale fading impacts radio performance, consider the IEEE 802.15.4 radio standard, in which receivers do not have equalizers. On the 2.4 GHz band 802.15.4-radios transmit 2 Mcps (chips per second) which translates to 2 MHz half-power bandwidth B. To assure that the signal sees frequency-non-selective channel, it is customary to require that the coherence bandwidth is at least twice the signal bandwidth. By setting $B_{C,50} = 2 \times 2$ MHz, we get $\sigma_\tau = 50$ ns. Based on this analysis, a RMS delay spread exceeding 50 ns would cause concern. Simulation studies [16] suggest that IEEE 802.15.4 can achieve packet error rates below $5 \cdot 10^{-3}$ as long as the RMS delay does not exceed 400 ns. However, this comes with the cost of reduced range as up to 10dB higher signal-to-noise ratio is needed to compensate the effect of inter-symbol interference. The simulations also suggest that half-rate at 915 MHz would tolerate RMS delay spreads up to 800 ns. Comparing the tolerance of the IEEE 802.15.4 receiver to the values measured in industrial settings, we can conclude that in most cases the radio can operate reliably but in some environments the range could be limited and it is even possible to find environments where reliable operation is not possible. The low transmit power of IEEE 802.15.4 also helps in reducing the number of multipath reflected components.

In the wireless channel, many echoed copies of the transmitted signal appear with almost equal delay. The multipath model groups these delay paths into discrete set of resolvable paths, called taps, by summing all the signals received with almost the same delay. By applying the central limit theorem, we can model the signal as complex Gaussian random variable (it is customary to model the base band signal as complex random variables - the modulated signal is then obtained by multiplying the signal with phasor $\exp(j2f_c t)$ rotating with the carrier frequency f_c and taking the real part of the resulting complex signal). The amplitude of the signal passing through a single radio path follows either Rician or Rayleigh distribution depending whether there is significant line-of-sight (LOS) component in the signal or not. That is, whether the complex Gaussian variables have non-zero mean or not. In both cases, the phase of the received signal is uniformly distributed random variable. In case of Rayleigh fading, the signal power follows chi-square distribution with two degrees of freedom which happens to coincide with the exponential distribution. In the Rician case, the received signal power follows non-central chi-squared distribution with two degrees of freedom. Yet another distribution that is often utilized to characterize the line-of-sight is the Nagagami-m distribution. For $m = 1$, the model coincides with Rayleigh distribution. The larger m is the less variations there are in the channel. Measurement results in industrial environment suggest that the firs tap in a line of sight channel, typically experience Nagagami-m distribution with m varying between 3 and 67 while the reflected paths follow Rayleigh distribution ($m = 1$) [63]. In case that m is an integer, the cumulative probability density of the Nakagami-m faded signal strength $\tilde{P}_{rx}$ having mean $\bar{P}_{rx}$ follows Erlang-m distribution and can be written as

$$\Pr\left\{\tilde{P}_{rx} \leq P\right\} = 1 - \sum_{k=0}^{m-1} \left(\frac{k}{\bar{P}_{rx}}\right)^k \frac{1}{k!} \exp\left(-\frac{P}{\bar{P}_{rx}}\right) \stackrel{def}{=} F_P(P)$$

Doppler spread and *coherence time* are parameters that describe the time dispersive nature of the wireless channel in the small-scale region. When a node or reflectors in its environment is moving, the velocity of the moving node and/or environment causes a shift in the frequency of the signal transmitted along each signal path. This phenomenon is known as the Doppler shift. Although most wireless sensor networks tends to be immobile, the industrial environment often have moving reflectors such as overhead cranes, forklifts *etc.* The Doppler spread is the maximum Doppler shift and is given by

$$B_D \approx \frac{v}{c} f_c$$

where v is the speed of the node, c m/s denotes the speed of light and f_c is the carrier frequency. Signals travelling along different paths can have different Doppler shifts, corresponding to different rates of change in phase. The difference in Doppler shifts between different signal components contributing to a single fading channel tap is known as the Doppler spread. The coherence time of the channel during which the channel stays essentially constant is approximately

$$T_c \approx \frac{0.423}{B_D}$$

In office environments, the Doppler spread varies from 2Hz to 20Hz which translates to coherence time from 20 to 200ms [32]. These values are longer than the typical packet lengths in the wireless sensor networks, implying that the channel stays essentially constant during the transmission of single packet.

Large-Scale Effects

Large-scale effects can be understood by simply observing how the wireless channel affects the received power. *Frii's equation* relates the transmit power P_{tx} to the received power P_{rx} and is given in logarithmic form as

$$P_{\mathrm{rx}}^{dBm} = P_{\mathrm{tx}}^{dBm} + G_{\mathrm{tx}}^{dBi} - L_{p}^{dB} + G_{\mathrm{rx}}^{dBi}$$

where G_{tx}^{dBi} and G_{rx}^{dBi} are antenna gains at the transmitter and receiver, respectively, and L_{p}^{dB} is the *path loss*. The unit of the power is typically dBm, while the antenna gains are stated in dBi where the 'i' refers to isotropic antenna, *i.e.* an antenna that that radiates equally in all directions. One of the most simplest path loss models is the *single-slope empirical propagation model* given by

$$L_{p}^{dB} = L_{0}^{dBm} + n10\log_{10}(d)$$

The parameter $d > 1$ m denotes the distance between transmitter and receiver (in meters), n is the path loss exponent and L_{0}^{dB} is the path loss at one meter distance. The experienced path loss depends on the environment and typically the model parameters need to be tuned to match the environment: the parameter L_{0}^{dB} depends on the carrier frequency used; reflection from smooth surfaces can cause the direct path and reflected path to interfere such that the path loss exponent becomes equal to 4. The path loss exponent can be even less than 2 in some cases due to a wave guiding effect of the environment. Measurements conducted in factory environment suggest that the path loss varies from 1.49 measured for line of sight paths in light clutter (small number of echoes) to 2.81 in case of obstructed path and heavy clutter (large number of echoes) [48]. Slightly larger path loss exponents have been reported in [63] for metal processing facilities where reflections from metallic bodies are likely to happen. The reported path loss exponents range from 2.56 to 4.24.

The following two slope model

$$L_{p}^{dB} = \begin{cases} 40.2 + 20\log_{10}(d) & \text{if } d \leq 8 \text{ m} \\ 58.5 + 33\log_{10}(d/8) & \text{if } d \geq 8 \text{ m} \end{cases}$$

has been utilized by IEEE in the studies for the co-existence of various radios on the 2.4 GHz Industry, Science, Medical frequency band [21]. The model assumes line of sight path in the close proximity of the transmitter ($d < 8$ m) and partially obstructed path for longer distances.

The impact of walls can be taken into account by increasing the pathloss. The attenuation cased by walls, floors and other objects which the radio wave can penetrate, depend on the dielectric properties of the materials of the object and frequency of the radio wave. The higher the frequency, the higher the pathloss. A majority of the empirical models available in literature consider building materials used in residential and office buildings, see *e.g.* [22], Obstacles shadowing the radio path cause diffraction losses to the signal, the impact of which is typically modelled as lognormal random variable $\tilde{S}^{dB}$ called shadowing that appears additively in Frii's equation. Shadow fading has zero mean and its standard deviation varies between 3 to 20 dB based on the environment. Values 4 - 9 dB have been reported for factory environments [63]. In a given link, the shadow fading term stays essentially constant, S^{dB}, unless there are large changes in the node position with respect to the shadowing screens.

Outage

An important metric of the radio channel quality is the *Signal-to-Noise Ratio* (SNR). The instantaneous received SNR at time instant t can be written as

$$\tilde{\gamma}(t) = \frac{\tilde{g}(t)P_{\text{tx}}}{N_0 B}$$

Here N_0 is the noise power density given in terms of Watts per Hertz (W/Hz) and B is the bandwidth of the signal. The random variable $\tilde{g}(t)$ is called the *link gain*. It is the ratio between received power and transmitted power. The link gain models both the short scale effects and large-scale effects. Assuming that the channel is *wide sense stationary* (*i.e.* the movement of the transmitters or receivers is slow compared to the time scale of interest), then the large scale effects are captured by the mean value of the channel gain

$$\bar{g}(t) = \mathbb{E}\{\gamma(t)\} = 10^{\left(G_{\text{tx}}^{dBi} - L_p^{dB} + S^{dB} + G_{\text{rx}}^{dBi}\right)/10}$$

while the short scale effects explain the distribution of $\tilde{g}(t)$, $F_g(g) = F_P(gP_{tx})$ and its correlation properties. According to Clarke's model, the correlation can be expressed as

$$\rho(\tau) = \mathbb{E}\left\{\left(\tilde{g}(t) - \bar{g}\right)\left(\tilde{g}(t+\tau) - \bar{g}\right)\right\} = J_0\left(2\pi B_D\right)$$

where J_0 denotes the Bessel function of the first kind. The distribution of the received SNR is then $F_\gamma(\gamma) = F_g\left(\frac{N_0 B}{P_{tx}}\right)$.

Assuming that the coherence time of the channel is large compared to the time it takes for the transmitter to transmit single information symbol (one symbol can consist of several bits depending on the modulation method used), the bit error probability can be expressed as monotonically decreasing function of the instantaneous SNR $p_{be}(\gamma)$. If the coherence time of the channel is large compared to the packet length, then SNR stays essentially constant during the transmission of a single packet of b bytes. Hence, the packet error probability $p_{pe}(\gamma)$ is given by

$$p_{pe}(\gamma) = 1 - (1 - p_{be}(\gamma))^{8b}$$

By fixing $p_{pe}(\gamma^{th}) = p_{pe}^{max}$, we can solve γ^{th} which guarantees that that as long as $\gamma \geq \gamma^{th}$, the packet error rate does not exceed p_{pe}^{max}. It is customary to express this threshold in terms of received power P_{rx} required to archive γ^{th}. This power value $P_s = \gamma^{th} N_0 B$ is called *receiver sensitivity*.

Example 2.1 (IEEE 802.15.4 O-QPSK PHY on 2.4 GHz band). According to the IEEE 802.15.4 standard [20], the bit error probability conditioned on SNR (before despreading) $\gamma(t) = \gamma$ is

$$p_{be}(\gamma) = \frac{1}{30}\sum_{j=2}^{16}\binom{16}{j}(-1)^j \exp\left(-20\left(1 - \frac{1}{j}\right)\gamma\right)$$

The resulting packet error probability PER is plotted in Figure 2.1 for 30 byte and 127 byte packets. It can be seen from the Figure, that the packet error rate is very high untill SNR exceeds certain threshold after which the PER drops quickly. Increasing physical layer packet size sifts the PER curve to the right. That is, for given SNR, the longer the packet, the higher the packet error probability - unless some measures are taken to protect the packet from bit errors. It can bee seen from Figure 2.1, that about 14 dB SNR is enough to obtain PER less than 0.001.

Due to the fact that the packet error probability has waterfall type of shape where after some threshold the packet drop probability falls down quickly, a very simple model for the effect of channel induced packet drops can be utilized. The simple *outage* model assumes that a packet is always successfully transmitted if the received power P_{rx} is above the receiver sensitivity; otherwise the packet is lost and we say that the link is in *outage*. Hence, in the average packet delivery ratio in the fading channel can be approximated as

$$p_{out} \approx F_P(P_s) \tag{2.1}$$

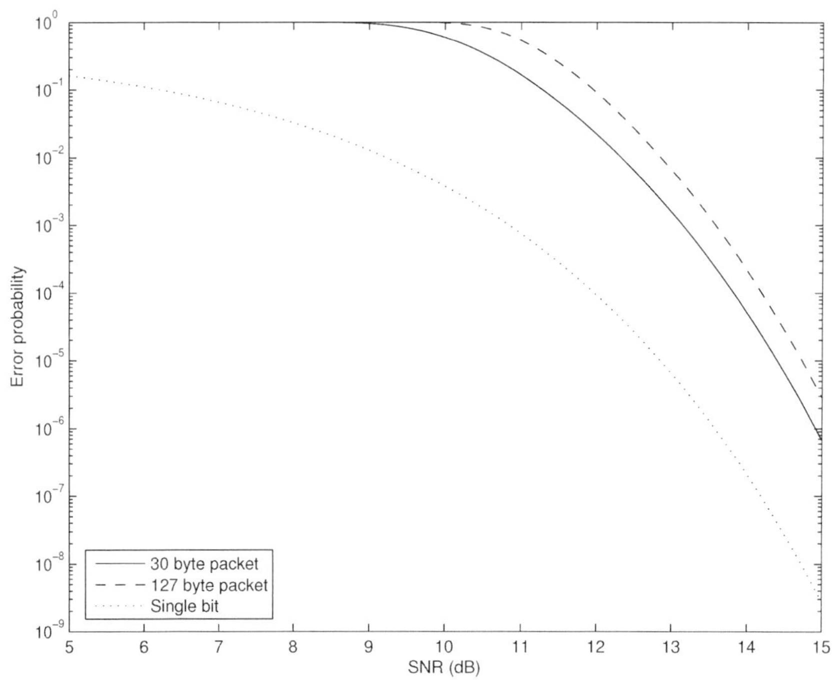

Fig. 2.1 Packet error probability as a function of SNR after despreading for IEEE 802.15.4 O-QPSK PHY operating on 915 MHz band

2.2.2 Markov Models for the Wireless Channel

In wireless channels, the errors typically come in bursts. The simple outage model described above is not enough to characterize the time correlation of the errors. Markov models have been utilized to model the evolution of the channel SNR in time. A Finite State Markov Chain (FSMC) model of the fading channel partitions the received SNR into finite amount of intervals. Each state corresponds to certain SNR interval. The transition probabilities between adjacent states are then selected to match the one sided level crossing rates of the fading process and the state probabilities simply refer to the probability of finding the received SNR in the corresponding interval [67]. There are, however, many ways of doing the partitioning and the choice will have an impact on the simulation results [23].

From application point of view, we are typically more interested in the correlation between consecutive packet drops rather than the evolution of the instantaneous SNR. Hidden Markov Models (HMM) have been suggested for this end. In regular Markov chains, the state (*e.g.* SNR) is directly visible to the observer, and therefore the state transition probabilities are the only parameters. In a hidden Markov model, the state is not directly visible, but output dependent on the state (packet received correctly or not) is visible. Each state has a probability distribution over the possible output values (packet error probability). The simplest HMM channel model is the Gilbert-Elliot (G-E) model [15, 12] . In the G-E model, the underlying discrete time Markov chain has only two states labeled as 0 or 'bad' and 1 or 'good'. In bad state, the packet is lost with probability $p_{pe,0}$ and in good state it is lost with probability $p_{pe,1}$. Let p_{01} and p_{10} denote the state transition probabilities between the states 0 and 1 and vice versa. Furthermore, let P_0 and P_1 denote the probabilities of finding the channel in bad and good state, respectively. In steady state, we have

$$
\begin{aligned}
P_0 &= \frac{1}{1+\frac{p_{01}}{p_{10}}} \\
P_1 &= \frac{p_{01}}{p_{10}} \frac{1}{1+\frac{p_{01}}{p_{10}}}
\end{aligned} \tag{2.2}
$$

The expected packet error rate of the channel is then $p_{pe} = P_0 p_{be,0} + P_1 p_{pe,1}$. Figure 2.2 illustrates the state transition diagram of the G-E model.

The error probabilities in the state and the transition probabilities can be fitted to measurement values. The state holding times of the Markov chain follow Geometric distribution with mean $1 - p_{ii}^{-1}, i = 0, 1$. The simplest, but somewhat artificial model is to have $p_{pe,0} = 0$ and $p_{pe,1} = 1$. In such case, the state holding time for state 1 is equivalent to the length of the error burst and the steady state probability P_1 gives directly the packet error probability. More realistic parameters can be obtained by fitting the model to the measurement data. The two-state G-E model can be viewed as first order approximation to the true channel in which only the mean length of the error burst is fitted to the data. If also higher moments need to be matched, then more states should be introduced. The G-E model is very popular among the researchers in

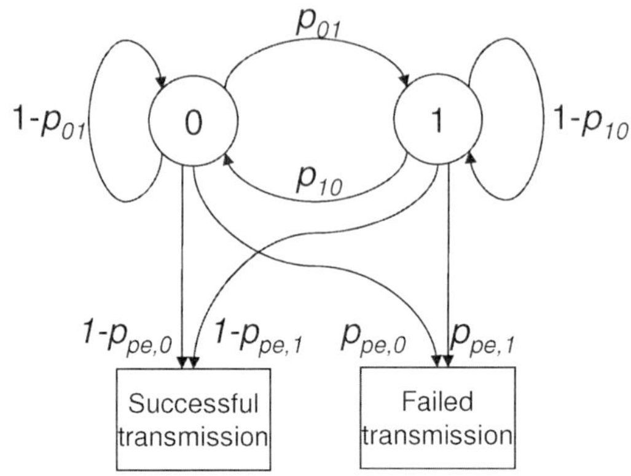

Fig. 2.2 State transition diagram of the G-E HMM model

the performance analysis of higher layer protocols (transport and application layer) due to its simplicity, see *e.g.* [23] and the references therein.

2.2.3 The ISM Band, Co-existence and Interference

Wireless devices are heavily regulated throughout the world. Most countries have allocated parts of the radio spectrum for open "license-free" use, while other parts of the spectrum can only be used with permission or "license". The license-free areas of the spectrum are also known as industrial, scientific, and medical (ISM) bands. The 2.4 GHz ISM band is available in most part of the world, while ISM bands in lower frequencies vary between continents (*e.g.* Europe provides license-free access in the 433 and 869 MHz bands, while the Americas have opened up the 915MHz band). Each frequency band is split into channels: higher frequency bands are wider, which allows for wider channels and higher data rates, but lower frequencies give greater operating distance at a given transmit power, and a less crowded spectrum ensures more reliable operation. Radio communication systems that use the ISM bands are typically free for use by anyone, but typically operate under a "license exempt" regime that sets limits on power, spectrum spreading techniques, or duty cycles to limit interference and make sure that no single device steals all bandwidth.

In the 2.4 GHz ISM band, at least three wireless technologies are likely to co-exist in current industrial deployments: wireless local area networks (WLAN, IEEE 802.11), Bluetooth (IEEE 802.15.1) and low-rate wireless personal area networks (LR-WPAN, IEEE 802.15.4). Figure illustrates the associated wireless landscape. The 802.15.4-2006 standard defines sixteen channels, numbered 11 to 26, in the 2.4-GHz band. Each channel has a bandwidth of 2MHz and channels are separated by 5MHz. The IEEE 802.11b and 802.11g standards operate in fourteen channels available in the 2.4-GHz band, numbered 1 to 14, each with a bandwidth of 22 MHz and a channel separation of 5 MHz. The most common WLAN configuration is to enable the mutually orthogonal channels 1, 6, and 11 as in Figure 2.3. Finally, the

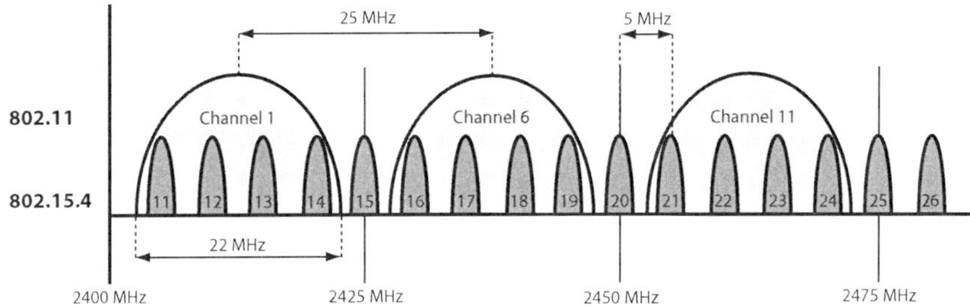

Fig. 2.3 ISM band channels for 802.11 and 802.15.4. Only a handful of LR-WPAN channels do not overlap with WiFi

IEEE 802.15.1 Bluetooth standard (not shown in the Figure) operates in 79 channels available worldwide in the 2.4-GHz band. Numbered 0 to 78, each channel has a bandwidth of one MHz and a channel separation of one MHz.

These three technologies operate using different output powers and offer communication over different ranges: WLAN output powers are typically around 20 dBm and operate within a 100m range; LR-WPAN decvices use a transmit power of 0dBm and operate within a 50 meter range; finally, Bluetooth output powers are generally less than 4 dBm and ranges below 10m for the more commonly used class 2 devices such as wireless headsets and keyboards. The less common class 1 devices can operate at up to 20 dBm and typically within a 100m range. The higher transmit powers used by WLAN and Bluetooth indicate that it could be difficult for 802.15.4-devices to coexist with these technologies. This has indeed been verified in practice *e.g.* in [56, 45]. Both report that a saturated WLAN practically kills the 802.15.4 connection unless the operating frequencies are off-set with at least 7-10 MHz. However, the IEEE 802.15.4 radio can achieve over 90% packet delivery ratio in the presence of 802.11b/g WLAN interference if the packet delivery ratio of the WLAN is low (less than 100 packets/s) or the signal-to-interference ratio at the 802.15.4 receiver exceeds 15 dB [31]. The upcoming IEEE 11n standard WLAN is expected to be an even more severe source of interference due to multipoint transmission and wider bandwidth. The influence of Bluetooth on 802.15.4 is more limited (packet error rates around 10% were reported in [56]) mainly due to the fact that it uses an agile frequency hopping scheme and relatively narrow channels. In the converse direction, 802.15.4 has virtually no impact on 802.11b performance unless the channels have the same center frequency and the 802.15.4 packets are very long [56].

If we are interested in modelling the impact of WLAN traffic on LR-WPAN communications, a reasonable first-order model (considering the power imbalance between the two technologies) is to assume that the WPAN packet is lost if there is a concurrent WLAN transmission on an overlapping channel. Simple and reasonably accurate semi-Markov models for WLAN interference, modelling the frequency and duration of transmit and idle periods, respectively can be found in [14, 61].

2.2.4 *Means for Increasing Reliability*

2.2.4.1 Error Control

In most of the applications it is important to detect whether the received packet contain errors or not. Hence, it is customary to use error detection coding. The most common error detection coding scheme is the Cyclic Redundancy Check (CRC) CRCs are popular because they are simple to implement in digital hardware using shift registers, are easy to analyze mathematically, and are particularly good at detecting common errors caused by noise in transmission channels. The IEEE 802.15.4 standard specifies 16 bit CRC to be utilized. The probability of an undetected error is bounded above by 2^{-16}. In a plant having 50 nodes each transmitting one packet per minute, the average number of undetected errors is approximately one per day. If 32 bit CRC is used, errors occur once per 200 years.

While CRC codes use redundant bits for detecting errors, Forward Error Correction (FEC) codes use the extra bits to correct some of the bit errors at the receiver hence increasing the robustness of the communications. The performance enhancement achievable by using FEC can be expressed in terms of *coding gain*. The coding gain expresses the difference in the required Signal-to-Noise with and without coding for a given bit error rate. The coding gain can be utilized to increase range, decrease transmitter power consumption or to increase robustness against noise and interference, see *e.g.* [65]. In the outage model discussed earlier, the coding gain simply translates to lower SNR threshold value γ^{th}.

Automatic Repeat reQuest (ARQ) algorithms are utilized to improve the transmission reliability at the cost of increased packet delay. The simplest and also most common ARQ method used in WSNs is Stop-and-Wait ARQ (SW-ARQ). If the receiver receives the packet successfully, it will transmit an acknowledge (ACK) packet. In case it detects an error, it can issue negative acknowledgment (NACK). It is also possible, that the packet gets entirely lost in the wireless channel and the receiver is not able to detect the transmission at all. In SW-ARQ, once a transmitter generates a packet is sets a timer. If the transmitter receivers ACK before its timer expires, it resets the timer and proceeds to the next packet. If it receives NACK, it resets the timer and retransmit the packet. If neither ACK or NACK is received before the timer expires, the transmitter deduces that the packet must have got lost and will try to retransmit the packet. Typically, the number of retransmission attempts is limited; after which the packet is considered lost.

The packet transmission using SW-ARQ can be modeled by unsing the Markov chain model and adding an absorbing state 'packet received'. In case of two state G-E model, we the number of transmission attempts $\tilde{a}$ needed to successfully transmit a packet follows discrete distribution

$$\Pr\{\tilde{a} \leq a\} = 1 - \tau \mathbf{T}^a \mathbf{1}$$

Here, $\tau = (P_0, P_1)$ denotes the probability of starting the process from state 0 and 1, respectively, $\mathbf{T}$ is the transition matrix

$$\mathbf{T} = \begin{bmatrix} (1-p_{01})\, p_{pe.0} & p_{01} p_{pe.0} \\ p_{10} p_{pe.1} & (1-p_{10})\, p_{pe.1} \end{bmatrix}$$

and $\mathbf{1} = (1,1)^T$ is a column vector of ones. Given that the maximum number of retransmission attempts in the system is a, we can can obtain the residual packet error probability p_{pe} after a transmission attempts assuming ideal error detection:

$$p_{pe}(a) = \Pr\{\tilde{a} > a\} = \tau \mathbf{T}^a \mathbf{1}$$

Example 2.2 (Performance of ARQ). The packet error characteristics for IEEE 802.15.4 [20] standard radio operating on the 2.4 GHz band were measured in an industry assembly hall. Measurement results obtained from a 30m link between ground level and overhead crane suggested that the mean packet error probability is $p_{pe} = 0.133$ for 119 byte packets send by the rate of 10 packets/second. A G-E model was fitted to the packet delivery traces suggesting $p_{pe.0} = 0.02$ and $p_{pe.1} = 0.74$.

The state holding times for good and bad state were 35.54 s and 6.58 s, respectively. Figure 2.4 shows the packet error probability as a function of the number of allowed transmission attempts per packet for both the G-E model and a simple model that assumes that packet errors are uncorrelated and error occurs with probability p_{pe}. It can be seen form Figure 2.4 that the bursty nature of the channel makes reliable communication much more difficult than what the simple geometric model assuming independent errors would suggest.

FEC can be combined with ARQ to obtain Hybrid ARQ (HARQ). HARQ schemes aim to exploit the advantages of both FEC and ARQ by incrementally increasing the error resiliency of the packet through retransmissions. The HARQ schemes are typically divided into two types. Type I-HARQ adds FEC on top of ARQ. The transmission starts with a low number of code bits or possibly with CRC only. Once the packet fails, the retransmitted packet is encoded with stronger code. In type II-HARQ, the first transmission uses an uncoded packet with CRC. Once the packet fails, the transmitter only sends code bits (incremental redundancy). The code bits are typically selected such that the packet is self-decodable and together with the original packet forms 1/2 rate code. The receiver can first try to check if the retransmitted packet was successful - if not, it then decodes the codeword composed by combining the original packet and the retransmitted one. Measurement resutls with MicaZ nodes employing IEEE 802.15.4 radio, indicate that type-II HARQ is the best to reduce latency and energy consumption [65].

2.2.4.2 Diversity Techniques

Due to frequency dependent multi-path fading certain frequency channels can be severely faded. In wireless sensor networks, the channel is typically slowly changing which indicates that once the channel is deeply faded, the situation is likely to remain the same for long time causing many consecutive packets to get dropped. The fading seen by two frequency channels that are spaced more than the

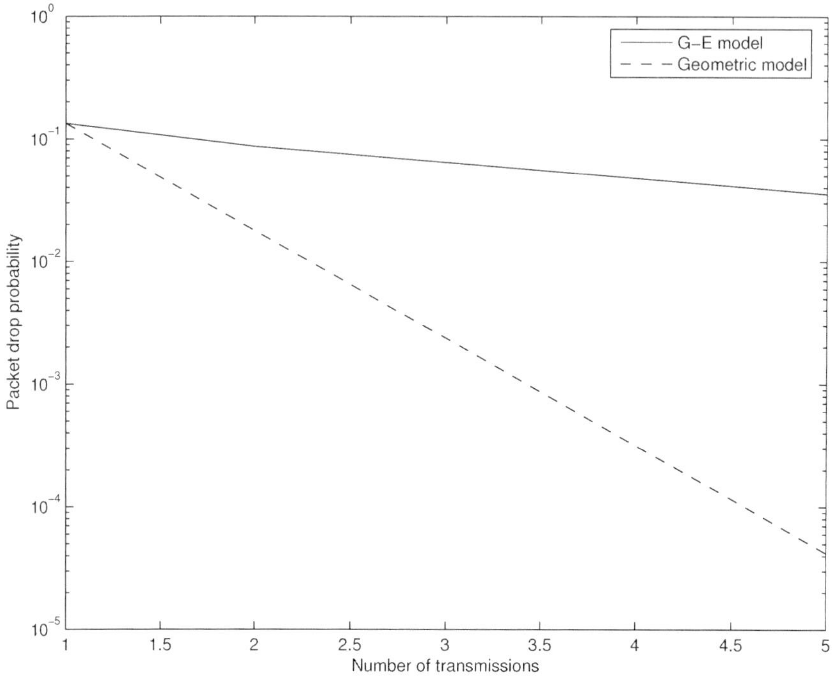

Fig. 2.4 Packet error probability as a function of allowed number of transmission attempts for channel with error burst (G-E model) and independent errors (Geometric).

coherence band from each other see independent fading. This fact can be exploited to decorrelate the transmission of the consecutive packets by utilizing different channel for each transmission attempt. *Frequency hopping* FH is a diversity technique, in which the synchronized transceivers changes channels according to some predefined *hopping pattern*. It is still possible that some channels are deeply faded or contain a lot of interference. In such situations, the system should avoid bad channels. To avoid these problems, adaptive FH techniques (AFH) have been developed. The basic idea implemented by these schemes is to remove from the hopping sequence frequencies experiencing bad channel conditions (for instance high packet error rate), consequently reducing the number of utilized frequency bands. To this purpose the resulting algorithms make use of a channel classification procedure that is basically required to identify channels unsuitable for packet transmissions. This procedure however introduces delays. Moreover, channels removed from the hopping sequence need to be periodically checked to verify if they still do not meet the specified performance requirements. An alternative to channel classification is to use probabilistic (pseudo random) frequency hopping where the probability of utilizing certain channel is constantly updated based on the estimated packet error rate for the channel. The use of FH in wireless sensor networks is discussed *e.g.* in [60].

In some applications, the number of available channels for sensor network operation can be very limited. For instance, in case of heavy WLAN traffic, there is only one channel available for IEEE 802.15.4 which is completely free from the WLAN. Spatial diversity can be exploited by utilizing multiple antennas on the sensor motes. If the antennas are at least half a wavelength ($\frac{\lambda}{2}$) apart from each other, and there is enough scattering in the environment, then each antenna would see independent fading. Just like in case of FH, we can decorrelate the transmission of the consecutive packets by utilizing different antenna pairs for each transmission attempt. In low mobility scenarios, simple electro-mechanical radio frequency switch can be utilized to select between multiple antennas. *Antenna switching* in WSN setting has been studied in [53] where it was demonstrated by laboratory scale tests.

If the receiver has multiple RF units, it is possible to utilize *receiver selection diversity*. In the simple outage model the packet error rate of single transmission was $p_{out} = F_P(P_s)$ with M-fold selection diversity (receiver has M radios), the outage becomes $p_{out}(M) = F_P^M(P_s)$ that is, the outage probability decreases exponentially as the number of receiver increase.

Example 2.3 (Spatial diversity). Example 2.2 illustrated G-E parameters for one particular 30 m link in an industry assembly hall. The measurements were done using two transmit antennas and four receive antennas. The antennas were placed half an wavelength from each other. The resulting 8 single input - single output spatial links were tested together. The resulting packet error rates (PER) for the links (labelled L1,L2,....L8) are plotted in Figure 2.5. The figure also shows result for random antenna switching (RS) that selects the transmit and receive antenna randomly for each transmission. The resulting PER is the average of individual link PERs. Ideal antenna switching (IS) selects the transmit - receiver antenna pair that minimizes PER. If there is separate receiver for each antenna at the sink node, we can utilize receiver diversity (D1 and D2 corresponding to the choice of transmitter antenna). In case receiver selection diversity is utilized, a packet is only lost if none of the four receivers were able to capture the packet. The receiver diversity scheme can be combined with random transmit antenna switching (RSD) and ideal transmit antenna selection (ISD).

The results indicate that in certain radio environments, the channel quality is very sensitive to the placing of transmit and receiver antennas. If the node size is not a limiting factor, then antenna diversity is a viable option to obtain diversity gains and increase the communication reliability.

2.3 Multiple Links: Medium Access Control

Aim: to understand various MAC protocols, their advantages and disadvantages, and how channel utilization and delay can be analyzed.

The wireless medium is inherently broadcast in the sense that a transmitter can be heard by multiple receivers and a receiver can hear many transmitters. If two transmitters transmit at the same time, their signals may interfere and become

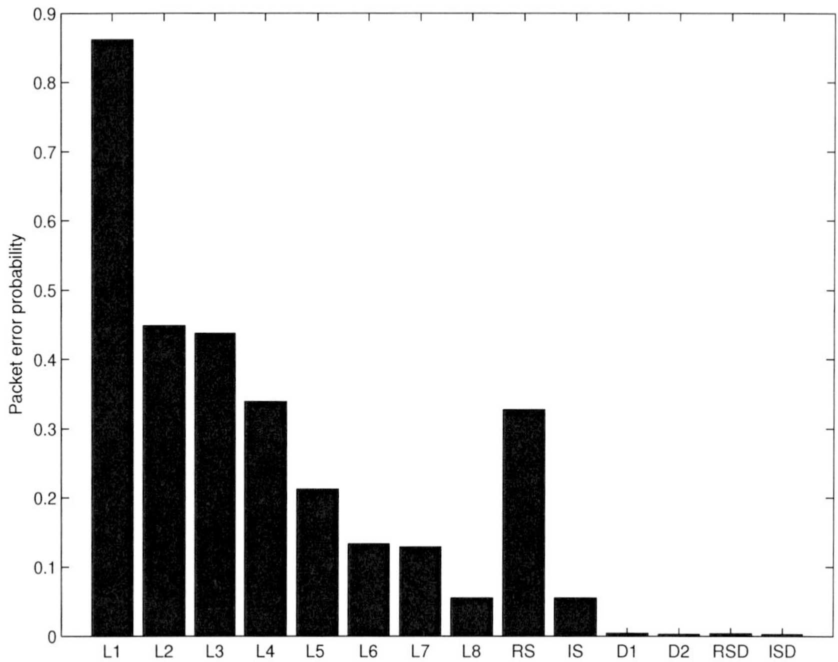

Fig. 2.5 Packet error probability for various transmit antenna and receiver antenna configurations

unrecoverable by the receivers, leading the associated packets to be lost. The role of the *medium access control* (MAC) protocol is to define a set of rules for how to share the medium between transmitters to avoid interference and communicate efficiently.

Although many MAC protocols exist, they can be broadly categorized as scheduled and contention-based. In scheduled approaches, the medium is divided into fixed non-overlapping portions (in frequency, time, etc) and assigned to individual transmitter pairs once and for all. In contention-based approaches, on the other hand, transmitters try to access the medium when they have traffic and release the channel when all data is sent. Roughly speaking, scheduled methods tend to work best when the traffic (the number of users and their traffic patterns) is predictable, while contention-based approaches are more efficient when the traffic is unpredictable.

To quantify this intuition, we need to specify the user traffic on the network. It is customary to assume that sensors generate packets according to a stationary ergodic process (*e.g.* that packet arrival times follow a Poisson distribution with a given fixed intensity). We say that the traffic is *light* when the probability that a node has multiple packets in its buffer is negligible and *saturated* when this probability approaches one. Classical analysis of MAC schemes assumes that the the arrival process of packets to nodes are mutually *independent*. However, in most sensor applications, sensor reading events tend to be *correlated*. An extreme situation, common in control scenarios, is when all sensors generate packets simultaneously. We will review some of the underlying principles of common MAC schemes next.

2.3.1 Scheduled Medium Access: TDMA and FDMA

The most common scheduled medium access techniques are time-division multiple access (TDMA) and frequency-division multiple access (FDMA). In TDMA, the time axis is divided into time slots of fixed length, and transmitter-receiver pairs are allocated a fixed number of time slots in which they can communicate. FDMA, on the other hand, divides the available communication bandwidth into disjoint frequency bands ("channels"), and allocates a fixed number of channels to each user. The two techniques can be combined into multi-channel TDMA protocols, where the communication channel is divided in frequency and space. Such an approach is natural in 802.15.4 networks, since the ISM band is channelized.

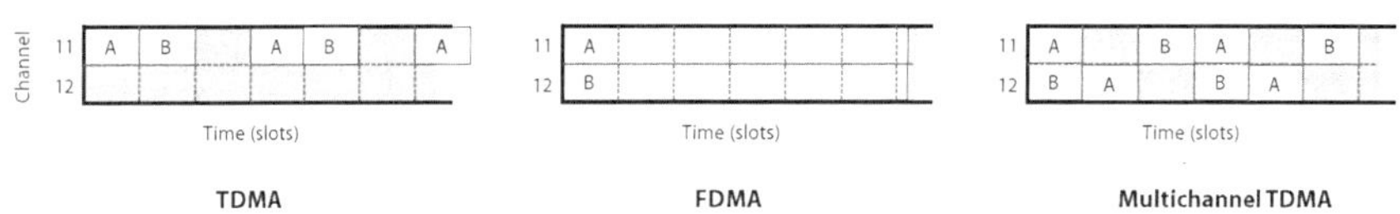

Fig. 2.6 TDMA, FDMA and multi-channel TDMA for two transmitters (A and B) on 802.15.4-channels 11 and 12 in the ISM band

Neither TDMA nor FDMA is able to use the full channel resource for communication. TDMA requires that all nodes are synchronized to a common clock. Typically, some bandwidth is lost to synchronization messages between nodes and time slots are made slightly longer than what is required to perform a single data transaction to allow for slight synchronization errors. Similarly, FDMA requires guard bands between channels to avoid co-channel interference. For example, the 802.15.4 channels only occupy 32 out of 100 MHz in the 2.4 GHz ISM band. Similarly, out of the 10ms slot time used by the WirelessHART multi-channel TDMA protocol (see § 2.4.4), only 5.088ms are used for data and acknowledgement transmissions.

Another performance loss occurs when there is a mismatch between the traffic characteristic and the deterministic transmission opportunities of scheduled access.

Example 2.4 (TDMA). To make this discussion more specific, consider a system with N sensors, each generating traffic according to a Poisson process with intensity g/N. The medium access is scheduled using TDMA with a frame of N slots of equal length T. Since a packet needs to wait until the next time slot allocated to that specific sensor appears in the schedule, the average delay is given by

$$D = T\left(1 + \frac{N}{2(1-gT)}\right) \tag{2.3}$$

if the throughput (or busy probability) gT does not exceed one [51]. As we will see shortly, contention-based MAC schemes are sometimes a better match with random traffic.

2.3.2 Contention-Based Medium Access: Aloha, CSMA and Beyond

While scheduled medium access such as TDMA provides deterministic latency (under perfect channel conditions), it does come with some implementation overhead: clocks need to be synchronized, topology information is needed for efficient transmission scheduling, and nodes joining and/or leaving the network dynamically typically means that the complete schedule needs to be recomputed and redistributed. Many of these issues are eliminated in contention-based medium access schemes.

The basic idea of contention-based medium access is the one of conflict resolution: users simply transmit whenever they have a packet to send and follow a specific set of rules to resolve conflicts when collisions occur. One of the simplest contention-resolution protocols is the Aloha-protocol [1]. In its slotted variant ("Slotted Aloha") [50], nodes are assumed to be synchronized and time is subdivided into slots, each of sufficient length to transmit a single packet. Whenever users have data to transmit, they will attempt to do so at the start of the next slot. If a collision occurs, then the node becomes backlogged. Backlogged nodes transmit in each slot with probability q until successful. The protocol performance can be optimized by tuning the access probability q. If the system is saturated, *i.e.* all N nodes have enough packets to attempt to access the medium in every slot, then one should make sure that $q < 1/N$ to avoid excessive collisions. In light traffic conditions, higher access probabilities might be advantageous, see *e.g.* [47].

In many multi-access channels, it is possible for a node to detect when other nodes are transmitting (after a slight propagation and detection delay). When this *carrier sensing* functionality is available, it is natural to check if the medium is free before transmitting. This leads to the so-called *p-persistent CSMA protocol*, which works as follows: whenever a node has data to send, it first performs carrier sense; if the channel is busy, the node does not transmit in the current slot; if the medium is idle, then the node transmits with probability p (and refrains from transmitting with probability $1-p$). An alternative CSMA variant is the *non-persistent CSMA*. Also here, nodes perform carrier sense when they have data to send and only attempt to transmit their data if the channel is considered idle. If the channel is busy, on the other hand, the node waits a random time before trying to access the channel again. The random wait mechanism is often implemented using a *back-off counter*: nodes initialize a counter by drawing a random number in the interval [0,CW] (CW for *contention window*) and decrement the counter in each slot. When the counter reaches zero, the node performs carrier sense again in hope of gaining channel access. To understand the relationship between the contention window size in non-persistent CSMA and the access probability in p-persistent CSMA, it is useful to consider the average waiting time after a detected collision. This waiting time is CW/2 for non-persistent CSMA and 1/p for p-persistent CSMA. Setting p=2/CW typically results in comparable performance of the two approaches [62]. Note, however, that the access probability distributions differ (uniform over time in non-persistent CSMA and geometrically distributed in p-persistent) and that this can influence the performance of certain applications [74].

Since the optimal access probability depends on the traffic load, it is natural to try to automatically adapt the contention window size based on the perceived congestion. The most common solution is the *binary exponential backoff* (BEB) algorithm. Here, the contention window is doubled each time a collision occurs (until it reaches a maximum allowed value) and reset to its minimum value at the beginning of each new transmission. Many successful wireless standards, such as 802.11 and 802.15.4, use medium access control mechanisms that are based on non-persistent CSMA with BEB. However, the actual implementation varies in many (sometimes subtle) ways from the basic description above. For example, in 802.11, the backoff timer only elapses when the medium is idle. Since listening consumes significant power, 802.15.4 decrements the backoff counter irrespectively of the state of the medium, but then performs carrier sensing in two consecutive slots to judge if the medium is idle. It is important to re-iterate that these mechanisms were introduced as a means to adapt to an unknown traffic situation. If the traffic is known, then the binary backoff offers few, if any, advantages over an appropriately tuned fixed contention window [34]. Naturally, one can apply similar ideas to p-persistent CSMA, *e.g.* halving the access probability at collision events [7]. Such approaches are sometimes called *predictive p-persistent CSMA*, and standardized in *e.g.* the LON Talk protocol of the ANSI/EIA 709.1 standard [37].

The following example illustrates the performance of contention-based MAC protocol under saturated random traffic.

Example 2.5 (Slotted ALOHA and CSMA delay under Poisson traffic). We assume that packets arrive at the rate g packets per time unit. The transmission time of the packet is T. Let $G = gT$ denote the traffic volume in terms of packets. In slotted ALOHA, transmissions can start only in the beginning of a slot. The slot length is equal to the transmission time of the packet T. Under the Poisson arrival assumption and large backoff range, the consecutive transmission attempts become independent of each other and follow a geometric distribution with mean $1/p_s$. The packet delay stays bounded as long as the probability of successful transmission fulfills $p_s = \exp(-G) > 0.5$ which implies upper bound for the offered traffic $G = gT < \ln 2$. Let ω denote the initial CW window size. The expected delay of the protocol has been derived in [73] and it is given by

$$D = \frac{T}{2}\left(\frac{3}{p_s} + \frac{\omega p_s}{1 - 2(1 - p_s)} - \omega\right) \tag{2.4}$$

In case of slotted CSMA, we assume that the time is divided into slots of length $\tau = aT$, $a \leq 1$. The slot length is selected to coincide with the carrier sensing time. As in the slotted ALOHA case, transmissions can only start in the beginning of a new slot. A sensor wishing to transmit must sense the channel to be free for the period of

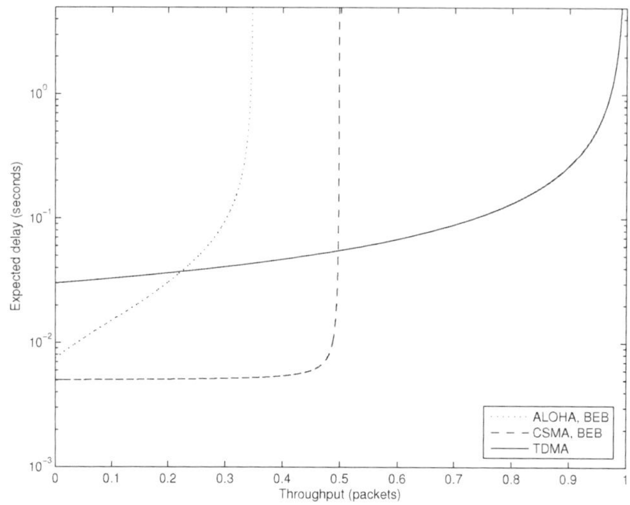

Fig. 2.7 Throughput-delay characteristics of TDMA (solid line), CSMA with BEB (dashed line) and ALOHA with BEP (dotted line)

one slot. If multiple sensors generated packets during the same slot, a collision will occur. The probability that the transmission is successful in case CSMA is utilized was derived in [27]

$$p_s = \frac{ae^{-aG}}{1+a-e^{-aG}}$$

For CSMA with BEB, the expected access delay stays bounded if $p_s > 0.5$ and has been derived in [73]:

$$D = \frac{T}{2}\left(\frac{a\omega p_s}{1-2(1-p_s)} + \frac{2+5a}{p_s} - \frac{2+4a}{p_s}p_b - a(4+\omega)\right) \quad (2.5)$$

The throughput for slotted ALOHA and CSMA is given by $S = p_sG < 1$.

The delay - throughput characteristics of the three protocols are shown in Figure 2.7. It can be seen that the maximum throughput of the ALOHA and CSMA is much less than what can be achieved with TDMA, but for low load, both contention based protocols can offer significantly shorter delay. It should be noted, however, that this conclusion is only valid for Poisson traffic. If all sensors generate packets at the same time, CSMA will result in multiple collisions and leading to significantly longer transmission times than scheduled transmission.

Example 2.6 (Delay under correlated traffic). In control systems, traffic is seldom completely random, but is typically correlated. For example, control loops often request synchronous samples from multiple sensors. To study this scenario, consider an idealized case where N sensors each generate a single packet at the same time

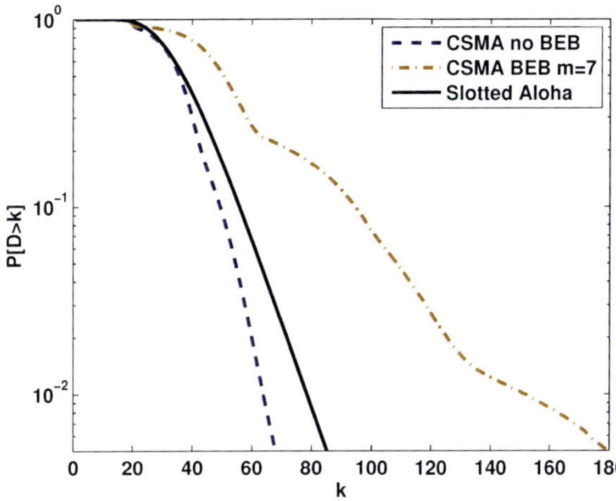

Fig. 2.8 Delay distribution for N sources that start contending for access at the same time (from [62])

and attempt to transmit these over a shared channel. For slotted Aloha, assuming that all sensor use the same transmit probability p, the delay distribution has been derived in [62] and is given by

$$\mathbf{Pr}(D = d) = \sum_{i=1}^{N} a_i \sum_{j=1}^{N} \frac{(a_j - 1)^N}{\prod_{k \neq j}(a_j - a_k)} (1 - a_j)^d$$

with $a_k = k(1-p)p^{k-1}$. While similar expressions can be derived also for CSMA with fixed contention-window, it appears hard to derive closed-form expressions for CSMA with BEB. Figure 2.8 demonstrates the latency distributions when $N = 10$ sensors simultaneously content for access using slotted Aloha (with $p = 0.1$), CSMA (with $CW = 20$) and CSMA with Binary Exponential Backoff, respectively.

Our discussion of CSMA protocols has assumed that carrier sensing is able to reliably detect conflicting transmissions. In reality, two phenomena known as the *hidden terminal* and *exposed terminal* problem, respectively, makes carrier sensing error-prone and degrades the performance of CSMA protocols. The hidden terminal problem occurs when two nodes which are outside each other's transmission range attempt to communicate with the same destination node. Since the two transmitter nodes cannot sense the carrier, they consider the medium to be idle and might transmit simultaneously, leading to collisions. The exposed terminal problem, on the other hand, occurs when two transmitters which are within transmission range of each other try to communicate with two different, well separated, receivers. Carrier sensing then blocks both transmissions although they would not have collided. The exposed and hidden terminal problems can be alleviated to some extent by implementing a two-way "request-to-send/clear-to-send" handshake between

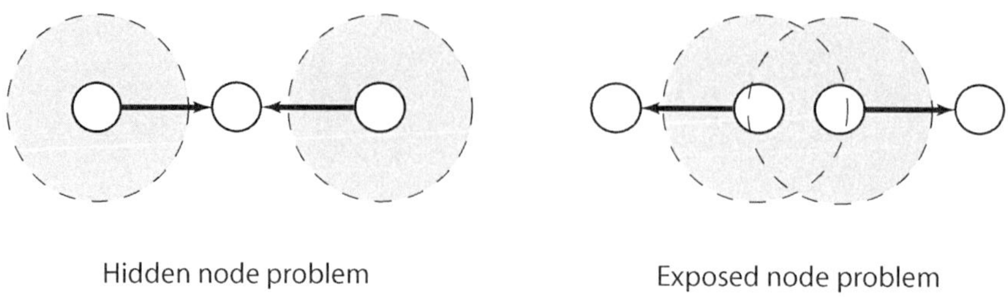

Fig. 2.9 Pictorial illustration of hidden and exposed node problem. Carrier sensing blocks nodes from transmitting within the carrier sensing range (marked in light grey) of an ongoing transmission. Hidden nodes may receive simultaneous transmissions despite collisions, and exposed nodes are block although simultaneous transmissions would be possible without collisions

transmitter and receiver prior to the actual data transmission (see, *e.g.*, [71] for a discussion).

2.3.3 Dynamic Access Scheduling via Polling and Reservation

While Aloha and CSMA allow nodes to arbitrate channel access under varying traffic conditions, the presence of collisions limits their performance at high traffic load. Better channel utilization can be achieved by scheduling nodes so that no collisions occur. In dynamically scheduled medium access schemes, nodes reserve the channel before transmitting. Channel access is coordinated by a central node and can be based on either polling (the coordinator asks nodes if they have data, as in the 802.11 point coordination function PCF) or reservation requests (nodes demand channel access from the coordinator, as in the 802.15.4 Guaranteed Time Slot service GTS). Dynamically scheduled medium access can be efficient if the reservation overhead is low, but is typically restricted to very simple node topologies. The following example provides some more details about the dynamically scheduled access in 802.15.4.

Example 2.7 (Dynamic scheduling in 802.15.4). The beacon-enabled mode of the 802.15.4 MAC supports a combination of contention-based and dynamically scheduled access. As illustrated in Figure 2.10, the MAC operates in a cyclic fashion where an initial beacon triggers a sequence of contention-based access, scheduled access and sleep periods. To reserve time slots in the contention-free period, nodes send reservation requests to the coordinator during the contention-access period. The requests can be for multiple time slots and extend over multiple superframes, until deallocated by the coordinator or the node itself.

The performance of the IEEE 802.15.4 beacon enabled mode has been analyzed *e.g.* in [36, 40]. The results indicate that the default parameter settings given in the standard lead to low performance and abrupt change from non-saturation to saturation regime, and that queue stability is limited by the amount of downstream traffic from coordinator to nodes. One observation, elaborated in [6], is that

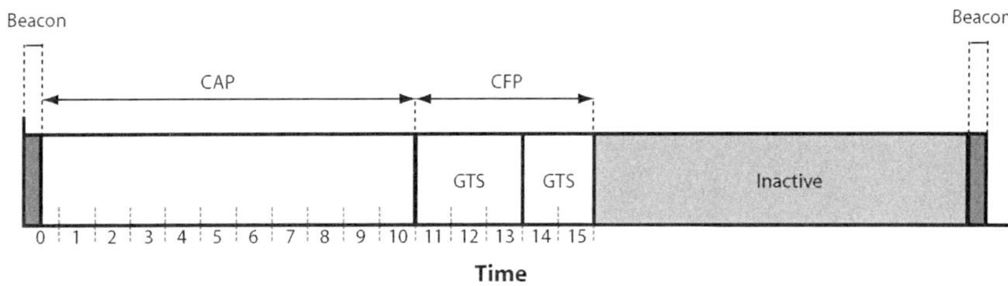

Fig. 2.10 Superframe structure of 802.15.4 in beacon-enabled mode. After the beacon, the active part of the superframe is divided into a contention access period and a contention free period

delay-sensitive control applications would have been better served if the CFP and CAP would have switched place in the superframe.

A noticeable alternative to the IEEE standards is the Z-MAC protocol [49], which combines TDMA with a mechanisms that allows backlogged nodes to detect and utilize vacant time slots. The basic idea is that the node assigned to a specific time slot has priority and starts its transmission at the slot boundary, while other backlogged nodes wait a short time before performing their carrier sense. In this way, the non-assigned nodes can sense the carrier of the prioritized node and refrain from transmission if the slot is occupied. Similar ideas for improving the efficiency of TDMA have been included in the ISA100 standard described later in this chapter.

2.3.4 *Energy-Efficient Medium Access Control*

Although the power consumption of wireless nodes can be reduced by avoiding collisions and unnecessary protocol overhead, the largest energy savings can be obtained by turning off the transceivers when they are not needed for communication. For example, in TDMA it is clear that nodes can go to sleep whenever they are not scheduled to transmit or receive. If further energy savings are needed and the traffic rate is not that high, one can introduce a *duty-cycle* where nodes are awake for communication D% of the duty cycle period and asleep the rest of the time. Many MAC protocols for sensor networks require nodes to synchronize their duty cycles, but rely on contention-based medium access in the access period [75]. Duty-cycling becomes more challenging when nodes are not synchronized, and is then typically based on *preamble sampling*: whenever a node has data to send, it starts transmitting a train of short preambles. When a receiver wakes up, it samples the medium. If no preamble is detected, the receiver goes back to sleep. If a preamble is detected, the receiver compares its ID with the one embedded in the preamble to see if it is the intended receiver of the pending data transmission. If it is the intended receiver, it sends an acknowledge message to the transmitter which aborts the preamble strobe and starts transmitting the actual data packet. Nodes that sample the preamble strobe but detect that they are not the intended receiver go back to sleep. Representative examples of such MAC protocols are B-MAC [46] and X-MAC [8].

2.4 From Single Links to Network: The Upper Networking Layers

2.4.1 Topologies and Multi-hop Communications

Two or more nodes that communicate form a network. The simplest network is the single receiver-transmitter pair (the single link). Although interesting phenomena occur already when multiple wireless links co-exist within transmission range, the first non-trivial network topology is the *star topology*; see Figure 2.11. Here, a single node maintains connection with all other nodes in the network. In many cases, the central node coordinates transmissions and might also work as the gateway to external networks. The star topology has at least two disadvantages: it has limited coverage (essentially limited by the maximum distance over which a transmitter-receiver pair can sustain reliable communication) and limited reliability (limited by the reliability of a single link). The range can be extended by forming a *cluster tree*. Here, the children of the central node in the initial star topology also have the option to act as cluster heads, maintaining a star topology with their own children, etc; see Figure 2.11. In this way, data might be routed over multiple hops, and coverage is increased (at the expense of increased latency). However, there is still only one path from each source to the destination and the failure of a single intermediate node destroys the communication for all its children. The *mesh topology* allows addressing both coverage and reliability issues. In a mesh topology, all nodes participate in the forwarding of data packets from sources to destinations. By letting nodes maintain communication with more than one neighbor, it is possible to set up multiple paths from each source to their destination, and quickly divert data to a new path if an intermediate router fails. The following set of examples illustrate key aspects of multi-hop networking: the energy consumption for delivering a packet is likely to increase; the latency and loss rate both increase as the number of hops between source and destination increases; retransmissions allow to improve reliability at the price of increased delay, but are less effective when packet erasures occur in burst; multi-path diversity allows to increase reliability; and the interaction between medium access control and multi-path routing can severely degrade the system performance.

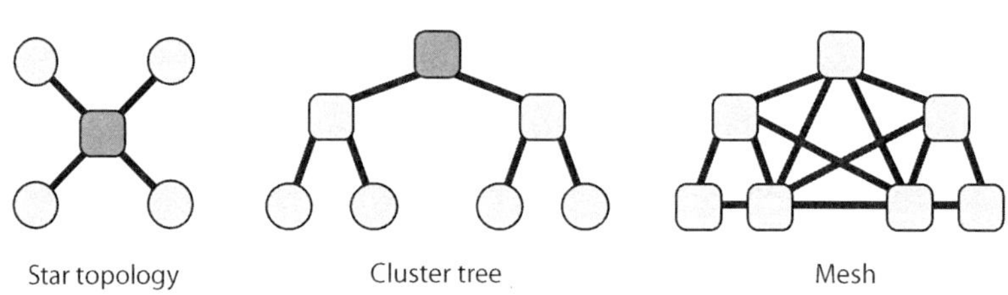

Fig. 2.11 Star, cluster tree and mesh topologies. Squares denote devices with routing functionality

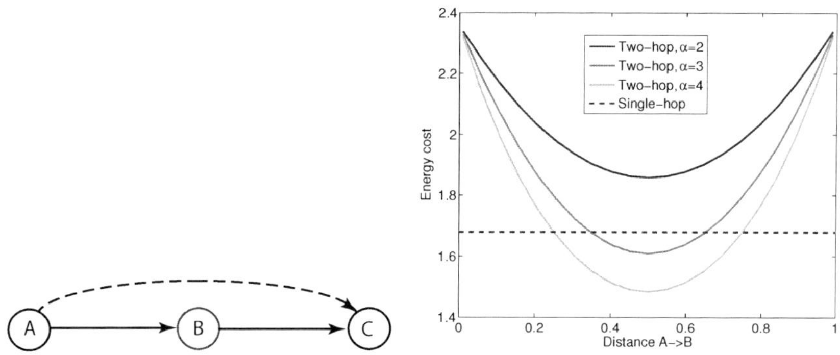

Fig. 2.12 Node A can transmit directly to C or send data via the intermediate node B (left). The right figure demonstrates the total energy cost for communication under various path loss coefficients α and for various placement of the intermediate node

Example 2.8 (Energy and latency cost of multi-hop networks [13]). Consider the three-node topology in Figure 2.12 (left). Transmitter A has the possibility to send data directly to the sink C, or forward data via the intermediate node B. The intermediate route requires two transmissions ($A \rightarrow B, B \rightarrow C$) which doubles the end-to-end delay compared to the direct transmission. However, since the path gain decays as $d^{-\alpha}$, the shorter links can reduce their transmit power by a factor $2^{-\alpha}$ while maintaining the same received power as the direct link, multi-hop transmissions could potentially result in energy savings. Unfortunately, the energy gains often disappear when we account for the fact that radios consume significant energy also in reception mode. In [13], measurements on a 802.15.4 radio revealed a receive energy of 0.68 times the maximal transmit energy $P_{\max}$. Even with a path loss of $\alpha = 4$ the difference between the energy cost of the direct path $P_{Tx} + P_{Rx} = 1.68 P_{\max}$ and that of the two-hop path $2 \cdot 2^{-4} P_{\max} + 2 \cdot 0.68 P_{\max} = 1.485 P_{\max}$ is very limited, and no energy gain is possible when $\alpha \leq 2.64$. As Figure 2.12(right) reveals the energy gains disappear even earlier when the intermediate B node is moved from its ideal position in the middle of the nodes.

Example 2.9 (Latency and reliability of multi-hop communications). To understand some of the latency and reliability issues of multi-hop networking better, consider the transmission of a single packet across the $(N+1)$-node topology shown in Figure 2.13. Assume a slotted system where a single time slot admits the transmission of one packet, and that link transmissions fail independently with probability p. The minimum latency is N time slots, and the probability of successful end-to-end communication in N time slots is $(1-p)^N$, which decays rapidly with N.

The reliability can be increased if we allow a longer latency, so that nodes can retransmit a message if it does not receive a packet acknowledgement from the intended receiver. If we allocate two consecutive time slots to each link (corresponding to primary transmission and retransmission, respectively), then the probability of reaching the sink in $2N$ time slots is $(1-p^2)^N$, *i.e.* the reliability has been

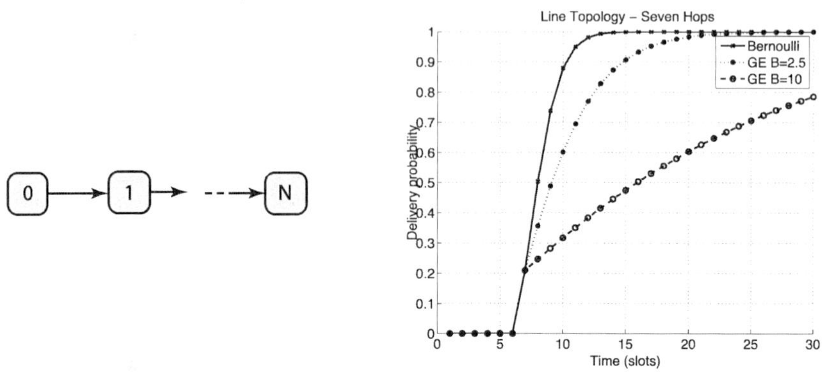

Fig. 2.13 Single packet transmissions over N hops (left). The latency distributions for a link loss probability of $p = 0.2$ for line of length 8.

increased a factor $(1+p)^N$. If we can allocate retransmission attempts dynamically (*e.g.* to let a node retransmit until successful) the probability of successful end-to-end communication with latency $D > N$ is given by [58]

$$(1-p)^N \sum_{r=0}^{D-N} \binom{N+r-1}{N-1} p^r$$

When losses are bursty, the effect of immediate retransmissions diminishes, since the probability that the retransmission will also fail is high. The latency-loss curves for dynamic retransmissions on a line where link losses are given by the Gilbert-Elliot model are shown in Figure 2.13. Note that a significant latency is needed to ensure high reliability.

Example 2.10 (The mesh forwarding advantage). As discussed earlier in this chapter, reliability can be improved by exploiting diversity. Rather than retrying a transmission on a bursty link, with high probability of continued loss, we could schedule the retransmission on another link (possibly also on another frequency). To illustrate the ideas, consider the "tube" topology shown in Figure 2.14(left) and assume that erasure events on links are dictated by independent (two-state) Markov chains, parameterized according to Gilbert-Elliot. The achievable loss-latency curve, shown in Figure 2.14(right) demonstrate an increased reliability compared to the line, with a significant reliability boost for low latencies [76].

Example 2.11 (Multiple data streams and interaction with MAC). To consider the interaction of medium access control and multi-hop networking, consider the problem of synchronized data collection from multiple sensors using the so-called *convergecast* operation. In convergecast, all nodes have a single packet that should be transmitted to the sink. Thus, if we consider a simple line with N sensors then convergecast on a single channel needs a total of

$$T_{\text{TDMA}} = \sum_{n=1}^{N} n = \frac{N(N+1)}{2}$$

time slots. The latency can be decreased to

$$T_{\text{MCH}} = 2(N-1)$$

time slots when we use multi-channel TDMA to allow for parallel transmissions [57]. Figure 2.15(right) demonstrates how the performance degrades significantly when nodes use slotted ALOHA to contend for channel access. Even using the optimal transmit probability, the average latency is 94.5 time slots compared with the 15 slots required for single-channel TDMA and 8 slots for multi-channel TDMA.

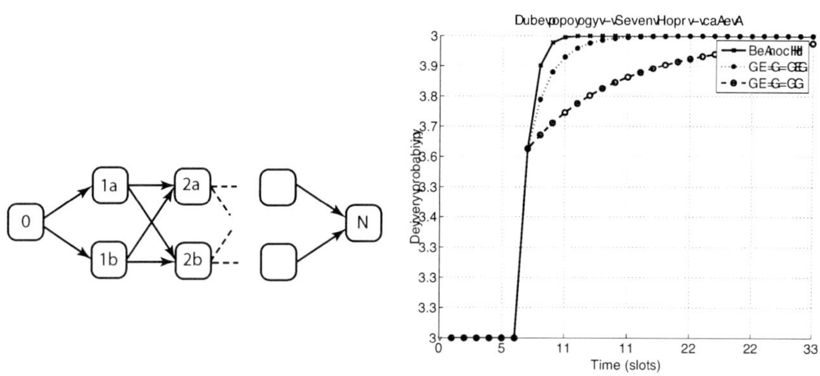

Fig. 2.14 Single packet transmissions over an N-hop tube (left). The latency distributions for a link loss probability of $p = 0.2$ for line of length 8.

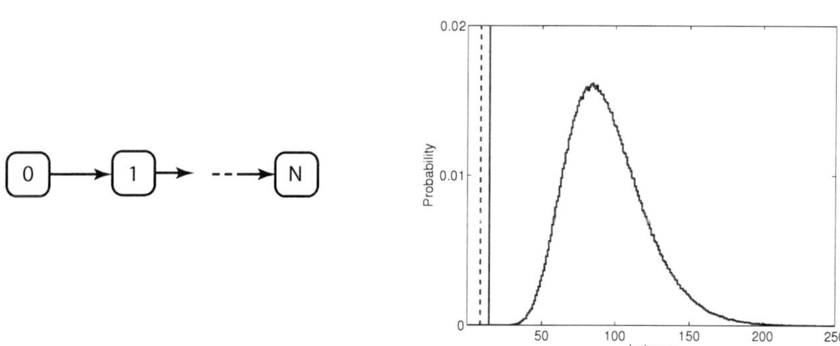

Fig. 2.15 Convergecast operation using N sensors in line topology (left). For a 5 node case, the right plot reveals the latency distribution for TDMA and multi-channel TDMA (deterministic at 8 and 15 time slots, respectively) and the latency distribution for completion of convergecast under the ALOHA protocol

2.4.2 *Routing*

Routing is the process of selecting paths along which to send data traffic. Data packets are then forwarded from sources to their final destinations via the intermediate nodes on the selected routes. Most routing algorithms use a single network path between sources and destinations, while multipath routing protocols maintain several alternative paths to improve reliability. Routing protocols are further classified as either reactive (on-demand) or pro-active (table-driven) [52]. Table-driven routing protocols attempt to maintain consistent up-do-date routing information from each node to every other node in the network, while on-demand routing protocols gather such information only when needed.

Routing on resource-constrained low-power radio nodes is challenging: constrained memory and processing power limits the size of routing tables that can be stored, low-power radios experience higher loss rates than high-end technologies, and limited power supply requires energy-optimized implementations and low overhead traffic. The wireless sensing community has developed many routing protocols that respect these design constraints and exploit the traffic and data characteristics of sensor networks to optimize performance (see, *e.g.*, [2] for a survey). We will not go into the details of the specific protocols, but rather focus on the resulting routing topology and its performance.

Link-Metrics and Shortest Path Routing

Paths are typically selected based on some quality metric, such as latency or reliability. These path metrics can often be written as the sum of the costs for using each link in the path. For example, the end-to-end latency of a path is the sum of the transmission times for its individual links. When link costs are additive, the best paths can be found by solving a shortest path problem using, for example, the algorithms due to Dijkstra or Bellman-Ford (see, *e.g.*, [5]). When path costs are multiplicative, a simple transformation allows for casting the optimal route calculation as a classical shortest path problem. The following example illustrates this idea.

Example 2.12 (Maximum reliable path as a shortest path problem). Consider the problem of finding the path with the maximum packet delivery probability. If links fail independently with probability p_l, then the reliability of path P, r_P, can be written as $r_P = \prod_{l \in P}(1 - p_l)$. Since the logarithm is monotone, the path that maximizes r_P will also maximize $log(r_P) = \sum_l \log(1 - p_l)$. Now, maximizing $\sum_l \log(1 - p_l)$ is the same as minimizing $\sum_l - \log(1 - p_l)$, so the most reliable path can be found by solving a shortest path problem with link weights equal to $-\log(1 - p_l)$.

Related to this model of link reliability is the expected transmission count, or ETX for short [11]. The aim of ETX is to increase overall network throughput by finding paths with the lowest expected end-to-end latency. Observing that to avoid link level retransmissions, both the data packet and its link-level acknowledgement must be successfully received, the expected number of transmissions for a given link is approximated by $\text{ETX} = 1/(d_f \cdot d_r)$. Here, d_f and d_r represent the delivery ratios in

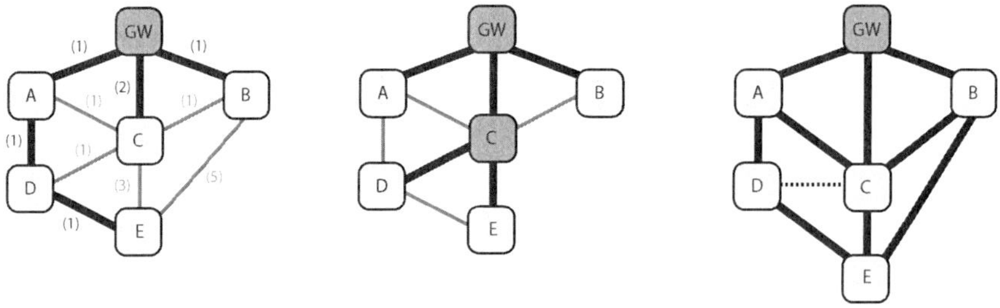

Fig. 2.16 Routing strategies. Numbers in parenthesis indicate ETX for the links. The leftmost figure demonstrates shortest-path routing in the ETX metric, where thicker lines indicate links selected for forwarding data towards the gateway (GW). The middle figure demonstrates constrained shortest path routing. Here, device C is the only battery operated field device allowed to forward data, which shifts the permissible shortest paths (again, indicated with thick lines). Finally, the rightmost figure demonstrates graph-based routing. Here, nodes are organized into "levels" according to their ETX distance to the sink, and additional links are introduced to neighbors on higher levels. The dashed line indicates a sibling link between devices on the same level, which may by used if care is taken to avoid routing loops

the forward (data) and reverse (acknowledgement) direction of the link. Note that under Bernoulli losses and symmetric links $d_f = d_r = (1 - p_l)$, in which case ETX and $-\log(1 - p_l)$ have a similar qualitative behavior.

Constraint-Based Routing

In some cases, it is useful to exclude certain links or nodes from the routing process. For example, we might want to make sure that battery operated nodes do not need to forward other nodes' packets. Such considerations are included in constrained shortest-path routing protocols. Here, nodes or links that do not satisfy certain constraints are removed from the routing topology, and the routes are found by computing the shortest paths in this reduced graph. For example, the most reliable path that does not use battery-operated devices can be computed by first removing the battery-driven devices from the graph representing the network topology, and then solving a shortest path problem with link weights equal to $-\log(1 - p_l)$. To balance reliability and latency, we could consider removing all unreliable links from the connectivity graph and then find the routing paths with minimum hop distance to the destination (for a comparison of this approach and the use of ETX, see [69]).

A more in-depth discussion about uses of constrained shortest paths in wireless sensor networks can be found in [77].

Graph-Based Routing

The routes computed by a constrained shortest path routing protocol form trees rooted in the destination. A drawback with this structure is if one node breaks, the full branch from this node to its leaves breaks. One way to come around this problem is to set up multiple (partially) disjoint trees rooted in the destination, and switch from one tree to another when a failure is detected. An alternative is to set-up a *directed acyclic graph* (DAG), in which nodes might have multiple parents, and forward messages along this graph. Since the graph is directed and does not contain any cycles, each successful message transmission guarantees forward progress towards the destination. Note that the forwarding policy now becomes a critical part of the routing algorithm: not only does the preferred parent matter, but also the decision about whether to retransmit on the same link or switch to a secondary route, *etc.* [76]. Setting up the graph is also non-trivial, but in many cases a reasonable graph can be built from the shortest path tree (in some metric) by connecting nodes in the DAG with neighbors that are closer to the shortest path tree root (in *e.g.* number of hops) than the node itself.

Example 2.13 (Markov-models for unicast forwarding). When link losses are independent, it is easy to construct Markov models for the latency of a single packet traversing a routing graph from source node s to destination node d under a given schedule. The basic idea is to construct a Markov chain whose n^{th} state corresponds to the packet being buffered at node n. The state transition depends on the schedule (and hence on time): if the packet is located at node n and one of its outgoing links l is scheduled for transmission at time t, then the packet is forwarded to the end-node of link l with probability $(1-p_l)$ and stay in node n with probability p_l.

To define the Markov chain in a compact way, consider network with N nodes and L links whose underlying topology is represented by a node-arc incidence matrix $A=[a_{nl}]\in\mathbb{R}^{N\times L}$ where, $a_{nl}=-1$ if link l is outgoing from node n, $a_{nl}=1$ if link l is incoming to node n and $a_{nl}=0$ otherwise. For convenient notation, we also define $A_m=\max(-A,0)$. Let $p\in\mathbb{R}^L$ be the vector of loss probabilities for links and let $P=\text{diag}(p)$. Introduce the diagonal matrix $S(t)=[s_{lm}(t)]\in 0,1^{L\times L}$ to encode the scheduling decision at time t: $s_{ll}(t)=1$ if link l is scheduled for transmission at time t while all other entries are zero. Let $\pi(t)$ be the probability distribution for the Markov chain at time t. Then,

$$\pi(t+1)=\left(A(I_{L\times L}-P)S(t)A_m^T+I_{N\times N}\right)\pi(t)$$

Since the packet is known to be at state s at time zero, we have that $\pi_n(0)=0$ for $n\neq s$ and $\pi_s(0)=1$. Similar models can also be developed for correlated losses and more complex communication patterns [9, 43, 44].

On-Demand Routing

The most well-known on-demand routing protocol is the Ad hoc on-demand distance vector routing protocol, AODV [42]. When a source node needs to send a packet to a destination to which it does not already have a route, it initiates a route discovery process: it broadcasts a route request packet to its neighbors, who forward the request to their neighbors, *etc.* The process continues until either the destination or an intermediate node with "fresh enough" route to the destination receives the request. The destination or the intermediate node then transmits a route reply message to the neighbor from which it received the first route request. The route reply message is then transmitted back along the reverse path, and nodes on this path set up forward route entries in their routing tables pointing to the node from which they received the route reply. Associated with each route is a timer which will cause a deletion of the entry if it is not used within the specified lifetime. By letting intermediate nodes wait in a controlled fashion before forwarding route requests and/or letting the destination wait for multiple route requests (that have arrived over different paths) and picking the best path, AODV can come close to generating (constrained or unconstrained) shortest paths [13].

2.4.3 Transport Layer Protocols and Traffic Patterns

The physical layer, data-link and network layers described so far provide all the essential functionalities required for moving information between remote hosts in the network. The role of the transport layer is now to "package" this functionality into an transparent end-to-end communication service for applications. While the wireless sensor networking community has developed a wide variety of transport protocols (see, *e.g.*, [66]), we will focus on the solutions that appear in the industrial wireless standards.

The most basic transport-layer protocol in the Internet is the *user datagram protocol (UDP)*. The UDP protocol simply adds information about the source and destination application process ("ports") along with a checksum bit for error detection, and passes the resulting packet to the network layer for forwarding to the end host. The protocol neither provides support for informing the source about lost or corrupted packets detected at the end-host, nor functionality for handling retransmissions. If such functionality is needed, it must be implemented by the application.

The *transmission control protocol (TCP)* adds reliability and flow control functionalities to the end-to-end communication primitive. In contrast to UDP, TCP is a connection-oriented protocol: before any communication can take place and end-to-end connection has to be established. In addition to a checksum, TCP packets also include sequence numbers to allow the end host to detect lost messages and to reorganize packets that are delivered out of order. The end-host informs the source about which packets it has received by returning acknowledgement messages. If the source does not receive an acknowledgement within a given time-out period, it assumes that the associated packet is lost and retransmits the data. Another important feature of TCP is flow control: the protocol automatically adjusts the rate at which packets are sent, so as to avoid congestion in the network. However, TCP's flow control mechanism was originally designed for fixed networks and has performance problems in case of wireless transmissions [18]. TCP treats lost packets as an indication of congestion and decreases the flow rate when no acknowledgements are returned. While optical networks are essentially free from transmission errors, most of the packet drops in wireless networks are due to fluctuating channel quality and not from congestion-induced buffer overflows. The flow control of TCP thus unnecessarily decreases the flow rate every time a packet gets lost in a wireless link, leading to poor utilization of the radio resources. Another known performance problem is related to the interaction between TCP protocol and BEB in multi-hop networks. Single-hop flows will capture a majority of the transmission opportunities leaving the multi-hop flows to starve. The problems of the TCP in multi-hop environments are discussed in, *e.g.*, [72]. The rich feature set of TCP makes it rather heavy-weight to implement and most real-time communication protocols rely on UDP for end-to-end communication.

Finally, note that contrary to fixed networks where point-to-point (unicast) communication is the most common traffic pattern, wireless sensor and actuator networks often rely on many-to-one (*e.g.* sensors to gateway) and one-to-many (*e.g.* controller to actuators) communication. A large body of work in the wireless sensor networks literature considers theory and tailored system solutions for convergecast (many-to-one) and multicast (one-to-many) solutions.

2.4.4 Standards and Specifications for Industrial Wireless Networking

Aim: To list some existing standards, and to describe which techniques/functionalities from above that they explore.

Zigbee PRO

Zigbee exists since 2004 and was one of the first attempts to provide a low-power radio standard for home automation and industrial control. Its focus was on providing a self-organizing scalable and secure short-range wireless solution with a battery life up to two years. However, Zigbee has some serious reliability flaws, which has stopped it from making its way into a true industrial standard. Zigbee remains, to date, in specification form.

Zigbee builds on the physical and MAC-layer of IEEE 802.15.4, and specifies the behavior of the higher protocol layers. There are two types of nodes in a Zigbee network: full-function devices and reduced-function devices. Full-function devices can route messages and act as network coordinator, while reduced-function devices can only communicate directly with a full-function device. The standard supports mesh networking among full-function devices and routing is performed using the AODV protocol. Hence, a node can send packets to any other node in the network, but need to execute the algorithms for route discovery and route maintenance as discussed earlier in this chapter. In a Zigbee network, all nodes share the same channel, and there is no frequency hopping. The only way to attempt to provide reliable operation is to scan the spectrum for a channel with low interference before deploying the network. Since AODV is a single-path routing protocol (unless we allow us the time to re-discover a route when we detect a breakage) neither frequency nor path diversity is supported in Zigbee and the overall reliability is typically low. Zigbee can operate in both beaconed and non-beaconed mode. Beaconing allows some degree of synchronization among nodes, but the medium access is generally contention-based

(CSMA/CA). There is an option to use guaranteed time slots, but the support for this is not mandatory and might break interoperability of nodes. As we have seen in Example 2.11, the lack of synchronized communication can seriously increase latency under correlated traffic. Moreover, the uncoordinated operation of nodes also means that devices need to stay awake practically all the time and that the energy consumption is higher than needed. The latency and reliability issues, together with the security concerns detailed in [29], are some of the key reasons for the limited industrial success of Zigbee.

WirelessHART

WirelessHART is an extension of wired HART, a transaction-oriented communication protocol for monitoring and control applications. As illustrated in Figure 2.17, the basic elements of a WirelessHART network include: *field devices* connected to the process equipment and able to source, sink and forward packets on behalf of other devices in the network; *gateways*, possibly equipped with multiple access points, enable communication between host applications and field devices; a *network manager* responsible for configuring the network, health monitoring, managing routing tables and scheduling communication between devices. WirelessHART networks may also include *adapters* for connecting to existing HART-compatible devices and handhelds to configure, maintain or control plant assets. An important restriction of WirelessHART is that device-to-device communication is not supported: all data must pass through the gateway.

Like Zigbee, WirelessHART is based on the 802.15.4-2006 physical layer, but it is restricted to operate on 15 channels (number 11-26) of the 2.4GHz ISM band. The medium access is controlled using a multi-channel TDMA MAC. A time slot is 10ms, which allows for a complete data packet and associated acknowledgement transaction. Transmissions are scheduled on one of the 15 logical channels, which are mapped onto the physical channel by a channel hopping sequence.

The transmission schedule is organized into multiple *superframes*. A superframe is simply a collection of links, each assigned to a specific time slot and a specific channel, see Figure 2.18 for an illustration. Superframes repeat in time, and should have periods that form a harmonic chain (*e.g.* $1, 2, 4, 8, 16, \ldots$ or any other period

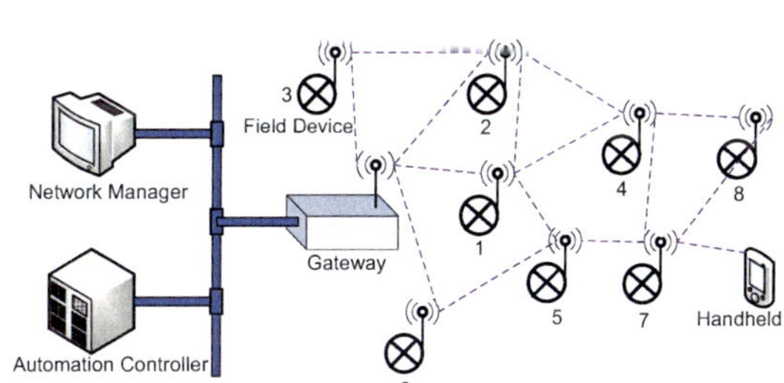

Fig. 2.17 Example of wirelessHART network infrastructure

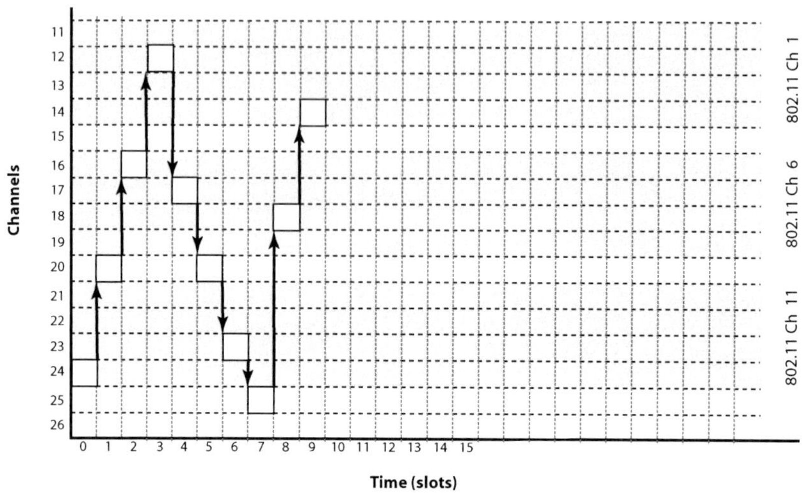

Fig. 2.18 WirelessHART schedules transmissions in time and frequency

that conforms to an expression on the form ab^n), so that all superframes are aligned at the start of the longest superframe. Each superframe is typically constructed to accommodate for communication among a specific group of devices and allows to run the whole network at different duty cycles. Additional superframes might be defined to support traffic at different scan rates, to handle alarm traffic, *etc.* At least one superframe is always enabled while additional superframes can be added and removed. The standard also specifies procedures for link arbitration to resolve conflicts that may appear if a network device participates in multiple superframes.

The schedule supports three different types of slots: dedicated slots, with a single transmitter-receiver pair; shared slots, where many transmitters but only a single receiver are scheduled; and broadcast slots, where one transmitter and multiple receivers are scheduled. Dedicated and shared slot transmissions are acknowledged, *i.e.*, the receiver transmits a short acknowledgement packet to the transmitter in the same time slot, while broadcast packets are unacknowledged. WirelessHART uses non-persistent CSMA to resolve conflicts in shared slots. If no acknowledgement is received, the backoff exponent of the link is incremented and the backoff counter is initialized with a random number in the associated backoff window. The backoff counter is decremented in every time the same link is scheduled in a shared slot, and the data is retransmitted when the backoff counter reaches zero. Note that this is a completely different time-scale than the back-off counter in the 802.15.4 CSMA and that back-off times in shared WirelessHART slots can be significant.

WirelessHART supports two routing paradigms: graph and source routing. In graph routing, network nodes store multiple DAG:s. Packets are tagged with the ID of the graph along which they should be forwarded. In source routing, on the other hand, the complete path is encoded into the packet. Source routing specifies a single path between source and destination, and is thus fragile to link failures, while graph routing supports multiple forwarding paths to improve reliability.

The transport layer provides both unacknowledged and acknowledged end-to-end communication. The unacknowledged service is, much like UDP, a basic protocol without acknowledgement packets and without guarantees of packet ordering at the destination node. The acknowledged service, on the other hand, provides end-to-end acknowledgements and guarantees packet ordering (but not advanced TCP features such as flow control).

The standard supports methods to maintain network-wide time synchronization, network formation and topology maintenance, security features on both link level and network level, and much more. Since these features do not have a direct influence on the end-to-end performance of the network, we do not describe these in detail here, but refer to the standard documents.

ISA 100

While WirelessHART formally only standardizes wireless communication with HART devices, the ISA standard was created with the more ambitious aim of providing a single wireless standard that supports multiple legacy application layer protocols. However, as the standardization work has proceeded, ISA100 and WirelessHART have evolved into rather similar specifications and there is now an ongoing effort in the standard committees to see if and how the two standards could "converge" into one. In this section, we will review ISA100 and highlight some of the features that differ from WirelessHART.

Like WirelessHART, ISA100 is based on the IEEE 802.15.4-2006 physical layer and the data link layer uses a multi-channel TDMA medium access control. However, in ISA100 different superframes can use different time slot lengths, ranging from the 10ms of WirelessHART up to 12ms. Longer time slot lengths give some extra margin against poor device synchronization, but perhaps more importantly this feature enables prioritized access in shared time slots and advanced transmission primitives beyond unicast and broadcast. Prioritized access in shared slots is implemented by transmitting high-priority data at the very beginning of time slots, while waiting a short time before performing CCA and attempting to transmit lower-priority data. The *duocast* primitive allows a field device to broadcast a data packet to multiple access points (backbone routers in the ISA100 terminology) and receive serial acknowledgements within the same time slot. The duocast transaction is considered successful if at least one link-level acknowledgement is received, and allows low-latency communication with high probability of first success. There is no restriction of individual superframe lengths, but the system realigns time slots of all superframes every 250ms by the insertion of short idle periods. Frequency hopping can be performed on a per time-slot basis ("slotted hopping", like in WirelessHART) or with longer intervals ("slow hopping") where the same channel offset is used for multiple consecutive transmissions. The slower hopping can be useful, for example at network discovery, since it allows for less accurate synchronization among nodes. Similarly to WirelessHART, ISA100 supports both source and graph routing. The network layer of ISA100 is based on the IETF 6LoWPAN standard, and the transport layer provides end-to-end encrypted communication through UDP datagrams as outlined in IETF RFCs RFC 768 and 2460.

Alternative and Emerging Standards

While WirelessHART and ISA100 appear to be able to provide the industrial-grade reliability and security that Zigbee was lacking, they still have a serious drawback: they both rely on centralized network configuration and resource management. The centralized management comes with large overhead for collecting network health information and disseminating routing graphs and transmission schedules to nodes, and the times required for a node to join a WirelessHART network can be substantial. It is thus natural to ask whether it is possible to create distributed routing and MAC protocols that could exploit diversity and offer the reliability required by control applications. No such standard exists at the writing of this book chapter, but (at least) two noticeable efforts are currently taking place towards that aim. The first one is the work on the Internet Engineering Task Force (IETF) on Routing over Low-power and Lossy Networks (ROLL). This effort attempts to develop and standardize a routing protocol that includes distributed DAG construction and maintenance. Complementing this work, the IEEE 802.15.4e attempts to amend the 802.15.4-2006 standard to better cater for industrial needs, and the current draft standard includes a time-slotted channel hopping medium access control.

2.5 Control Relevant Models of Latency and Loss

Aim: show how the existing models used in the control literature apply to various scenarios above.

The purpose of this section is not to provide an extensive overview of design methodologies for networked control (such expositions can be found elsewhere in this book and in several survey papers, *e.g.* [17, 64]). Rather, our aim is to provide the link between the communication network models derived earlier and abstractions that have been used in the theoretical control community.

The Network as a Delay

The initial work on networked control, *e.g.* [30, 39], which focused on wired control loops, advocated to model the network as a time-varying delay. The general set-up, shown in Figure 2.19, includes delays from sensors to controller $\tau_{sc}(t)$ and from controller to actuator $\tau_{ca}(t)$.

If we assume that

$$\tau_{sc}(t) + \tau_{ca}(t) \leq h$$

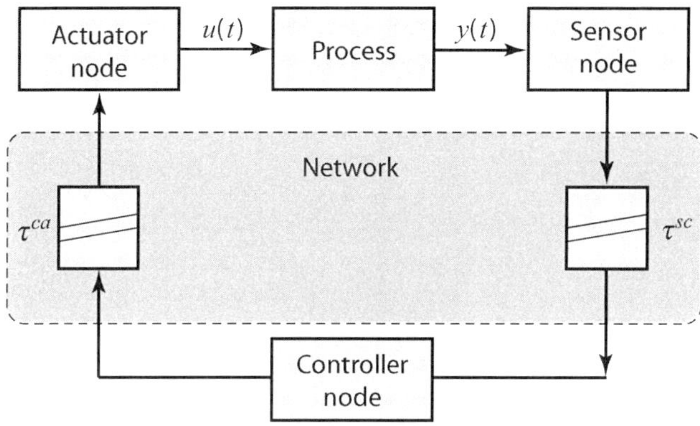

Fig. 2.19 The network abstracted as a time-varying delay

Then, a discrete-time controller with sampling period h can be developed that does not need to consider the possibility of packet loss. The time-varying delays can be dealt with using varying degrees of sophistication.

The simplest approach is to let the sensor and controller work synchronously with sampling period h, and insert a buffer in the actuator node that waits until the end of the sampling interval before applying the control [30]. In this way, all stochasticity is removed and replaced by a fixed one-sample information delay in the control loop. Hence, a wealth of classical control design techniques apply.

As the control performance generally degrades with increasing information delay, it is natural to let the controller be event-driven, in the sense that it computes the control signal as soon as senor data arrives and immediately transmits the command to the actuator [39]. Since the sensor-controller actuator delay is known at the controller, a time-varying Kalman filter allows to optimally estimate the process state. In this case, we do not need to know the latency distribution

$$\mathbf{Pr}(\tau_{\text{sc}} = \kappa)$$

(*e.g.* using one of the models developed earlier in this chapter to design the algorithm, only to evaluate its performance. Suboptimal schemes that use a fixed gain Kalman filter, however, will need access to latency distributions in their design [10]. The latency distributions are also needed for computing the appropriate controller gains, since the delays between controller and actuator are not known to the controller (see, *e.g.* [24] for alternative solutions).

When it is not possible to guarantee a communication delay less than the sampling interval, it is natural to consider designs that allow

$$\tau_{\text{sc}}(t) + \tau_{ca}(t) \leq Kh$$

for some fixed $K > 1$. A new problem that appears is that packets might now arrive out of order. Similarly as above, one can use buffers to remove the stochasticity in the control loop. Although the advantage of acting as quickly as possible when new data arrives increases when K increases, dealing with out-of-order delivery increases implementation complexity. The optimal filter, for example, now needs buffers to store the last K sensor packets received and the covariance matrix of the time-varying Kalman filter at time $t - Kh$, and it needs to run K iterations of the Riccati equations when new data arrives [54].

When the latencies are upper-bounded, then one can also consider the networked control problem from a robust control perspective. One useful result here is the jitter-margin [25] which relates the maximum tolerable network delay to the "bandwidth" of the complementary sensitivity function (see also [35] for less conservative results).

As we have seen, however, natural models for wireless control systems are stochastic and do not allow any deterministic guarantees on latencies. The possibility of packet loss should thus be considered.

The Network as an Erasure Link

One other class of models disregard communication delay and only consider whether or not packets arrive at their destination; see Figure 2.20.

The most common model in the theoretical control literature is the Bernoulli loss process, which assumes that losses occur independently with a fixed probability p_{loss}. The loss probability can be computed from the latency distributions derived earlier, as

$$p_{\text{loss}} = \int_{\kappa=h}^{\infty} \mathbf{Pr}(\tau = \kappa)\, d\kappa \tag{2.6}$$

For this type of model, bounds on the tolerable loss probability can be computed and some optimal controllers and estimators can be derived [55]. Co-design of network and controllers are also possible [44].

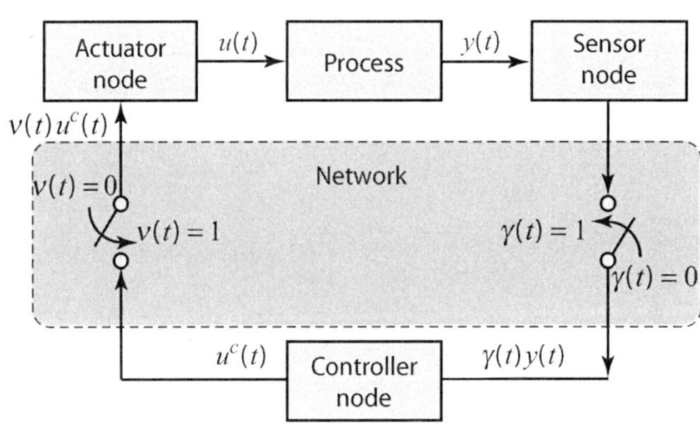

Fig. 2.20 The network abstracted as an erasure link

The independent loss model is reasonable if the sampling times are longer than the coherence time of the wireless channels, but not appropriate for modelling single links at the short time-scale. In the shorter time-scale of a few consecutive transmissions, the loss process on a given link tends to be correlated [68] and more complex loss models, such as higher-order Markov models should be used. In this case, the natural modeling framework is the one of jump-linear systems [10, 70]. Some specific results for estimation over Gilbert-Elliot models also exist [19, 3].

Models which assume a bounded delay of hK also applies if one can guarantee that the network produces no more than K consecutive losses. However, once again, no such deterministic guarantees can be derived from the natural stochastic models of loss described earlier in this chapter.

Combined Latency-Loss Models

In many cases, the most natural models for control is to pay careful attention to the shorter delays, in order to act quickly when the network incurs a short delay, but then put an upper bound on the tolerable latency after which the packet is considered as lost. Such a mechanism is easy implemented by retransmission and buffering policies in the single link and using time-to-live counters for in the multi-hop scenario. Also in this case, the appropriate modelling framework is one of jump-linear systems [70]. Once again, the latency distributions using various technologies can be derived using the techniques described earlier in this document and the loss probability calculation (2.6).

It is natural to relate the time-to-live not only to the network but also to the sampling interval and characteristics of the process at hand. Networking-controller co-design procedures following this philosophy are presented in, *e.g.* [47, 44].

2.6 Conclusions

This chapter has given an overview of technologies and models for wireless networking for control and monitoring applications. Starting with physical signal transmission of radio signals, we have described theoretical models for propagation, fading, and outage. These models describe how packets on a single isolated link can get corrupted and impossible to decode at the receiver. We shortly discussed various diversity techniques for minimizing the probability of outage, and coding techniques for correcting occasional bit errors. We have also discussed how the co-existence with other technologies on the license-free ISM band creates interference for low-power radios. Next, we described various methods for sharing the spectrum among several transmitters. These medium access control methods could broadly be characterized as scheduled or contention-based: scheduled medium access gives predictable channel access for predictable traffic, but tends to be inefficient when the traffic itself is random and uncorrelated. Contention-based medium access, on the other hand, is more efficient in dealing with random traffic but creates random and sometimes unnecessarily long latencies when traffic is correlated and predictable.

From the single link, we moved to multi-hop networks to illustrate several critical issues such as increased end-to-end latency and decreased reliability. We demonstrated how multi-path diversity improved reliability, especially in the case of correlated losses. We also described various routing paradigms, including single path and multi-path (graph-based) routing and relevant link metrics. With this basic understanding, we described several standards and specifications for low-power industrial wireless communication. The chapter was concluded with a short section describing how control-relevant models of latency and loss could be developed based on the individual models that we have described.

Acknowledgements. Luca Stabellini kindly let us use Figure 2.8 from his paper [62]. Zhenhua Zou generated the Figures 13 and 14 based on our recent work. Joonas Pesonen, Pablo Soldati, Haibo Zhang, Olaf Landsiedel and Maurice Heemels gave several constructive comments that helped to improve the quality of the text.

References

1. Abramson, N.: The ALOHA system – another alternative for computer communications. In: Proc. of the Fall Joint Computer Conference, pp. 281–285 (1970)
2. Akkaya, K., Younis, M.: A survey on routing protocols for wireless sensor networks. Ad Hoc Networks 3, 325–349 (2005)
3. Almstrom, P., Rabi, M., Johansson, M.: Networked state estimation over a Gilbert-Elliot type channel. In: Proc. IEEE Conference on Decision and Control, Shanghai, China (December 2009)
4. Antsaklis, P., Baillieul, J.: Special issue on technology of networked control systems. Proceedings of the IEEE 95(1) (January 2007)
5. Bertsekas, D., Gallager, R.: Data Networks. Prentice Hall, Englewood Cliffs (1987)
6. Bhatti, G., Mehta, A., Sahinoglu, Z., Zhang, J., Viswanathan, R.: Modified beacon-enabled IEEE 802.15.4 MAC for lower latency. In: Global Telecommunications Conference, IEEE GLOBECOM 2008, pp. 1–5. IEEE, Los Alamitos (2008)
7. Buchholz, P., Plonnigs, J.: Analytical analysis of access-schemes of CSMA type. In: Proc. IEEE International Workshop on Factory Communication Systems, Vienna, Austria (2004)
8. Buettner, M., Yee, G.V., Anderson, E., Han, R.: X-MAC: a short preamble mac protocol for duty-cycled wireless sensor networks. In: SenSys 2006: Proceedings of the 4th international conference on Embedded networked sensor systems, pp. 307–320 (2006)
9. Chen, P., Sastry, S.: Latency and connectivity analysis tools for wireless mesh networks. In: ROBOCOMM, p. 33 (2007)
10. Costa, O.L.V., Fragoso, M.D., Marques, R.P.: Discrete-Time Markov Jump Linear Systems. Springer, Heidelberg (2005)
11. De Couto, D.S.J.: High-Throughput Routing for Multi-Hop Wireless Networks. PhD thesis, MIT (2004)
12. Elliot, E.O.: Estimates of error rates for codes on burst-noise channels. Bell SystemTechnology Journal 39, 1253–1264 (1963)
13. Gao, C.: Performance and energy efficiency in wireless self-organized networks. PhD thesis, University of Vaasa (November 2009)

14. Geirhofer, S., Tong, L., Sadler, B.M.: Dynamic spectrum access in WLAN channels: Empirical model and its stochastic analysis. In: Proceedings of the First International Workshop on Technology and Policy for Accessing Spectrum, Boston, MA (2006)
15. Gilbert, E.N.: Capacity of a burst-noise channel. Bell Systems Technology Journal 39, 1253–1264 (1960)
16. Gorday, P.: 802.15.4 multipath. IEEE 803.15-4-0337-00-004b (2004)
17. Hespanha, J.P., Naghshtabrizi, P., Xu, Y.: A survey of recent results in networked control systems. Proceedings of the IEEE 95(1), 138–162 (2007)
18. Holland, G., Vaidya, N.: Analysis of TCP performance over mobile *ad hoc* networks. Wireless Networks 8(2), 275–288 (2002)
19. Huang, M., Dey, S.: Stability of Kalman filtering with Markovian packet losses. Automatica 43, 598–607 (2007)
20. IEEE standard for information technology - telecommunications and information exchange between systems - local and metropolitan area networks specific requirements part 15.4: Wireless medium access control (mac) and physical layer (phy) specifications for low-rate wireless personal area networks (lr-wpans). IEEE Std 802.15.4-2003 (2003)
21. IEEE recommended practice for information technology - telecommunications and information exchange between systems - local and metropolitan area networks - specific requirements part 15.2: Coexistence of wireless personal area networks with other wireless devices operating in unlicensed frequency bands. IEEE Std 802.15.2-2003 (2003)
22. ITU-R. Propagation data and prediction methods for the planning of indoor radiocommunication systems and radio local area networks in the frequency range 900 mhz to 100 ghz. Recommendation ITU-R P.1238-6 (2009)
23. Krishnamurty, P., Arauz, J., Labrador, M.A.: Discrete rayleigh fading channel modelling. Wireless Communications and Mobile Computing 4, 413–423 (2004)
24. Johansson, M., Xiao, L.: On optimal control over packet-oriented communication links. In: 38th Annual Allerton Conference on Communication, Control, and Computing, Monicello, vol. II (October 2000)
25. Kao, C.-Y., Lincoln, B.: Simple stability criteria for systems with time-varying delays. Automatica 40(8), 1429–1434 (2004)
26. Kemp, A., Bryant, E.: Channel sounding of industrial sites in the 2.4 GHz ISM band. Wireless Personal Communications 31, 235–248 (2004)
27. Kleinrock, L., Tobagi, F.: Packet switching in radio channels: Part I–carrier sense multiple-access modes and their throughput-delay characteristics. IEEE Transactions on Communications 23(12), 1400–1416 (1975)
28. Koumpis, K., Hanna, L., Andersson, M., Johansson, M.: Wireless industrial control and monitoring beyond cable replacement. In: Proc. 2nd PROFIBUS International Conference, Coombe Abbey, UK (June 2005)
29. Lennvall, T., Svensson, S., Hekland, F.: A comparison of WirelessHART and Zigbee for industrial applications. In: IEEE International Workshop on Factory Communication Systems, WFCS 2008, pp. 85–88 (May 2008)
30. Luck, R., Ray, A.: An observer-based compensator for distributed delays. Automatica 25(6), 903–908 (1990)
31. Mahmood, A., Hossain, M.M.A., Jantti, R.: Channel ranking algorithms for cognitive coexistence of IEEE 802.15.4. In: Proc. IEEE PIMRC 2009, Tokyo, Japan (2009)
32. MacLeod, H., Loadman, C., Chen, Z.: Experimental studies of the 2.4-GHz ISM wireless indoor channel, pp. 63–68 (May 2005)
33. Marrón, P.J., Minder, D.: Embedded WiSeNts Consortium. Embedded WiSeNts research roadmap. Logos Verlag, Berlin (2006)

34. Medepalli, K., Tobagi, F.A.: Towards performance modeling of IEEE 802.11 based wireless networks: a unified framework and its applications. In: Proc. IEEE Infocom, Bercelona, Spain (April 2006)
35. Mirkin, L.: Some remarks on the use of time-varying delay to model sample-and-hold circuits. IEEE Trans. Automat. Control 52(6), 1109–1112 (2007)
36. Misic, J., Shafi, S., Misic, V.B.: Performance of a beacon enabled IEEE 802.15.4 cluster with downlink and uplink traffic. IEEE Trans. Parallel Distrib. Syst. 17(4), 361–376 (2006)
37. Miskowicz, M., Sapor, M., Zych, M., Latawiec, W.: Performance analysis of predictive p-persistent CSMA protocol for control networks. In: 4th IEEE International Workshop on Factory Communication Systems, pp. 249–256 (2002)
38. Morse, J.: Market pulse: wireless in industrial systems: cautious enthusiasm. Industrial Embedded Systems 2(7), 10–11 (2006)
39. Nilsson, J.: Real-time control systems with delays. PhD thesis, Lund Institute of Technology, Lund, Sweden (1998)
40. Park, P., Fischione, C., Johansson, K.H.: Performance analysis of GTS allocation in beacon enabled IEEE 802.15.4. In: 6th Annual IEEE Communications Society Conference on Sensor, Mesh and Ad Hoc Communications and Networks, SECON 2009, pp. 1–9 (June 2009)
41. Pérez-Yuste, A.: Early developments of wireless remote control: The Telekino of Torres-Quevedo. Proceedings of the IEEE 96(1), 186–190 (2008)
42. Perkins, C.E., Royer, M.E.: Ad-hoc on demand distance vector routing. In: Proc. 2nd IEEE Workshop on Mobile Computing Systems and Applications, New Orleans, LA, pp. 90–100 (1999)
43. Pesonen, J.: Stochastic estimation and control over Wireless HART networks: Theory and implementation. Master's thesis, Royal Institute of Technology (February 2010)
44. Pesonen, J., Zhang, H., Soldati, P., Johansson, M.: Methodology and tools for controller-networking codesign in Wireless HART. In: Proc. 14th IEEE International Conference on Emerging Techonologies and Factory Automation, Palma di Mallorca, Spain (2009)
45. Petrova, M., Riihijarvi, J., Mahonen, P., Labella, S.: Performance study of IEEE 802.15.4 using measurements and simulations, vol. 1, pp. 487–492 (2006)
46. Polastre, J., Hill, J., Culler, D.: Versatile low power media access for wireless sensor networks. In: SenSys 2004: Proceedings of the 2nd international conference on Embedded networked sensor systems, pp. 95–107 (2004)
47. Rabi, M., Stabellini, L., Proutiere, A., Johansson, M.: Networked estimation under contention-based medium access. International Journal of Robust and Nonlinear Control (2010)
48. Rappaport, T.S.: Characterization of UHF multipath radio channels in factory buildings. IEEE Transactions on Antennas and Propagation 37(8), 1058–1069 (1989)
49. Rhee, I., Warrier, A., Aia, M., Min, J., Sichitiu, M.L.: Z-MAC: a hybrid mac for wireless sensor networks. IEEE/ACM Trans. Netw. 16(3), 511–524 (2008)
50. Roberts, L.G.: Dynamic allocation of satellite capacity through packet reservation. In: Computer communication networks, Noordhoff Internat Publishing, Groningen (1972)
51. Rom, R., Sidi, M.: Multiple access protocols – performance and analysis. Springer, Heidelberg (1989), `http://webee.technion.ac.il/people/rom/PDF/MAP.pdf`
52. Royer, E.M., Toh, C.-K.: A review of current routing protocols for *ad hoc* mobile wireless networks. IEEE Personal Communications, 46–55 (April 1999)
53. Jantti, R., Nethi, S., Nassi, V.: Time and antenna diversity in wireless sensor and actuator networks. In: Proc. IEEE MICC 2009, Kuala Lumpur, Malaysia (2009)

54. Schenato, L.: Optimal estimation in networked control systems subject to random delay and packet drop. IEEE Transactions on Automatic Control 53(5), 1311–1317 (2008)
55. Schenato, L., Sinopoli, B., Franceschetti, M., Poolla, K., Sastry, S.S.: Foundations of control and estimation over lossy networks. Proceedings of the IEEE 95(1), 163–187 (2007)
56. Sikora, A., Groza, V.F.: Coexistence of IEEE 802.15.4 with other systems in the 2.4 GHz ISM band, vol. 3, pp. 1786–1791 (2005)
57. Soldati, P., Zhang, H., Johansson, M.: Deadline-constrained transmission scheduling and data evacuation in WirelessHART networks. In: Proc. European Control Conference, Budapest, Hungary (September 2009)
58. Soldati, P., Zhang, H., Zou, Z., Johansson, M.: Optimal routing and scheduling of deadline-constrained traffic over lossy networks. In: IEEE Globecom 2010, Miami, Florida (2010)
59. Kemp, A., Bryant, E.: Channel sounding of industrial sites in the 2.4 GHz ISM band. Wireless Personal Communications 31, 235–248 (2004)
60. Stabellini, L.: Design of reliable communication solutions for wireless sensor networks, Licentate Thesis. Technical report, Royal Institute of Technology, KTH (2009)
61. Stabellini, L.: Quantifying and modeling spectrum opportunities in a real wireless environment. In: Proc. IEEE WCNC 2010, Sydney, Australia (April 2010)
62. Stabellini, L., Proutiere, A.: Evaluating delay and energy in sensor networks with sporadic and correlated traffic. In: The 7th Scandinavian Workshop on Wireless Ad-hoc and Sensor Networks, Stockholm, Sweden (2009)
63. Tanghe, E., Joseph, W., Verloock, L., Martens, L., Capoen, H., Van Herwegen, K., Vantomme, W.: The industrial indoor channel: large-scale and temporal fading at 900, 2400, and 5200 MHz. IEEE Transactions on Wireless Communications 7(7), 2740–2751 (2008)
64. Tipsuwan, Y., Chow, M.-Y.: Control methodologies in networked control systems. Control Engineering Practice, Special Section on Control Methods for Telecommunication 11(10), 1099–1111 (2003)
65. Vuran, M.C., Akyildiz, I.F.: Error control in wireless sensor networks: A cross layer analysis. IEEE/ACM Transactions on Networking 17(4), 1186–1199 (2009)
66. Wang, C., Sohraby, K., Li, B., Daneshmand, M., Hu, Y.: A survey of transport protocols for wireless sensor networks. IEEE Network 20(3), 34–40 (2006)
67. Wang, H.S., Moayeri, N.: Finite-state Markov channel-a useful model for radio communication channels. IEEE Transactions on Vehicular Technology 44(1), 163–171 (1995)
68. Willig, A.: A new class of packet- and bit-level models for wireless channels. In: The 13th IEEE International Symposium on Personal, Indoor and Mobile Radio Communications, vol. 5, pp. 2434–2440 (September 2002)
69. Woo, A., Tong, T., Culler, D.: Taming the underlying challenges in reliable multihop routing in sensor networks. In: ACM Sensys, Los Angeles, CA (2003)
70. Xiao, L., Hassibi, A., How, J.P.: Control with random communication delays via a discrete-time jump system approach. In: Proceedings of the 2000 American Control Conference, vol. 3, pp. 2199–2204 (2000)
71. Xu, K., Gerla, M., Bae, S.: How effective is the IEEE 802.11 RTS/CTS handshake in *ad hoc* networks. In: Global Telecommunications Conference. IEEE GLOBECOM 2002, vol. 1, pp. 72–76 (November 2002)
72. Xu, S., Saadawi, T.: Does the IEEE 802.11 MAC protocol work well in multihop wireless *ad hoc* networks. IEEE Communications Magazine (June 2001)
73. Yang, Y., Yum, T.-S.P.: Delay distributions of slotted ALOHA and CSMA. IEEE Transactions on Communications 51(11), 1846–1857 (2003)

74. Jamieson, K., Tay, Y.C., Balakrishnan, H.: Collision-minimizing CSMA and its applications to wireless sensor networks. IEEE Journal on Selected Areas in Communications 22(6), 1048–1057 (2004)
75. Ye, W., Heidemann, J., Estrin, D.: Medium access control with coordinated adaptive sleeping for wireless sensor networks. IEEE/ACM Transactions on Networking 12(3), 493–506 (2004)
76. Zou, Z., Soldati, P., Zhang, H., Johansson, M.: Delay-constrained maximum-reliability routing over lossy links. In: IEEE Conference on Decision and Control, Atlanta, GA (December 2010)
77. Winter, T., Thubert, P. (eds.), The ROLL Team: RPL. IPv6 Routing Protocol for Low power and Lossy Networks. Internet Draft draft-ietf-roll-rpl-06 (work in progress)

Chapter 3
A Survey on Distributed Estimation and Control Applications Using Linear Consensus Algorithms

Federica Garin and Luca Schenato

Abstract. In this chapter we present a popular class of distributed algorithms, known as linear consensus algorithms, which have the ability to compute the global average of local quantities. These algorithms are particularly suitable in the context of multi-agent systems and networked control systems, *i.e.* control systems that are physically distributed and cooperate by exchanging information through a communication network. We present the main results available in the literature about the analysis and design of linear consensus algorithms,for both synchronous and asynchronous implementations. We then show that many control, optimization and estimation problems such as least squares, sensor calibration, vehicle coordination and Kalman filtering can be cast as the computation of some sort of averages, therefore being suitable for consensus algorithms. We finally conclude by presenting very recent studies about the performance of many of these control and estimation problems, which give rise to novel metrics for the consensus algorithms. These indexes of performance are rather different from more traditional metrics like the rate of convergence and have fundamental consequences on the design of consensus algorithms.

3.1 Introduction

In the past decades we have being witnessing the growth of engineering systems composed by a large number of devices that can communicate and cooperate to achieve a common goal. Although complex large-scale monitoring and control systems are not new, as for example nuclear plants and air traffic control, a new

---

Federica Garin
INRIA Grenoble Rhône-Alpes, France
e-mail: `federica.garin@inrialpes.fr`

Luca Schenato
Department of Information Engineering, University of Padova, Italy
e-mail: `schenato@dei.unipd.it`

A. Bemporad, M. Heemels, and M. Johansson: Networked Control Systems, LNCIS 406, pp. 75–107.
springerlink.com

architectural paradigm is emerging, mainly due to the adoption of smart agents, *i.e.*, devices that have the ability to cooperate and to take autonomous decisions without any supervisory system. In fact, traditional large-scale systems have a centralized or at best a hierarchical architecture, which has the advantage to be relatively easy to be designed and has safety guarantees. However, these systems require very reliable sensors and actuators, are generally very expensive, and do not scale well due to communication and computation limitations. The recent trend to avoid these problems is to substitute costly sensors, actuators and communication systems with a larger number of devices that can autonomously compensate potential failures and computation limitations through communication and cooperation. Although very promising, this new paradigm brings new problems into the picture, mainly due to the lack of analysis and design tools for such systems. In particular, there are only few tools for predicting the global behavior of the system as a whole starting from the local sensing and control rules adopted by the smart sensors and actuators. As a consequence, there has been a strong effort in past years by many engineering areas to develop such tools.

One of the most promising tools are the linear consensus algorithms, which are simple distributed algorithms which require only minimal computation, communication and synchronization to compute averages of local quantities that reside in each device. These algorithms have their roots in the analysis of Markov chains [53] and have been deeply studied within the computer science community for load balancing [61, 42] and within the linear algebra community for the asynchronous solution of linear systems [30, 56]. More recently they have been rediscovered and applied by the control and robotics communities for cooperative coordination of multi-agent systems, as surveyed in [52, 51] and in the recent book [12].

The spirit of this chapter is mostly tutorial. We start in Section 3.2 by presenting a coherent description of the linear consensus algorithms and by surveying the most important results. No prior knowledge is required except for standard linear algebra and control systems theory. A special attention has been placed on the design of such algorithms, which, in our opinion, is one of the most relevant aspects for a control engineer. In Section 3.3 we illustrate through some examples how these algorithms can be applied to relevant estimation and control problems such as least squares, sensor calibration, and vehicle coordination, just to name a few. Section 3.4 presents some more recent research directions. More precisely, starting from the analysis of control applications of consensus algorithms, such as those described in Section 3.3, we show that the performance indexes to be considered are different from the traditional index given by rate of convergence, *i.e.* the essential spectral radius of the consensus matrix, and in general this index depends on all the eigenvalues of the consensus matrix. This observation has relevant consequences in terms of analysis and design of consensus algorithms, which goes beyond the current results and opens up new research directions, which we believe are particularly relevant for the control community.

3.2 Linear Consensus Algorithms: Definitions and Main Results

In this section, we review some of the main results on the analysis and design of consensus algorithms and we also provide references for more recent developments under different scenarios and assumptions. In particular, we will concentrate on linear discrete-time consensus algorithms. However we will give some references to continuous time and nonlinear consensus. We start by introducing some mathematical preliminaries. Let us consider the following linear update equation:

$$x(t+1) = Q(t)x(t) \tag{3.1}$$

where $x(t) = [x_1(t)\ x_2(t)\ \cdots\ x_N(t)]^T \in \mathbb{R}^N$ and, for all t, $Q(t) \in \mathbb{R}^{N\times N}$ is a *stochastic matrix*, *i.e.* $[Q(t)]_{ij} = q_{ij}(t) \geq 0$ and $\sum_{j=1}^N q_{ij} = 1, \forall i$, *i.e.* each row sums to unity. Equation (3.1) can be written as

$$\begin{aligned} x_i(t+1) &= \sum_{j=1}^{N} q_{ij}(t)x_i(t), \quad i = 1,\dots,N & (3.2)\\ &= x_i(t) + \sum_{j\neq i} q_{ij}(t)(x_j(t) - x_i(t)) & (3.3) \end{aligned}$$

where the local updates of each component of the vector x is written explicitly.

A stochastic matrix Q is said *doubly-stochastic* if also $\sum_{i=1}^N q_{ij} = 1, \forall j$, *i.e.* each column sums to unity. Clearly if a stochastic matrix is symmetric, *i.e.* $Q = Q^T$, then it is also doubly-stochastic. An important class of doubly-stochastic matrices is given by the class of stochastic matrices which are also circulant. A matrix $Q = \mathrm{circ}(c_1, c_2, \dots, c_N)$ is a *circulant matrix* if

$$Q = \begin{bmatrix} c_1 & c_2 & c_3 & \cdots & c_N \\ c_N & c_1 & c_2 & \cdots & c_{N-1} \\ \vdots & & & \ddots & \vdots \\ c_2 & c_3 & c_4 & \cdots & c_1 \end{bmatrix} \tag{3.4}$$

All eigenvalues λ_i of a stochastic matrix Q are included in the unit circle, *i.e.* $|\lambda_i| \leq 1$, and the vector $\mathbf{1} = [1\ 1 \cdots 1]^T \in \mathbb{R}^N$ is an eigenvector for Q and its eigenvalue is equal to one, i.e. $Q\mathbf{1} = \mathbf{1}$. The *essential spectral radius* $\mathtt{esr}(Q)$ of a stochastic matrix Q is defined as the second largest eigenvalue in modulus of the matrix Q, *i.e.* if we consider the ordered eigenvalues in modulus $1 = |\lambda_1| \geq |\lambda_2| \geq \cdots \geq |\lambda_N|$, then $\mathtt{esr}(Q) = |\lambda_2|$.

Many important results about convergence of consensus algorithms can be reframed as graph properties. Therefore we provide some useful preliminary definitions. We define the (directed) graph associated with a stochastic matrix Q as $\mathscr{G}_Q = (\mathscr{N}, \mathscr{E}_Q)$, where the nodes are $\mathscr{N} = \{1, 2, \dots, N\}$ and the edges are $\mathscr{E}_Q = \{(j,i)\,|\,q_{ij} > 0\}$, *i.e.* $(j,i) \in \mathscr{E}$ implies that node i can receive information from node j. A graph is *undirected* if $(i,j) \in \mathscr{E}$ implies that also $(j,i) \in \mathscr{E}$.

We also say that a matrix Q is *compatible* with the graph $\mathscr{G} = (\mathscr{N}, \mathscr{E})$ if its associated graph $\mathscr{G}_Q = (\mathscr{N}, \mathscr{E}_Q)$ is such that $\mathscr{G}_Q \subseteq \mathscr{G}$, *i.e.*, is a subgraph of $\mathscr{G}$. We denote with $\mathbb{G}_{\text{sl}}$ the set of graphs which include all self-loops, *i.e.* $\mathscr{G} \in \mathbb{G}_{\text{sl}}$ if and only if $(i,i) \in \mathscr{E}, \forall i \in \mathscr{N}$. The *in-degree* of a node i is defined as $d_{in}(i) = |\mathscr{V}_{in}(i)|$, where $\mathscr{V}_{in}(i) = \{j \,|\, (j,i) \in \mathscr{E}, i \neq j\}$ is the set of neighbors that can send information to i and $|\cdot|$ indicates the cardinality of a set. Similarly, the *out-degree* of a node i is defined as $d_{out}(i) = |\mathscr{V}_{out}(i)|$ and $\mathscr{V}_{out}(i) = \{j \,|\, (i,j) \in \mathscr{E}, i \neq j\}$. For an undirected graph, in-neighbors and out-neighbors of a node i coincide and they are simply denoted by the set $\mathscr{V}(i)$ whose degree is $d(i) = |\mathscr{V}(i)|$.

The *adjacency matrix* $A \in \{0,1\}^{N \times N}$ of a graph $\mathscr{G} = (\mathscr{N}, \mathscr{E})$ is defined as $[A]_{ij} = 1$ if $(i,j) \in \mathscr{E}$ and $i \neq j$, and $[A]_{ij} = 0$ otherwise. The *Laplacian matrix* L of a undirected graph is defined as $L = D - A$, where $D = \text{diag}\{d(1), d(2), \ldots, d(N)\}$ is diagonal and $d(i)$ is the degree of node i. The Laplacian L is positive semidefinite and $L\mathbf{1} = 0$.

A graph is *rooted* if there exists a node $k \in \mathscr{N}$ such that for any other node $j \in \mathscr{N}$ there is a unique path from k to j. A graph is *strongly connected* if there is a path from any node to any other node in the graph. Clearly a strongly connected graph implies that it is also rooted for any node. The *diameter* of a graph is defined as the length of the longest among all shortest paths connecting any two nodes in a strongly connected graph. A graph is *complete* if $(i,j) \in \mathscr{E}, \forall i,j \in \mathscr{N}$. The *union* of two graphs $\mathscr{G}_1 = (\mathscr{N}, \mathscr{E}_1)$ and $\mathscr{G}_2 = (\mathscr{N}, \mathscr{E}_2)$ is defined as the graph $\mathscr{G} = (\mathscr{N}, \mathscr{E}) = \mathscr{G}_2 \cup \mathscr{G}_1$ where $\mathscr{E} = \mathscr{E}_1 \cup \mathscr{E}_1$.

3.2.1 Analysis

In this section we describe three main frameworks for modeling consensus algorithms. The first is related to static synchronous implementation, where updates at each node are performed simultaneously, thus being well-represented by constant matrices. The second and the third are both more suitable for modeling asynchronous implementations, where information exchanges and local variable updates are not necessarily coordinated, thus being well-represented by time-varying matrices. The second framework addresses the problem of finding the weakest sufficient conditions that guarantee convergence to consensus from a worst-case point of view, thus being able to characterize a wide class of consensus implementations. The drawback of this approach is that it is very hard to estimate performance indexes such as the rate of convergence and, when possible, the predictions are often over-pessimistic. The third framework considers randomized asynchronous implementations which has three main advantages as compared to the second approach. The first advantage is that randomized communication and updates require almost no coordination among nodes and are easy to implement in practice. The second advantage is that this approach naturally models stochastic nature of the environment, such as communication losses, communication noise and quantization. The third advantage is that the estimation of performance such as rate of convergence is closer to the experimental performance observed through simulations and experiments.

Let us consider the following consensus problem definitions:

Definition 3.1. *Let us consider Eqn. (3.1). We say that $Q(t)$ solves the* ***consensus problem*** *if* $\lim_{t\longrightarrow\infty} x_i(t) = \alpha$, $\forall i = 1,\ldots,N$, *where $x_i(t)$ is the i-th component of the vector $x(t)$. We say that $Q(t)$ solves the* ***average consensus problem*** *if in addition to the previous condition we have $\alpha = \frac{1}{N}\sum_{i=1}^{N} x_i(0)$. If $Q(t)$ is a random variable, then we say that Q solves the* ***probabilistic (average) consensus problem*** *if the limit above exists almost surely.*

These definitions include a wide class of consensus strategies: strategies with a time–invariant matrix $Q(t) = Q$, deterministic time-varying strategies $Q(t)$, and randomized strategies where $Q(t)$ is drawn from a set of stochastic matrices $\mathscr{Q}$ according to a probability distribution. The next theorem describes some sufficient conditions which guarantee deterministic and probabilistic (average) consensus.

Theorem 3.1. *Let us consider the sequence of constant matrices $Q(t) = Q$. If the graph $\mathscr{G}_Q \in \mathbb{G}_{\mathrm{sl}}$ and is rooted, then Q solves the consensus problem, and*

$$\lim_{t\longrightarrow\infty} Q^t = \mathbf{1}\eta^T$$

where $\eta \in \mathbb{R}^N$ is the left eigenvector of Q for the eigenvalue one and has the properties $\eta_i \geq 0$ and $\mathbf{1}^T\eta = 1$. If $\mathscr{G}_Q$ is strongly connected, then $\eta_i > 0, \forall i$. If in addition Q is doubly-stochastic, then $\mathscr{G}_Q$ is strongly connected and Q solves the average consensus problem, i.e. $\eta = \frac{1}{N}\mathbf{1}$. Moreover, in all cases the convergence is exponential and its rate is given by the essential spectral radius $\mathtt{esr}(Q)$.

This theorem is well known and can be found in many textbooks on Markov chains such as in [53]. The assumption that $\mathscr{G}_Q \in \mathbb{G}_{\mathrm{sl}}$ is not necessary to achieve consensus; for example consider $Q = \begin{bmatrix} 1 & 0 \\ 1 & 0 \end{bmatrix}$, for which $x(t) = x_1(0)\begin{bmatrix} 1 \\ 1 \end{bmatrix}$ for each $t \geq 1$ and $x(0) = [x_1(0)\ x_2(0)]^T$. However, some additional assumption besides $\mathscr{G}_Q$ being rooted is actually needed in order to guarantee consensus: for example $Q = \begin{bmatrix} 0 & 1 \\ 1 & 0 \end{bmatrix}$ is such that $\mathscr{G}_Q$ is rooted, but it gives $x(2t) = \begin{bmatrix} x_1(0) \\ x_2(0) \end{bmatrix}$ and $x(2t+1) = \begin{bmatrix} x_2(0) \\ x_1(0) \end{bmatrix}$ for all t. In this chapter, for the sake of simplicity, we will use the assumption that $\mathscr{G}_Q \in \mathbb{G}_{\mathrm{sl}}$, also noting that this is a very mild requirement since it means that any agent can communicate to itself; however in some cases, such as in the de Bruijn graphs [24], it is useful to consider also graphs not in $\mathbb{G}_{\mathrm{sl}}$.

Besides the results on constant matrices Q, much research has been devoted to the analysis of time-varying linear consensus which is addressed by the next theorem.

Theorem 3.2. *Consider the deterministic sequence of stochastic matrices $\{Q(t)\}_{t=0}^{+\infty}$ and the corresponding associated graphs $\mathscr{G}(t) = \mathscr{G}_{Q(t)}$. Suppose $\mathscr{G}(t) \in \mathbb{G}_{\mathrm{sl}}, \forall t$. Then the sequence $Q(t)$ solves the consensus problem if and only if there exists a finite positive integer number T such that the graphs $\overline{\mathscr{G}}(\tau)$ obtained from the* *union of the graphs $\mathscr{G}(\tau)$ in the following way: $\overline{\mathscr{G}}(\tau) = \mathscr{G}(\tau) \cup \mathscr{G}(\tau+1) \cup \ldots \cup \mathscr{G}(\tau+T-1)$ with $\tau = 0, 1, \ldots$ are all rooted. If in addition the matrices $Q(t)$ are all doubly-stochastic, then they solve the average consensus problem.*

A simple proof of the previous theorem can be found in [41], but its roots can be tracked back at least to [61], and it has been rediscovered several times in the past years [33, 50, 8, 13]. The previous theorem states that it is not necessary for graphs associated to the matrices $Q(t)$ to be connected at all time, but only over a time window. This assumption basically guarantees that information travels, possibly with some delay, from at least one node to all other nodes infinitely many times. What is particularly remarkable in this theorem and also in Theorem 3.1, is that convergence is completely characterized by connectivity properties of the graphs $\mathscr{G}_{Q(t)}$, regardless of the specific values of the entries of the matrices $Q(t)$. On the other hand, the negative side is that the rate of convergence is hard to estimate since it is based on worst-case analysis. Therefore in general it is over-pessimistic and of little practical use. Recent work has tried to address this problem by finding tighter bounds on the rate of convergence while adding only general constraints on the topological properties of the graphs $\mathscr{G}_{Q(t)}$ and on the numerical values for the entries of $Q(t)$ [2].

A more recent approach to consensus is to model time-varying consensus in term of randomized strategies. The advantage of a randomized approach is to preserve simple convergence conditions based on graph properties while obtaining good estimates for the rate of convergence of typical realizations. The next theorem provides convergence conditions for the randomized linear consensus.

Theorem 3.3. *Consider a random i.i.d. sequence of stochastic matrices $\{Q(t)\}_{t=0}^{+\infty}$ drawn according to some probability distribution from the set $\mathscr{Q}$, and the stochastic matrix $\overline{Q} = \mathbb{E}[Q(t)]$. If the graphs $\mathscr{G}(t) = \mathscr{G}_{Q(t)} \in \mathbb{G}_{\mathrm{sl}}, \forall t$ and if $\mathscr{G}_{\overline{Q}}$ is rooted, then the sequence $Q(t)$ solves the probabilistic consensus problem. The rate of convergence in mean square sense defined as $\rho = \sup_{x(0)} \limsup_{t\to\infty} (\mathbb{E}[||x(t) - x(\infty)||^2])^{1/t}$ is bounded by*

$$(\mathtt{esr}(\overline{Q}))^2 \leq \rho \leq \mathtt{sr}(\mathbb{E}[Q^T(t)\Omega Q(t)])$$

where $\Omega := I - \frac{1}{N}\mathbf{1}\mathbf{1}^T$ and $\mathtt{sr}(P)$ indicates the spectral radius of the matrix P, i.e. its largest eigenvalue in absolute value. If in addition $Q(t)$ are all doubly-stochastic, then they solve the probabilistic average consensus problem.

The proof of this theorem can be found in [26]. Similarly to the previous two theorems, even in a randomized scenario the convergence conditions are characterized in terms of graphs connectivity properties. In particular, it states that convergence is guaranteed if the graph is connected on average. However, differently from Theorem 3.2, the randomized framework provides tighter bounds on the rate of convergence. Another advantage of considering a randomized framework is the ability to model scenarios subject to random communication links or nodes failure.

There is a rich literature on randomized consensus that extends the results of the previous theorem. One direction is to find weaker convergence conditions, more specifically by relaxing the hypothesis of i.i.d. sequences to ergodicity only [58]. Another direction is to add additional hypotheses on the matrices $Q(t)$ or on the set $\mathscr{Q}$ in order to improve the convergence bounds. For example, in [11] it was shown that if $Q(t)$ are symmetric and idempotent. *i.e.* $Q(t) = Q^T(t)$ and $Q^2(t) = Q(t)$, then the upper bound is given by $\mathtt{sr}(\mathbb{E}[Q^T(t)\Omega Q(t)]) = \mathtt{esr}(\overline{Q})$.

There is also a rich literature on the analysis of consensus under different scenarios. For example, there is an equivalent version of the consensus problem in continuous time given by

$$\dot{x} = A(t)x \tag{3.5}$$

where A is a *Metzler matrix*, *i.e.* a matrix whose off-diagonal elements are nonnegative and the row-sum is null, *i.e.* $A\mathbf{1} = 0$. This types of systems have been well characterized by Moreau [40]. For example, the opposite of a Laplacian matrix is a Metzler matrix, which implies that $\dot{x} = -Lx$ achieves consensus under general connectivity properties of the associated graph. The continuous time framework is particularly suitable for modeling flocking and vehicle dynamics [28, 52, 59].

Another research direction is concerned with convergence conditions for consensus with delayed information, *i.e.* for consensus whose dynamics can be written as

$$x_i(t+1) = \sum_{j=1}^{N} q_{ij} x_j(t - \tau_i(t)), \quad i = 1, \ldots, N$$

where the delay $\tau_i(t)$ can be unknown and time-varying [46, 8, 7, 60, 54, 62]. The main finding is that consensus is very robust to delay, which is particularly important in networked systems where delay is unavoidable. This comes from the observation that the convex hull of the points $x_i(t)$ can only shrink or remain constant, and delay only marginally affects this property [41, 8].

Also much interest has been generated from consensus subject to quantization and in particular to quantized communication. In this context the dynamics can be written as

$$x_i(t+1) = \sum_{j=1}^{N} q_{ij} \mathfrak{q}_d(x_j(t)), \quad i = 1, \ldots, N$$

where $\mathfrak{q}_d(\cdot) : \mathbb{R} \to \mathfrak{Q}_d$ and $\mathfrak{Q}_d$ is a finite or countable set. A typical example is $\mathfrak{q}_d(x) = \lfloor x \rfloor$, where $\lfloor x \rfloor$ indicates the largest integer smaller than x. This problem is particularly challenging due to the fact that quantization acts similarly to noise, thus being particularly harmful since the consensus matrices $Q(t)$ are not strictly stable but always have an eigenvalue in one and convergence might not be guaranteed. Therefore, much effort has been given in finding quantization strategies and quantization functions that still guarantee consensus [37, 18, 29, 38, 36, 43].

Another interesting aspect is related to consensus subject to lossy communication , *i.e.* a scenario where communication scheduled between two nodes fails due to random interference or noise. This scenario naturally fits the randomized framework of Theorem 3.3, however it also requires the design of a compensation mechanism when a packet is lost. Different strategies have been proposed and studied [35, 27, 47]. For example a natural scheme is to compensate for the lost packets by replacing the the lost value x_j from the transmitting node j with the self value x_i of the receiving node i, more formally:

$$x_i(t+1) = \Big(q_{ii} + \sum_{j=1,i\neq j}^{N} (1-\gamma_{ij}(t))q_{ij}\Big)x_i + \sum_{j=1,i\neq j}^{N} \gamma_{ij}(t)q_{ij}x_j(t), \quad i=1,\ldots,N$$

where $\gamma_{ij}(t)$ is a random variable such that $\gamma_{ij}(t) = 1$ if transmission at time t from node j to node i was successful, and $\gamma_{ij}(t) = 0$ otherwise [27]. These works show that packet loss in general does not affect convergence to consensus, but it can reduce convergence rate and change the final consensus value as compared to ideal scenario with perfect communication, *i.e.* $\gamma_{ij}(t) = 1, \forall i, j, t$.

A different setting is studied in [64], where additive noise is included in the consensus dynamics, *i.e.*

$$x(t+1) = Qx(t) + v(t)\,.$$

Note that, in all cases described above, noise affects the speed of convergence and the final value obtained (which is not the desired average), but does not prevent convergence. Differently, in the case when there is noise in the transmissions among nodes (without feedback), so that the messages sent by an agent are received by its neighbors corrupted by noises which might be different, and which are unknown to the sender, then convergence itself is an issue. The difficulty is in the design of a modified consensus algorithm capable of avoiding noise accumulation. Algorithms dealing with variations on this setting have been designed and analyzed by various authors, *e.g.* [49, 32, 34] (using time-varying weights in the consensus algorithm, to decrease the effect of neighbors' noise) and [16] (using error-correcting codes of increasing length to decrease the communication noise).

3.2.2 Design

Up to now, we provided a short overview of the properties of consensus algorithms under different scenarios and assumptions. However, in many engineering applications it is also very important to be able to design such algorithms. From a consensus design perspective, the design space is given by the communication graph $\mathscr{G} = \{\mathscr{N}, \mathscr{E}\}$ of a network of $N = |\mathscr{N}|$ agents, and the design problem consists in finding suitable $Q(t)$ compatible with $\mathscr{G}$ that achieve consensus or average consensus. We assume that the graph $\mathscr{G}$ includes self-loops, *i.e.* $\mathscr{G} \in \mathbb{G}_{\mathrm{sl}}$, and that it is at least rooted.

There are two main approaches to design. The first focuses on local design methods which require only local information, *i.e.* each node can design its communication and consensus updates weights almost independently of the other nodes. Obviously, with this approach optimality with respect to some performance index is not guaranteed. The second approach focuses on methods which try to optimize some global performance index. As a consequence, this often leads to a centralized optimization problem that strongly depends on the topology and might be suitable if the network static and has small size. We start by presenting these two approaches first within the context of static consensus, *i.e.* $Q(t) = Q$ and then in the context of time-varying consensus strategies.

3.2.2.1 Matrix Design – Static Consensus: Q

If only consensus is required then a simple local strategy to design the matrix Q is given by:

$$q_{ij} = \frac{1}{d_{in}(i)+1}, \quad (j,i) \in \mathscr{E}$$

Clearly $\mathscr{G}_Q = \mathscr{G}$, and Q is stochastic, thus satisfying hypotheses of Theorem 3.1.

Differently, if average consensus is required, various solutions are possible. If the graph is undirected a possible solution is to choose:

$$q_{ij} = \begin{cases} \varepsilon & \text{if } (j,i) \in \mathscr{E} \text{ and } i \neq j \\ 1 - \varepsilon d(i) & \text{if } i = j \end{cases} \tag{3.6}$$

where $\varepsilon < \frac{1}{\max_i d(i)}$. This matrix is clearly symmetric since the non-zero off-diagonal terms are all equal and positive $q_{ij} = q_{ji} = \varepsilon, \forall i, j$. The condition on ε is necessary to guarantee that all diagonal terms are positive. As a consequence, Q is a stochastic symmetric matrix, therefore it is also doubly-stochastic. Moreover $\mathscr{G}_Q = \mathscr{G}$ and by hypothesis $\mathscr{G}$ is rooted[1], thus satisfying hypotheses of Theorem 3.1. Note that this matrix is strongly related to the Laplacian matrix L of the graph $\mathscr{G}$. In fact, consider the discretized dynamics of Eqn. (3.5) where $A = -L$ with time step ε, *i.e.* $x(t+1) = e^{-\varepsilon L}x(t) = Q_d x(t)$, then the first order expansion of Q_d, *i.e.* $Q_d = I - \varepsilon L + O(\varepsilon)$, has the same structure of the Q given by Eqn. (3.6).

Another possible strategy for undirected graphs is based on the Metropolis-Hastings weights:

$$q_{ij} = \begin{cases} \frac{1}{\max(d(i),d(j))+1} & \text{if } (j,i) \in \mathscr{E} \text{ and } i \neq j \\ 1 - \sum_{j=1,i\neq j}^{N} q_{ij} & \text{if } i = j \end{cases} \tag{3.7}$$

Clearly the matrix Q is symmetric and the diagonal elements are strictly positive since $q_{ii} = 1 - \sum_{j=1,i\neq j}^{N} q_{ij} \geq 1 - \sum_{j=1,i\neq j,(i,j)\in\mathscr{E}}^{N} \frac{1}{d(i)+1} = 1 - \frac{d(i)}{d(i)+1} = \frac{1}{d(i)+1} > 0$, therefore Q is doubly-stochastic and $\mathscr{G}_Q = \mathscr{G}$ which are sufficient conditions to guarantee average consensus. As compared to the strategy based on the Laplacian of Eqn. (3.6), the strategy based on the Metropolis weights of Eqn. (3.7) is local, *i.e.* each node requires only the knowledge of local information, namely the degrees of its neighbors, while the former requires the knowledge of an upper bound on the degree of all nodes of the network. Moreover, the Metropolis-based consensus matrix has in general faster convergence rate than the Laplacian-based consensus matrix.

If the communication graph $\mathscr{G}$ is directed, then the design of a consistent doubly-stochastic matrix is not trivial. A possible strategy is based on the design of a doubly-stochastic matrix based on a convex combination of permutation matrices, where a permutation matrix P is defined as $P \in \{0,1\}^{N\times N}, \mathbf{1}^T P = \mathbf{1}^T, P\mathbf{1} = \mathbf{1}$. Note that a permutation matrix is doubly-stochastic. This procedure is basically an application

[1] If an undirected graph is rooted, then it is also strongly connected.

of the Birkhoff's Theorem [39]. We start from the assumption that the graph is strongly connected. This implies that for each edge $e=(j,i)\in\mathscr{E}$ there exists a path connecting node i to node j, which in turns implies there exists at least one simple cycle $\mathscr{C}$ in the graph including the edge e, *i.e.* there exists a sequence of non repeated vertices $\ell_1,\ell_2,\dots,\ell_L\in\mathscr{N}$ such that $\ell_1=i,\ell_L=j$, $(\ell_i,\ell_{i+1})\in\mathscr{E}$ for $i=1,\dots,L-1$ and $(\ell_L,\ell_1)\in\mathscr{E}$. Associated to this cycle it is possible to define a permutation matrix P_e as follows:

$$\begin{array}{ll}[P_e]_{\ell_r\ell_{r+1}}=1 \text{ for } r=1,\dots,L-1 & \\ [P_e]_{\ell_L\ell_1}=1 & \\ [P_e]_{kk}=1 & \text{for } k\neq\ell_r,\ r=1,\dots,L \\ [P_e]_{hk}=0 & \text{otherwise}\end{array}$$

Clearly $\mathscr{G}_{P_e}\subseteq\mathscr{G}$. According to this procedure it is always possible to find M cycles in the graph $\mathscr{G}$ and permutation matrices $P_i, i=1,\dots,M$ constructed as above, that includes all edges of the graphs. Let us consider now the matrix $Q=a_0I+\sum_{i=1}^M a_iP_i$ where $a_i>0,\forall i=0,\dots,M$ and $\sum_{i=0}^M a_i=1$, then Q is still doubly-stochastic since it is a convex combination of doubly-stochastic matrices. Also since all edges of $\mathscr{G}$ are included in Q, then $\mathscr{G}_Q=\mathscr{G}$. These two facts guarantee that Q achieves average consensus.

However, this procedure is rather tedious and requires global knowledge of the graph topology. There is an elegant alternative solution to achieve average consensus [1], which requires only local knowledge of the graph topology. Let us consider the matrix Q designed as follows:

$$q_{ij}=\frac{1}{d_{out}(j)+1},\quad (j,i)\in\mathscr{E}$$

This matrix is column-stochastic, *i.e.* its transpose is stochastic ($Q^T\mathbf{1}=\mathbf{1}$), and $\mathscr{G}_Q=\mathscr{G}$ is strongly connected. This implies by Theorem 3.1 that $\lim_{t\to\infty}Q^t=\lim_{t\to\infty}((Q^T)^t)^T=(\mathbf{1}\rho^T)^T=\rho\mathbf{1}^T$ where $\rho_i>0,\forall i$. Now let us consider $z(t+1)=Qz(t)$ and $w(t+1)=Qw(t)$ where the initial condition are $z(0)=x(0)$ and $w(0)=\mathbf{1}$, and the $x(t)$ such that $x_i(t)=\frac{z_i(t)}{w_i(t)}$. From $\lim_{t\to\infty}Q^t=\rho\mathbf{1}^T$, it follows that $\lim_{t\to\infty}z(t)=\left(\sum_{i=1}^N z_i(0)\right)\rho=\left(\sum_{i=1}^N x_i(0)\right)\rho$ and $\lim_{t\to\infty}w(t)=\left(\sum_{i=1}^N w_i(0)\right)\rho=N\rho$, therefore $\lim_{t\to\infty}x_i(t)=\frac{\rho_i\left(\sum_{i=1}^N x_i(0)\right)}{\rho_iN}=\frac{1}{N}\sum_{i=1}^N x_i(0)$ as desired. Note that average consensus is achieved through a nonlinear algorithm that uses two parallel linear iterative updates very similar to standard consensus. The weak point of this approach is that perfect communication is required since the algorithm can become unstable if lossy links are considered.

So far, we just considered design strategies to achieve consensus or average consensus, but we did not discuss about their rate of convergence. Design of consensus algorithms with fast rate of convergence is not a trivial task. If simple consensus is required, there is a simple strategy that achieves in a finite number of steps. Given a rooted graph, it is always possible to find a tree that connects one node, namely the root, to all other nodes. Without loss of generality, assume that the root is node $i=1$, and let us consider only the set of directed edges associated with this tree, *i.e.* $\mathscr{E}_{\text{tree}} \subseteq \mathscr{E}$. Note that $\mathscr{E}_{\text{tree}}$ does not contain self-loops. Let us consider the matrix Q designed as follows:

$$q_{11} = 1, \quad q_{ij} = 1 \quad (j,i) \in \mathscr{E}_{\text{tree}}, j \neq 1$$

Clearly the matrix is stochastic and it is not difficult to see that $Q^t = \mathbf{1}[1 \ 0 \ \cdots \ 0]$ for $t \geq \ell$, *i.e.* $x_i(t) = x_1(0)$ for $t \geq \ell$, where ℓ is the maximum distance of all nodes from the root. This implies that $\mathtt{esr}(Q) = 0$. In other words, each node sets the value of its variable $x_i(t)$ to the value received from its parents, therefore after a finite number of steps all nodes will have a copy of the initial condition of the root. This gives very fast convergence rate even for very large networks, as long as the diameter, *i.e.* the largest path distance within any two nodes, is small.

If average consensus is required, then the previous strategy is obviously not suitable. Optimal design of Q in terms of fast rate of convergence is not trivial in directed graph. If the graph is undirected, then it has been shown by Xiao et al. [63] that finding a symmetric stochastic matrix consistent with the graph with smallest $\mathtt{esr}$ is a convex problem. *i.e.*

$$\begin{aligned} \min_{Q} \quad & \mathtt{esr}(Q) \\ \text{s.t.} \quad & Q = Q^T, Q\mathbf{1} = \mathbf{1}, [Q]_{ij} \geq 0, \mathscr{G}_Q = \mathscr{G} \end{aligned}$$

Actually the non-negativeness constraint on the elements of Q is not necessary to have a convex problem, and therefore can be removed, thus providing a matrix Q with possible negative entries which can lead to an even smaller $\mathtt{esr}$. On the other hand, this is a centralized optimization problem, and the whole topology of the network is needed to find the optimal solution. Local optimization strategies to minimize the $\mathtt{esr}$ are still an open area of research.

3.2.2.2 Matrix Design – Dynamic Consensus: $Q(t)$

Now, we address the problem of designing dynamic consensus strategies where the consensus matrix is not constant but can change over time. The major drawback of static consensus is that it requires some sort of synchronization among all nodes of the network. In fact, between one iteration and the subsequent iteration, nodes need to exchange information and then update their local variables simultaneously. This can be difficult to enforce or simply too costly. Therefore, there is much interest in designing consensus strategies that require little coordination and synchronization among nodes. These algorithms are also referred as *asynchronous algorithms*. Some of the most popular asynchronous strategies are motivated by practical consideration based on the communication schemes that can be implemented on networks. These include broadcast [3], asymmetric gossip [25] and symmetric gossip [11].

In the *broadcast* scheme , one node i transmits its information to all its neighbors $\mathscr{V}_{out}(i)$, and each receiving node updates its local variable using consensus. More formally, given a possibly directed graph $\mathscr{G} = (\mathscr{N}, \mathscr{E})$, then $Q(t) \in \mathbb{Q}_B = \{Q_1, Q_2, \ldots, Q_N\}$, where $N = |\mathscr{N}|$ and

$$Q_i = I - w \sum_{j \in \mathscr{V}_{out(i)}} e_j (e_j - e_i)^T$$

where $w \in (0,1)$, I is the identity matrix of dimension N, and $e_i \in \mathbb{R}^N$ is a vector of all zeros except for the i-th entry which is set to one. Clearly all Q_i are stochastic, have self-loops, and $\mathscr{G}_{Q_i} \subseteq \mathscr{G}$.

Differently, in the *asymmetric gossip* one node i selects only one of its possible neighbors $\mathscr{V}_{out}(i)$, which after receiving the message updates its local variable. More formally, given a possibly directed graph $\mathscr{G} = (\mathscr{N}, \mathscr{E})$, then $Q(t) \in \mathbb{Q}_{AG} = \{Q^{ij} \mid (i,j) \in \mathscr{E}, i \neq j\}$, where

$$Q^{ij} = I - w e_j (e_j - e_i)^T$$

where $w \in (0,1)$ and e_i are defined as above. Clearly all Q^{ij} are stochastic, have self-loops, and $\mathscr{G}_{Q^{ij}} \subseteq \mathscr{G}$. Note that even if the graph $\mathscr{G}$ is undirected, than the matrices Q^{ij} are only stochastic and do not guarantee average consensus. The same consideration applies to the broadcast matrices Q_i defined above.

The *symmetric gossip* is applicable only to undirected graphs. In this scheme, one node i transmits its information to only one of its neighbors j, which in turn transmits back to the node i another message with its local value. Only after the completion of this procedure the two nodes update their local values using a consensus scheme based on the same weight w. More formally, given the undirected graph $\mathscr{G} = (\mathscr{N}, \mathscr{E})$, then $Q(t) \in \mathbb{Q}_{SG} = \{Q^{ij} \mid (i,j) \in \mathscr{E}, i \neq j\}$, where

$$Q^{ij} = I - w(e_j - e_i)(e_j - e_i)^T$$

Clearly all Q^{ij} are doubly-stochastic, are idempotent (*i.e.*, $(Q^{ij})^2 = Q^{ij}$), have self-loops, and $\mathscr{G}_{Q^{ij}} \subseteq \mathscr{G}$. Although symmetric gossip is somewhat more complex from a communication point of view, differently from broadcast and asymmetric gossip, it has the advantage to preserve the average at any time instant, therefore convergence to consensus automatically guarantees convergence to average consensus.

At this point, the design problem is how to select a sequence of $Q(t)$ from the sets defined above for the broadcast, asymmetric gossip and symmetric gossip, and how to choose the consensus weight w. In general the consensus weight is set to $w = 1/2$ and more attention is paid on the drawing of matrices $Q(t)$. One approach is to deterministically select these matrices according to some sequence, however this still requires some sort of coordination and synchronization. A more natural approach is to select these matrices randomly, possibly according to some i.i.d. distribution on the sets $\mathbb{Q}$. This distribution can be represented by a vector $p \in \mathbb{R}^N$, such that $p \geq 0$ and $\mathbf{1}^T p = 1$ for the broadcast model, where $p_i = \mathbb{P}[Q(t) = Q_i]$. Similarly, the probability distribution in the symmetric and asymmetric gossip can be represented by a matrix $P \in \mathbb{R}^{N \times N}$ which is nonnegative, *i.e.* $[P]_{ij} \geq 0$, is consistent with the graph, *i.e.* $\mathscr{G}_P \subseteq \mathscr{G}$, and sum to unity, *i.e.* $\mathbf{1}^T P \mathbf{1} = 1$, where $[P]_{ij} = \mathbb{P}[Q(t) = Q^{ij}]$. In this case, the design space corresponds to the probability distribution of these sets, *i.e.* the vector p or the matrix P. The proper framework to analyze these strategies is given by Theorem 3.3. Many results about exact rate of convergence and its optimal design are available for communications graphs $\mathscr{G}$ that present special symmetries like complete graphs, circulant graphs, hypercubes, and tori [17, 26]. Differently, for general undirected graphs, Boyd et al. [11] showed that under the randomized symmetric gossip schemes with weight $w = 1/2$, the rate of convergence can be bound by $\rho \leq \mathtt{esr}(\overline{Q})$ thus suggesting the following optimization criteria for maximizing the rate of convergence:

$$\begin{aligned} \min_{P} \quad & \mathtt{esr}(\overline{Q}) \\ \text{s.t.} \quad & \overline{Q} = \sum_{i=1}^{N}\sum_{j=1}^{N}[P]_{ij}Q^{ij},\ [P]_{ij} \geq 0, \mathbf{1}^T P\mathbf{1} = 1,\ \mathscr{G}_P \subseteq \mathscr{G} \end{aligned}$$

which turns out to be a convex problem. This optimization problem is a centralized problem, however the authors in [11] suggested also suboptimal decentralized optimization schemes. Fagnani et al. [25] studied the asymmetric gossip for general undirected graphs and showed that rate of convergence can be bound by $\rho \leq \mathtt{sr}([Q^T(0)\Omega Q(0)]) = 1 - 2w((1-w) - wN^{-1})\mu$, where μ is the smallest positive eigenvalue of the positive semidefinite matrix $S = \text{diag}(P\mathbf{1}) - (P+P^T)/2$, where $\text{diag}(x) : \mathbb{R}^n \to \mathbb{R}^{n\times n}$ indicates a diagonal matrix whole diagonal entries are the entries of the vector x. Therefore in this scenario a possible optimization criterium for minimizing the rate of convergence is to minimize ρ which is minimized by setting $w = \frac{1}{2}\frac{N+1}{N} \approx \frac{1}{2}$ and by maximizing μ. If we restrict to symmetric probability matrices $P = P^T$, maximizing μ is equivalent to the following convex optimization problem:

$$\begin{aligned} \max_{P,\varepsilon} \quad & \varepsilon \\ \text{s.t.} \quad & \text{diag}(P\mathbf{1}) - P \geq \varepsilon I,\ P = P^T,\ [P]_{ij} \geq 0, \mathbf{1}^T P\mathbf{1} = 1,\ \mathscr{G}_P \subseteq \mathscr{G} \end{aligned}$$

Similarly to [11] also this optimization problem is centralized and therefore might not be suitable for fully distributed optimization.

3.2.2.3 Graph Design

In the previous sections, we focused on the issue of how to design the coefficients of the matrix Q for a given communication graph $\mathscr{G}$. However, there are scenarios for which also the communication graph can be designed, therefore it is useful to understand the effect of the graph topology on the performance and how it scales as the number of nodes increases. Also, it is important to note that, in many cases, the effect of the graph topology on performance is much more relevant than the actual choice of the weights, *i.e.* of the non-zero entries of Q. In fact, for example, Xiao et al. [64] studied consensus over random geometric graphs [48] and compared optimal design with suboptimal decentralized strategies like the consensus based on the Metropolis matrix, showing that performance difference was not so drammatically different and seemed to scale similarly with the graph size.

In this context, let us consider the static consensus $x(t+1) = Qx(t)$. Asking what graph allows for the fastest convergence, without any further constraint, is trivially answered (the complete graph, *i.e.* every pair of nodes is connected by an edge) and is not very meaningful: the complete graph corresponds to centralized computation. A more interesting question is asked by Delvenne et al. [23, 24]: what is the best graph, under the constraint that each agent receives at most ν messages at each iteration (*i.e.*, $\mathscr{G}_Q$ has bounded in-degree)? The answer is given by a family of graphs known as *de Bruijn graphs*, well-known in the computer science literature for their expansion properties, and capable of giving the exact average in finite time (not only $\lim_{t\to\infty} x(t) = \frac{1}{N}\mathbf{1}^T x(0)$, but also $x(\bar{t}) = \frac{1}{N}\mathbf{1}^T x(0)$ for some $\bar{t}$), and moreover the time $\bar{t}$ is the smallest possible with the constraint on the in-degree.

The very good performance of de Bruijn graphs is surprising if compared with a family of graphs, *Abelian Cayley graphs* [17], which are grids on d-dimensional tori (a circle for $d = 1$), and whose algebraic structure (a generalization of circulant matrices) allows to compute the eigenvalues and to prove that $\mathtt{esr}(Q) \geq 1 - cN^{\frac{1}{\nu+1}}$, where ν is the degree of the nodes and c is a positive scalar independent of the graph. This proves that, when $N \to \infty$, $\mathtt{esr}(Q) \to 1$, *i.e.*, convergence is considerably slowed down by the size of the network. However, this is not always the case: in addition to de Bruijn graphs, there are other significant classes of graphs, known as *expander graphs*, such that $\mathtt{esr}(Q)$ is bounded away from 1 when $N \to \infty$ (see [45] for the study of such graphs in the context of consensus algorithms). A particular family of graphs which allow fast information transfer (having a small diameter despite the small degree of each node) are the so-called *small-world* graph, which are considered as a reasonable model for many social interactions (*e.g.*, the collaboration graph for scientific authors, or the spread of some diseases) and for the world-wide web; they have been studied in the consensus literature by Olfati-Saber [44] and Tahbaz-Salehi et al. [57].

All such graphs have good properties in terms of fast convergence, despite the small (average) number of neighbors of each node, and as opposed to Abelian Cayley graphs (roughly speaking: grids) where convergence is very slow for large networks. The key fact that makes this difference is that in grids not only the number of neighbors is little, but also their position is forced to be local, in a somehow geometrical sense. In many practical deployments of sensor networks, geometrical constraints are indeed present, and thus the very structured and symmetrical Abelian Cayley graphs can be thought as an idealized version of realistic settings, and are important in that they underline the strong limitations that such locality constraint has on performance and gives guidelines for the design of the number of nodes in the network, in the case when the topology is bound to have such a given structure and the size only is the objective of design. A step towards a more realistic, less structured family of graphs where geometrical bounds are enforced is the study of *random geometric graphs* [48]. Random geometric graphs are undirected graphs which are widely used to model wireless sensor networks, and they are obtained by randomly generating points in the Euclidean space (usually, in the plane) according to a Poisson point process (the number of points in any bounded region is a Poisson random variable with average proportional to the area, and the position of points is uniformly distributed in the region) and then drawing an edge between two nodes if and only if their relative distance is smaller than a predefined communication radius r.

The analysis of the effect of the graph topology on performance has been considered also for time-varying consensus algorithms, and particularly for randomized algorithms (as opposed to the previously-mentioned results, where families of random graphs were considered in the sense that the one time-invariant graph is randomly selected before starting the algorithm). An early work by Hatano et al. [31] studies the case where, at each time step, the graph is chosen randomly according to the Erdős-Rényi model, *i.e.*, the presence or absence of edges between any pair of nodes are given by i.i.d. Bernoulli random variables. A more recent research line has studied convergence of various randomized gossip algorithms, when the random activation of a node or of an edge is restricted to an underlying graph smaller than the complete graph. In this context, a relevant result by Fagnani et al. [26] concerns the rate of convergence of various algorithms (including symmetric, asymmetric and broadcast gossip) when the underlying graph is an Abelian Cayley graph. Another very interesting result can be found in [11], where the rate of convergence of symmetric gossip is found for random geometric graphs and compared to the faster convergence in the preferential connectivity model (a popular model for the graph of the world wide web, and an example of small-world graph).

3.3 Estimation and Control Problems as Average Consensus

In this section we illustrate with few examples that some estimation and control problems can be reframed as the computation of the average of some quantities, which therefore can be efficiently computed in a distributed fashion using average consensus algorithms.

3.3.1 Parameter Estimation with Heterogeneous Sensors

Let us consider N sensors that measure a noisy version of the true parameter $\theta \in \mathbb{R}$ as follows:

$$y_i = \theta + v_i, \quad v_i \sim \mathcal{N}(0, \sigma_i^2), \quad i = 1, \dots, N$$

where v_i are independent zero-mean random variable with covariance σ_i^2, *i.e.* sensors have different accuracy. The minimum-variance estimate of the parameter θ, given all the measurements, is given by:

$$\hat{\theta}_{\text{MV}} = \sum_{i=1}^{N} \alpha_i y_i, \qquad \alpha_i = \frac{\frac{1}{\sigma_i^2}}{\sum_{j=1}^{N} \frac{1}{\sigma_j^2}}$$

i.e. it is a convex combination of the measurements. It is easy to see that the previous estimator can be written as:

$$\hat{\theta}_{\mathrm{MV}} = \frac{\frac{1}{N}\sum_{i=1}^{N}\frac{1}{\sigma_i^2}y_i}{\frac{1}{N}\sum_{j=1}^{N}\frac{1}{\sigma_j^2}}$$

i.e. it is the ratio of two averages. Therefore, it can be asymptotically computed in a distributed fashion using two average consensus algorithms in parallel whose initial condition are set to $x_i^y(0) = \frac{1}{\sigma_i^2}y_i$ and $x_i^\sigma(0) = \frac{1}{\sigma_i^2}$, so that

$$\lim_{t\to+\infty} \hat{\theta}_i(t) := \frac{x_i^y(t)}{x_i^\sigma(t)} = \hat{\theta}_{\mathrm{MV}}, \ \forall i.$$

3.3.2 Node Counting in a Network

In many applications it is important to know how many nodes there are in a network. This can be easily computed via an average consensus algorithm, by setting all the initial conditions to zero except for one node, *i.e.* $x_1(0) = 1$ and $x_i(0) = 0, i = 2,\ldots,N$. Since average consensus guarantees converge to the average of initial conditions, an asymptotically correct estimator of the total number of node N is given by:

$$\hat{N}_i(t) := \frac{1}{x_i(t)},$$

because

$$\lim_{t\to+\infty} \hat{N}_i(t) = \lim_{t\to\infty}\frac{1}{x_i(t)} = \frac{1}{\frac{1}{N}\sum_{j=1}^{N}x_i(0)} = N, \ \forall i.$$

3.3.3 Generalized Averages

Besides the common arithmetic average it is also possible to compute other types of averages such as

$$z_\alpha = \sqrt[\alpha]{\frac{1}{N}\sum_{i=1}^{N} y_i^\alpha}$$

where $\alpha = 1$ gives rise to the usual arithmetic average, $\alpha = 2$ the mean square, $\alpha = -1$ the harmonic mean. Also note that $z_\infty := \lim_{\alpha\to+\infty} z_\alpha = \max_i y_i$ [6, 21]. These generalized averages can be computed using average consensus by setting the initial condition $x_i(0) = y_i^\alpha$ and computing an estimate of the desired average as follows:

$$\lim_{t\to+\infty} \hat{z}_i(t) := \sqrt[\alpha]{x_i(t)} = z_\alpha, \ \forall i$$

Another important average is the geometric mean defined as:

$$z_g = \sqrt[N]{\prod_{i=1}^{N} y_i}$$

The geometric mean can be written as $z_g = \exp(\log z_g) = \exp\left(\sum_{i=1}^{N} \log y_i\right)$, therefore it can be computed using average consensus by setting the initial conditions to $x_i(0) = \log y_i$ and the following estimator:

$$\lim_{t \to +\infty} \hat{z}_i(t) := \exp(N x_i(t)) = z_g, \quad \forall i$$

Note, however, that in this case the number of nodes N needs to be known in advance.

3.3.4 Vehicle Rendezvous

An important example of vehicle formation control is the rendezvous problem (see *e.g.* [12]), where all vehicles are required to meet at a common location using only relative position information for all initial conditions. In its simplest formulation, the vehicle dynamics is given by

$$x_i(t+1) = x_i(t) + u_i(t)$$

and the goal is to find a linear control strategy which uses only relative distance information, *i.e.*

$$u_i(t) = \sum_{j=1}^{N} q_{ij}(t)(x_j(t) - x_i(t))$$

such that $\lim_{t \to +\infty} x_i(t) = \bar{x}$ for some $\bar{x}$. This is indeed a consensus problem that can be solved by choosing the weights $q_{ij}(t)$ that guarantees convergence[2]. Besides convergence, it is also relevant to compute performance of the rendezvous strategy. A natural approach is to consider a linear quadratic (LQ) measure given by:

$$J_{\mathrm{LQ}} = J_x + \varepsilon J_u = \sum_{t=0}^{\infty} ||x(t) - x(\infty)||^2 + \varepsilon \sum_{t=0}^{\infty} ||u(t)||^2$$

where $x = [x_1\ x_2\ \cdots\ x_N]^T$, $u = [u_1\ u_2\ \cdots\ u_N]^T$, and ε is a positive scalar that trades off the integral square error of all vehicles from the rendezvous location $x(\infty) = \bar{x}\mathbf{1}$, namely J_x, versus the integral energy of all vehicles required to achieve consensus, namely J_u.

3.3.5 Least Squares Data Regression

Least squares are one of the most popular estimation techniques in data regression, where the objective is to estimate a function $y = f(x)$, from a noisy data set $\mathscr{D} = \{(x_i, y_i)\}_{i=1}^{N}$. A standard approach is to propose a parametrized function

[2] In realistic scenarios the gains q_{ij} are a function of vehicle location, *i.e.* $q_{ij} = q_{ij}(x)$. A typical model is to consider limited communication range $r > 0$, *i.e.* $q_{ij} = 0$ if $|x_i - x_j| > r$. This gives rise to nonlinear dynamics which is not captured by the model presented in Section 3.2. The analysis of these systems is beyond the scope of this work and we refer the interested reader to [22] an references therein.

$f_\theta(x) := \sum_{j=1}^{M} \theta_i g_i(x)$, where $g_i(x)$ are known functions, often called basis functions, and $\theta_i, i = 1\ldots, M$ are unknown parameters to be determined based on the data set $\mathscr{D}$. The least squares estimate of the parameter vector $\theta = [\theta_1\ \theta_2\ \cdots\ \theta_M]^T$ is defined as

$$\hat{\theta}_{\text{LS}} = \arg\min_\theta \sum_{i=1}^{N} (y_i - f_\theta(x_i))^2$$

If we define the vectors $g_i = [g_1(x_i)\ g_2(x_i) \cdots\ g_M(x_i)]^T \in \mathbb{R}^M, i = 1,\ldots,N$, $y = [y_1\ y_2\ \cdots\ y_M]^T \in \mathbb{R}^N$, and the matrix $G = [g_1\ g_2\ \cdots\ g_M]^T \in \mathbb{R}^{N\times M}$, then we have

$$\begin{aligned}\hat{\theta}_{\text{LS}} &= \arg\min_\theta \|y - G\theta\|^2 = (G^T G)^{-1} G^T y = \left(\sum_{i=1}^{N} g_i g_i^T\right)^{-1} \left(\sum_{i=1}^{N} g_i y_i\right) \\ &= \left(\frac{1}{N}\sum_{i=1}^{N} g_i g_i^T\right)^{-1} \left(\frac{1}{N}\sum_{i=1}^{N} g_i y_i\right)\end{aligned}$$

under the implicit assumption that $(G^T G)^{-1}$ exists. From last equation it is clear that the estimate can be computed as a nonlinear combination of two averages, therefore a consensus based strategy is to run two average consensus algorithms with initial conditions $x_i^{gg}(0) = g_i g_i^T \in \mathbb{R}^{M\times M}$ and $x_i^{gy}(0) = g_i y_i \in \mathbb{R}^M$, and then asymptotically computing the least square estimate as:

$$\lim_{t\to+\infty} \hat{\theta}_i(t) := \left(x_i^{gg}(t)\right)^{-1} x_i^{gy}(t) = \hat{\theta}_{\text{LS}}, \ \ \forall i$$

Note that in this scenario x_i^{gg} are matrices and x_i^{gy} are vectors, therefore they are not scalar as usually considered in Eqn. (3.1), however all results of Section 3.2 still apply by considering the local updates rules of Eqn. (3.2) or Eqn. (3.3) [65, 9].

3.3.6 Sensor Calibration

Often inexpensive sensors might be affected by unknown offsets due to fabrication imperfections or aging. A common example is given by the sensor that measures the signal strength, the RSSI, in the radio chip of commercial wireless sensor nodes [9]. The RSSI is often used to estimate the relative distance between two of these wireless nodes for localization and tracking applications. More precisely the signal strength y_{ij} measured by node i from node j can be modeled as:

$$y_{ij} = f(\xi_i, \xi_j) + o_i$$

where ξ_i and ξ_j are the locations of the receiving node i and the transmitter node j, respectively, and o_i is the offset of the receiving node. Typically, $f(\xi_i, \xi_j)$ is a function of the distance $\|\xi_i - \xi_j\|$ only, but in indoor environments this cannot be the case. However, it still holds that

$$f(\xi_i, \xi_j) = f(\xi_j, \xi_i),$$

i.e. the function f is symmetric in terms of nodes locations. The objective of calibration is to estimate the offset o_i for each node in order to remove it from the measurements. This is clearly impossible, unless at least one node is calibrated or if the function f and the node locations ξ are known. A less demanding requirement is to find offset estimates $\hat{o}_i$ such that $o_i - \hat{o}_i = \bar{o}$ for all i, *i.e.* to be able to have all nodes with the same offset $\bar{o}$. This can be interpreted as a consensus problem on the variable $x_i(t) = o_i - \hat{o}_i(t)$. However, this is still an undetermined problem since $\bar{o}$ is arbitrary. One solution to remove this ambiguity is to choose one node as a reference, for example node $i = 1$, *i.e.* $\bar{o} = o_1$. A less arbitrary choice is to find $\bar{o}$ such that

$$\arg\min_{\bar{o}} \sum_{i=1}^{N} \hat{o}_i^2 = \arg\min_{\bar{o}} \sum_{i=1}^{N} (o_i - \bar{o})^2 = \frac{1}{N}\sum_{i=1}^{N} o_i = \frac{1}{N}\sum_{i=1}^{N} x_i(0)$$

where the last equality is obtained by setting $\hat{o}_i(0) = 0$. This strategy, which aims at minimizing the magnitude of offset compensation terms $\hat{o}_i$, implies that average consensus is to be sought. By substituting $x_i(t) = o_i - \hat{o}_i(t)$ into Eqn. (3.3) we get:

$$\begin{aligned}
o_i - \hat{o}_i(t+1) &= o_i - \hat{o}_i(t) + \sum_{j=1}^{N} q_{ij}(t)\big(o_j - \hat{o}_j(t) - (o_i - \hat{o}_i(t))\big) \\
\hat{o}_i(t+1) &= \hat{o}_i(t) - \sum_{j=1}^{N} q_{ij}(t)\big(f_{ji} + o_j - \hat{o}_j(t) - (f_{ij} + o_i - \hat{o}_i(t))\big) \\
&= \hat{o}_i(t) + \sum_{j=1}^{N} q_{ij}(t)\big(\hat{o}_j(t) - \hat{o}_i(t) + y_{ij} - y_{ji}\big)
\end{aligned}$$

where we used the notation $f(\xi_i, \xi_j) = f_{ij}$ and the assumption that $f_{ij} = f_{ji}$. From average consensus we have that:

$$\lim_{t\to+\infty} \hat{o}_i(t) = o_i - \frac{1}{N}\sum_{j=1}^{N} o_i$$

From this expression, it is clear that if the offset are normally distributed, *i.e.* $o_i \sim \mathcal{N}(0, \sigma^2)$, then $\lim_{N\to+\infty} |\hat{o}_i(\infty) - o_i| = 0$ almost surely, *i.e.* if the number of nodes is large, then the offset estimate is very close to the true offset.

3.3.7 Kalman Filtering

Estimation of dynamical systems is another important area. Let us consider the following dynamical systems observed by N sensors:

$$\begin{aligned}
\xi(t+1) &= A\xi(t) + w(t) \\
y_i(t) &= C_i\xi(t) + v_i(t), \quad i = 1, \ldots, N
\end{aligned}$$

where $w(t) \sim \mathcal{N}(0,Q)$ and $v_i(t) \sim \mathcal{N}(0,R_i)$ are uncorrelated white Gaussian noises. If we define the new vectors $y(t) = [y_1(t)\ y_2(t)\ \cdots\ y_N(t)]^T$ and $v(t) = [v_1(t)\ v_2(t)\ \cdots\ v_N(t)]^T$. The minimum error covariance estimate is given by $\hat{\xi}(h|t) := \mathbb{E}[\xi(h)\,|\,y(t),y(t-1)\dots y(1)]$ and its error variance is $P(h|t) := \mathrm{Var}(\xi(h)-\hat{\xi}(h|t))$. The optimal estimator is known as the Kalman Filter, whose equations are given by:

$$\begin{aligned}
\hat{\xi}(t|t-1) &= A\hat{\xi}(t-1|t-1)\\
P(t|t-1) &= AP(t-1|t-1)A^T+Q\\
\hat{\xi}(t|t) &= \hat{\xi}(t|t-1)+P(t|t-1)C^T(CP(t|t-1)C^T+R)^{-1}(y(t)-C\hat{\xi}(t|t-1))\\
P(t|t) &= P(t|t-1)-P(t|t-1)C^T(CP(t|t-1)C^T+R)^{-1}CP(t|t-1)
\end{aligned}$$

The first two equations are known as the prediction step, while the last two equations are known as the correction step. Using the matrix inversion lemma, the correction step can be written as

$$\begin{aligned}
\hat{\xi}(t|t) &= P(t|t)(P(t|t-1)\hat{\xi}(t|t-1)+C^TR^{-1}y(t))\\
&= P(t|t)(P(t|t-1)\hat{\xi}(t|t-1)+\sum_{i=1}^{N}C_i^TR_i^{-1}y_i(t))\\
&= P(t|t)(P(t|t-1)\hat{\xi}(t|t-1)+z(t))\\
P(t|t) &= (P(t|t-1)+C^TR^{-1}C)^{-1} = (P(t|t-1)+\sum_{i=1}^{N}C_i^TR_i^{-1}C_i)^{-1}\\
&= (P(t|t-1)+Z)^{-1}
\end{aligned}$$

which are also known as the inverse covariance filter. From these equations it is evident that the sufficient statistics necessary to recover the centralized Kalman filter are the quantities $z(t) = N(\frac{1}{N}\sum_{i=1}^{N}C_i^TR_i^{-1}y_i(t))$ and $Z = N(\frac{1}{N}\sum_{i=1}^{N}C_i^TR_i^{-1}C_i)$ which are averages of local quantities. Therefore, a possible strategy to run a local filter on each local node, which, between two measurements $y(t-1)$ and $y(t)$, runs m iterations of the average consensus algorithm to recover $z(t)$ and Z, and then updates its estimate using the centralized Kalman gain. If m is sufficiently large and if the total number of nodes N is known to each sensor, then each local filter coincides with the centralized Kalman filter [55]. If m is not sufficiently large to guarantee that the consensus has converged, then performance of the local filters needs to evaluated and also the consensus algorithms design should be designed accordingly to improve it. In this context [14], if scalar dynamics is considered, *i.e.* $\xi \in \mathbb{R}$ where $A = C_i = 1, \forall i$, $Q = q$, and $R = r$, then the equations for the consensus-based Kalman filter can be written as

$$\begin{cases}\hat{x}(t|t-1) = Q^m\hat{x}(t-1|t-1)\\ \hat{x}(t|t) = (1-\ell)\hat{x}(t|t-1)+\ell y(t)\end{cases} \tag{3.8}$$

where $\hat{x} = [\hat{x}_1(t)\ \hat{x}_2(t)\ \cdots \hat{x}_N(t)]^T \in \mathbb{R}^N$ is the vector of the local estimators of the true state ξ at each node and $\ell \in (0,1)$ is the Kalman gain.

3.4 Control-Based Performance Metrics for Consensus Algorithms

The performance analysis of consensus algorithms presented in Sect. 3.2, which exploits results from Markov chains literature, is focused on predicting the speed of convergence to the average. This is very useful, but however it is not the whole story. In fact, when convergence to the average is not an objective per se, but is used to solve an estimation or control problem, it is important to consider different performance measures, more tightly related to the actual objective pursued. Also, the introduction of other performance indices allows a better understanding of large-scale networks, because for some very relevant families of communication graphs, *e.g.*, for grids (lattices), the essential spectral radius goes to one when the number of agents N grows, so that it is not clear whether $\texttt{esr}(Q)^t$ will go to zero or not, if both N and t tend to infinity. In this section, we will present examples of some alternative performance indices, and references to the relevant literature; however, this research topic is very recent and presently active, so that very likely new papers will appear in the next years.

For the sake of simplicity, we restrict our attention to constant Q, instead of studying all the (randomly)-time-varying schemes introduced in the previous sections. Moreover, we will always assume that $\mathscr{G}_Q$ is rooted and has all self-loops, so that Thm. 3.1 holds true. Additional assumptions that we will often use are that Q is doubly-stochastic, so that $\eta = \frac{1}{N}\mathbf{1}$, and that Q is normal, *i.e.*, $Q^TQ = QQ^T$; under these assumptions, all the costs we consider can be re-written as simple functions of the eigenvalues of Q.

3.4.1 Performance Indices

In this sections we give some examples of performance metrics arising in the use of consensus algorithm for estimation or control tasks. This is not a comprehensive list of all indices presented in the recent literature on distributed estimation and networked control; for example, we do not present here the interesting results related to estimation from relative measurements [5], to the costs arising from vehicle formation control [4], and clock synchronization [15].

3.4.1.1 LQ Cost

As discussed in Sect. 3.3.4, an interesting performance metric is the LQ cost $J_{\text{LQ}} = J_x + \varepsilon J_u$, where $J_x = \frac{1}{N}\sum_{t\geq 0}\mathbb{E}\left(\|x(t)-x(\infty)\|^2\right)$ is related to the speed of convergence, while a second term $J_u = \frac{1}{N}\sum_{t\geq 0}\mathbb{E}\left(\|x(t+1)-x(t)\|^2\right)$ takes into account the energy of the input sequence.

Let us see how to obtain easier expressions for J_x and J_u [23, 20]. Let us focus on the case when Q is doubly-stochastic, so that $x(\infty) = \frac{1}{N}\mathbf{1}^T x(0)$. Under this assumption, the following equalities hold true[3]:

$$J_x = \frac{1}{N}\sum_{t\geq 0}\left\|Q^t - \tfrac{1}{N}\mathbf{1}\mathbf{1}^T\right\|_{\mathrm{F}}^2 \quad \text{and} \quad J_u = \frac{1}{N}\sum_{t\geq 0}\left\|Q^{t+1} - Q^t\right\|_{\mathrm{F}}^2, \tag{3.9}$$

where $\|\cdot\|_{\mathrm{F}}$ the Frobenius norm of a square matrix, *i.e.*, $\|A\|_{\mathrm{F}} = \sqrt{\mathrm{tr} A^T A}$.

If in addition Q is normal, then the expression furtherly simplifies to:

$$J_x = \frac{1}{N}\sum_{\substack{\lambda\in\Lambda(Q)\\ \lambda\neq 1}} \frac{1}{1-|\lambda|^2} \quad \text{and} \quad J_u = \frac{1}{N}\sum_{\substack{\lambda\in\Lambda(Q)\\ \lambda\neq 1}} \frac{|1-\lambda|^2}{1-|\lambda|^2} \tag{3.10}$$

where $\Lambda(Q)$ denotes the set of all eigenvalues of Q (with their multiplicity).

The proof —as all proofs in this section— repeatedly uses linearity of expectation and of trace, plus the observation that for any scalar $a \in \mathbb{R}$ we have $a = \mathrm{tr}\, a$, and the property $\mathrm{tr}(ABC) = \mathrm{tr}(CAB)$ where A, B, C are matrices of suitable size.

The first expression in Eqn. (3.9) is obtained as follows:

$$\begin{aligned}
J_x &= \frac{1}{N}\sum_{t\geq 0}\mathbb{E}\left[x(0)^T(Q^t - \tfrac{1}{N}\mathbf{1}\mathbf{1}^T)^T(Q^t - \tfrac{1}{N}\mathbf{1}\mathbf{1}^T)x(0))\right] \\
&= \frac{1}{N}\sum_{t\geq 0}\mathbb{E}\left[\mathrm{tr}\left(x(0)^T(Q^t - \tfrac{1}{N}\mathbf{1}\mathbf{1}^T)^T(Q^t - \tfrac{1}{N}\mathbf{1}\mathbf{1}^T)x(0)\right)\right] \\
&= \frac{1}{N}\sum_{t\geq 0}\mathrm{tr}\left((Q^t - \tfrac{1}{N}\mathbf{1}\mathbf{1}^T)^T(Q^t - \tfrac{1}{N}\mathbf{1}\mathbf{1}^T)\mathbb{E}\left[x(0)x(0)^T\right]\right).
\end{aligned}$$

where we assume uniform distribution of initial conditions, *i.e.* $\mathbb{E}[x(0)x(0)^T] = I$. The second expression is easily obtained by the same techniques.

In order to prove Eqn. (3.10), we recall that normality of Q implies that all powers of Q, as well as Q^T and $Q^T Q$ are diagonalized with the same change of basis. Moreover, by stochasticity and primitivity of Q, also $Q - \frac{1}{N}\mathbf{1}\mathbf{1}^T$ (and all its powers, and its transpose) are diagonalized in that same basis and, denoting the eigenvalues of Q by $\lambda_1 = 1, \lambda_2, \ldots, \lambda_N$, we have that the eigenvalues of $Q^t - \frac{1}{N}\mathbf{1}\mathbf{1}^t$ are $\lambda_1 - 1 = 0, \lambda_2 - 0 = \lambda_2, \ldots, \lambda_N - 0 = \lambda_N$, so that $\|Q^t - \frac{1}{N}\mathbf{1}\mathbf{1}^t\|_{\mathrm{F}}^2 = \sum_{h=2}^N \|\lambda_h\|^{2t}$, and finally $J_x = \frac{1}{N}\sum_{h=2}^N \sum_{t\geq 0}(\|\lambda_h\|^2)^t = \frac{1}{N}\sum_{h=2}^N \frac{1}{1-\|\lambda_h\|^2}$.

[3] J_x and J_u might be infinite for some choices of Q. A sufficient condition for convergence of both costs is that Q is doubly-stochastic, $\mathcal{G}_Q$ is rooted and $\mathcal{G}_Q \in \mathbb{G}_{\mathrm{sl}}$. This is easily proved from Eqn. (3.9) using the following property of Frobenius norm: $\|AB\|_{\mathrm{F}} \leq \|A\|_{\mathrm{F}}\|B\|_{\mathrm{F}}$. Thus, $J_x \leq \frac{1}{N}\sum_{t=0}^{\infty}\left\|\left(Q - \frac{1}{N}\mathbf{1}\mathbf{1}^T\right)\right\|_{\mathrm{F}}^{2t} = \frac{1}{N}\mathrm{tr}\sum_{t=0}^{\infty}\left(Q^T Q - \frac{1}{N}\mathbf{1}\mathbf{1}^T\right)^t$, where the convergence of the last series is ensured by the fact that $Q^T Q$ is stochastic (Q being doubly-stochastic) and $\mathcal{G}_{Q^T Q}$ is a subgraph of $\mathcal{G}_Q$ (thanks to the self-loops in $\mathcal{G}_Q$) and thus inherits its properties. A similar proof can be given also for J_u, after noting that $J_u = \sum_{t\geq 0}\|(Q - \frac{1}{N}\mathbf{1}\mathbf{1}^T)^t(Q - I)\|_{\mathrm{F}}^2$.

For the second part of Eqn. (3.10), note $Q^{t+1} - Q^t$ is normal and has eigenvalues $\lambda_h^t(\lambda_h - 1)$ for $h = 1, \dots, N$, and then conclude with the same technique as above.

3.4.1.2 Steady-State Performance for Noisy or Quantized Consensus

For the consensus algorithm of Eqn. (3.1), Thm. 3.1 tells everything about steady-state performance: when $t \to \infty$, $x(t) \to x(\infty) := \eta^T x(0)\mathbf{1}$, and if Q is doubly-stochastic, then $\eta = \frac{1}{N}\mathbf{1}$. However this is no longer true if there is noise in the consensus process, or quantization in the exchanged messages.

In the presence of noise within the successive iterations of the consensus algorithm, the steady state can be different from the average of the initial values, despite Q being doubly-stochastic. Here we present a case analyzed in [64], where the noise is additive.

Consider the following consensus algorithm affected by noise:

$$x(t+1) = Qx(t) + v(t),$$

where $\{v_i(t)\}$ are noises uncorrelated w.r.t. both i and t, with zero mean and unit variance. Consider the case when Q is doubly-stochastic, so that, for any initial condition $x(0)$, $\mathbf{1}^T\mathbb{E}[x(t)] = \mathbf{1}^T x(0)$ for all t, and $\mathbb{E}[x(t)] \to \frac{1}{N}\mathbf{1}^T x(0)$. However, it is clear that the average-preserving property, and the convergence to $\frac{1}{N}\mathbf{1}^T x(0)$ are true only in expectation, and not for all realizations of the noise process. Thus, it is more reasonable to define the error as the distance from current average $\delta(t) = x(t) - \frac{1}{N}\mathbf{1}\mathbf{1}^T x(t))$ rather than distance from average consensus, which might not even exist. Hence, the relevant average quadratic cost is here defined as

$$J_{\text{noisy}} := \lim_{t\to\infty} \frac{1}{N}\mathbb{E}\left[\|x(t) - \tfrac{1}{N}\mathbf{1}\mathbf{1}^T x(t)\|^2\right]$$

Notice that J_{noisy} turns out to be the same as the cost J_x introduced when studying the LQ-cost. In fact, note that

$$x(t) = Q^t x(0) + \sum_{s=0}^{t-1} Q^s v(t-1-s),$$

so that $\delta(t) = (Q^t - \frac{1}{N}\mathbf{1}\mathbf{1}^T)x(0) + \sum_{s=0}^{t-1}(Q^s - \frac{1}{N}\mathbf{1}\mathbf{1}^T)v(t-1-s)$. Thus, by the statistical assumptions on the noises (zero-mean, uncorrelated, unit variance):

$$\begin{aligned}
\mathbb{E}\left[\|\delta(t)\|^2\right] &= \|(Q^t - \tfrac{1}{N}\mathbf{1}\mathbf{1}^T)x(0)\|^2 \\
&\quad + 2\sum_{s=0}^{t-1} x(0)^T(Q^t - \tfrac{1}{N}\mathbf{1}\mathbf{1}^T)^T(Q^s - \tfrac{1}{N}\mathbf{1}\mathbf{1}^T)\mathbb{E}[v(t-1-s)] \\
&\quad + \sum_{r,s=0}^{t-1} \operatorname{tr}\left\{(Q^r - \tfrac{1}{N}\mathbf{1}\mathbf{1}^T)^T(Q^s - \tfrac{1}{N}\mathbf{1}\mathbf{1}^T)\mathbb{E}[v(t-1-r)v(t-1-s)]\right\} \\
&= \|(Q^t - \tfrac{1}{N}\mathbf{1}\mathbf{1}^T)x(0)\|^2 + \sum_{s=0}^{t-1}\operatorname{tr}(Q^s - \tfrac{1}{N}\mathbf{1}\mathbf{1}^T)^T(Q^s - \tfrac{1}{N}\mathbf{1}\mathbf{1}^T)
\end{aligned}$$

When $t \to \infty$, the first term goes to zero, while the sum becomes an infinite sum, thus ending the proof.

A similar cost has been considered in [29], where however the noise was used as a model for **quantization** error, and thus noise appears in the equation in a different way, as follows:

$$x(t+1) = Q[(x(t)+v(t)] - v(t)$$

The fact that noise is multiplied by Q takes into account that the quantization error is within all messages passed to neighbors, while the substraction $-v(t)$ is possible, as every agent knows its own quantization error, and is useful for avoiding accumulation of errors over time: in this way, the average $\frac{1}{N}\mathbf{1}^T x(t)$ is kept constant.

As in the previous case, the assumption is that $v_i(t)$'s are uncorrelated with respect to both i and t, and have zero-mean and unit variance, and Q is doubly-stochastic, so that $\mathbb{E}x(t) \to \frac{1}{N}\mathbf{1}^T x(0)$. Again the relevant cost is the variance of the distance from consensus $\delta(t) = x(t) - \frac{1}{N}\mathbf{1}\mathbf{1}^T x(t)$, in the limit of infinite number of iterations:

$$J_{\text{quantiz}} := \lim_{t\to\infty} \frac{1}{N}\mathbb{E}\left(\|x(t) - \tfrac{1}{N}\mathbf{1}\mathbf{1}^T x(t)\|^2\right)$$

Clearly, due to the different update equation for $x(t)$, this will result in an expression for J_{quantiz} different from the one for J_{noisy}; it turns out that J_{quantiz} is equal to the cost J_u defined when dealing with the LQ-cost.

To prove this, notice that

$$x(t+1) = Q^t x(0) + \sum_{s=0}^{t-1} Q^s (Q-I) v(t-1-s)$$

so that $\delta(t+1) = (Q^t - \frac{1}{N}\mathbf{1}\mathbf{1}^T)x(0) + \sum_{s=0}^{t-1} (Q^{s+1} - Q^s)v(t-1-s)$.

By exploiting linearity of expectation and of trace, and the fact that arguments of the trace can be cyclically permuted, together with the assumptions on the noise, we get

$$\mathbb{E}\left(\|\delta(t)\|^2\right) = \|(Q^t - \tfrac{1}{N}\mathbf{1}\mathbf{1}^T)x(0)\|^2 + \sum_{s=0}^{t-1} \operatorname{tr}\left\{(Q^{s+1}-Q^s)^T(Q^{s+1}-Q^s)\right\}$$

By taking the limit for $t \to \infty$, the first term goes to zero, while the summation becomes an infinite sum, giving $J_{\text{quantiz}} = \frac{1}{N}\sum_{t\geq 0}\|Q^{t+1} - Q^t\|_{\text{F}}$ and thus ending the proof.

3.4.1.3 Performance of Static Estimation Algorithm

Consider the static estimation problem described in Sect. 3.3.1, but in the simplest case, when all sensors have the same variance $\sigma^2 = 1$. In this case, the best estimate is the average, $\hat{\theta}_{\text{MV}} = \frac{1}{N}\mathbf{1}^T x(0)$ and the sensors can compute it in a distributed way by simply using a consensus algorithm $x(t+1) = Qx(t)$, for some stochastic matrix

Q. What is peculiar to this setting, is that the focus is not on how precisely the average is computed, but on how good the estimate of θ is. In fact, knowing that $x(t)$ converges to $x(\infty) = \eta^T x(0)\mathbf{1}$ does not answer the questions on how well is θ estimated by $x(\infty)$ if the matrix Q is not doubly-stochastic and on how well is θ estimated after t iterations of the algorithm.

To address these questions, we consider the estimation error $e(t) := x(t) - \theta\mathbf{1}$. To answer the first question, let us first notice that, if Q is doubly-stochastic, then $x(\infty)$ is the average of the measurements, *i.e.*, $x(\infty) = \hat{\theta}_{\mathrm{MV}}\mathbf{1}$, and $\hat{\theta}_{\mathrm{MV}}$ has zero-mean and variance $\frac{1}{N}$. If Q is not doubly-stochastic, then it is interesting to study the error; it is easy to see that $e(\infty) = \mathbf{1}\eta^T v$, and so $\mathbb{E}[e(\infty)] = 0$, while its covariance matrix is

$$\mathbb{E}\left[e(\infty)e(\infty)^T\right] = \mathbf{1}\eta^T\mathbb{E}\left[vv^T\right]\eta\mathbf{1}^T = \mathbf{1}\|\eta\|^2\mathbf{1}^T = \|\eta\|^2\mathbf{1}\mathbf{1}^T,$$

i.e., each sensor's final estimate has variance $\|\eta\|^2$. Notice that $1/N \leq \|\eta\|^2 \leq 1$, since $\|\eta\|_1 = 1$.

Now let us turn our attention to the more interesting problem of understanding how well θ is estimated after a finite number of iterations, t, studying $e(t)$. More precisely, the relevant performance measure is the average quadratic error, defined as

$$J_{\mathrm{estim}}(t) := \tfrac{1}{N}\mathbb{E}\left[\|x(t) - \theta\mathbf{1}\|^2\right]$$

This cost can be re-written as:

$$J_{\mathrm{estim}}(t) = \tfrac{1}{N}\,\mathrm{tr}[(Q^T)^t Q^t]$$

and, if Q is normal, the expression simplifies as follows:

$$J_{\mathrm{estim}}(t) = \tfrac{1}{N}\sum_{\lambda\in\Lambda(Q)} |\lambda|^{2t}. \tag{3.11}$$

To prove the first claim, note that

$$J_{\mathrm{estim}}(t) = \tfrac{1}{N}\mathbb{E}\left[\|Q^t x(0) - \theta\mathbf{1}\|^2\right] = \tfrac{1}{N}\mathbb{E}\left[\|(Q^t - I)\theta\mathbf{1} + Q^t v\|^2\right] = \tfrac{1}{N}\mathbb{E}\left[v^T(Q^t)^T Q^t v\right]$$

from which the claim follows by taking the trace and cyclically permuting its arguments, recalling that $\mathbb{E}[vv^T] = I$. Then the simplified expression for normal Q is immediate

3.4.1.4 Distributed Kalman Filter

Consider the distributed Kalman filter presented in Sect. 3.3.7, and in particular its scalar version described in Eqn. (3.8). There are different ways of analyzing performance of such algorithm. One interesting performance index is the asymptotic quadratic estimation error, defined as:

$$J_{\mathrm{K.est}} := \tfrac{1}{N}\lim_{t\to\infty}\mathbb{E}\left[\|\hat{x}(t|t) - x(t)\mathbf{1}\|^2\right]$$

This cost can be re-formulated as follows:

$$J_{\text{K,est}} = \frac{q(1-\ell)^2}{1-(1-\ell)^2} + \tfrac{1}{N}\sum_{s=0}^{\infty} \|(1-\ell)^s Q^{sm}\|_{\text{F}}^2$$

and, in the case when Q is normal, the following easier characterization holds true:

$$J_{\text{K,est}} = \frac{q(1-\ell)^2}{1-(1-\ell)^2} + \frac{r\ell^2}{N} \sum_{\lambda \in \Lambda(Q)} \frac{1}{1-(1-\ell)^2|\lambda|^{2m}}.$$

Another relevant performance metric is the asymptotic quadratic prediction error

$$J_{\text{K,pred}} := \tfrac{1}{N} \lim_{t\to\infty} \mathbb{E}\left[\|\hat{x}(t|t-1) - x(t)\mathbf{1}\|^2\right],$$

which can be re-written as

$$J_{\text{K,pred}} = \frac{q}{1-(1-\ell)^2} + \frac{r\ell^2}{N}\sum_{s=0}^{\infty} \left\|(1-\ell)^s Q^{(s+1)m}\right\|_{\text{F}}^2$$

and, for normal Q, is also equal to

$$J_{\text{K,pred}} = \frac{q}{1-(1-\ell)^2} + \frac{r\ell^2}{N} \sum_{\lambda \in \Lambda(Q)} \frac{|\lambda|^{2m}}{1-(1-\ell)^2|\lambda|^{2m}}.$$

The techniques used for obtaining the simplified expressions are similar to those shown for the costs previously presented and details can be found in [14].

3.4.2 *Evaluation and Optimization of Performance Indices*

Clearly any performance index can be numerically computed for a given matrix Q, and gives a way of comparing the quality of different choices for Q. However, there are two research lines which lead to interesting results using some performance index. A first line concerns optimization of a chosen cost among all matrices Q consistent with a given communication graph. A second interesting direction is the study of the different costs for some relevant families of graphs and matrices, in particular for large-scale graphs. The more classical results in this two directions when the performance index is the essential spectral radius are discussed in Section 3.2.2.

Providing a comprehensive summary of the results is beyond the scope of this chapter: we give here some examples, so as to illustrate some curious or unexpected results and motivate the need for different performance metrics, and then we give pointers to some relevant literature, with the disclaimer that —this being a very recent and still active research area— our reference list will surely turn out to be incomplete.

An interesting work on design of the entries of Q for a given graph by optimization of a cost different from $\texttt{esr}(Q)$ is Xiao et al. [64]. Here noisy consensus is analyzed, so that the relevant metric is $J_{\text{noisy}} = J_x$. The authors show that the problem of finding, for a given graph and among all symmetric choices of weights, the weights minimizing J_{noisy}, is a convex optimization problem, and they provide efficient (although centralized) algorithms for its solution. They also compare numerically, for various graphs topologies, the three costs J_{noisy} obtained with the optimal Q, with the Q minimizing $\texttt{esr}Q$ and the simple Metropolis rule; for some topologies the difference is significant, while for other graphs the three results are very similar.

Another example is Carli et al. [14], where the problem of optimizing $J_{\text{K,pred}}$ for a given graph among normal matrices is examined. The first interesting result is that symmetric matrices are indeed optimal, and then, the authors prove that, for fixed ℓ, the optimization problem among symmetric matrices is convex in Q; however, despite the problem being also convex in ℓ, it is not jointly convex in Q and ℓ. Then simplified problems (under the limit for infinite communication or for small measuerement noise) are studied more in detail.

The optimality of de Bruijn graphs with respect to convergence speed, among all graphs with bounded in-degree, is confirmed, at least asymptotically in N and for small ε, also when the LQ cost is considered [23].

Another approach which is receiving much attention is the study of asymptotic performance in large-scale graphs. The idea is to consider families of graphs of increasing size, sharing the same common properties (in some sense that will be specified in the examples, having the same shape), and to analyze how the cost scales with the number of nodes. This is more an analysis than a design problem, but it gives useful hints on the number of nodes. Here we present a simple example.

Example 3.1 (Circle). Consider a graph $\mathscr{G}_N$ consisting of a circle o N nodes, where each node has a self-loop and an outgoing edge towards its neighbor on the right. Consider a coefficient $1/2$ on each edge, so that $Q_N = \text{circ}(1/2, 1/2, 0, \ldots, 0)$ is a circulant matrix. Because $\mathscr{G}_N$ is circulant, we know that it is normal, and we can easily compute its eigenvalues: $\Lambda(Q_N) = \{\frac{1}{2} + \frac{1}{2}e^{i\frac{2\pi}{N}h}, h = 0, \ldots, N-1\}$ [17]. Thus, the essential spectral radius is

$$\texttt{esr}(Q_N) = |\lambda_1^2| = \sqrt{\tfrac{1}{2}(1+\cos(\tfrac{2\pi}{N}))} = 1 - \tfrac{\pi^2}{N^2} + O(\tfrac{1}{N^4}) \quad \text{for } N \to \infty.$$

Now we can plug the expression for the eigenvalues in Equations (3.10) and (3.11). Then, an explicit computation (see *e.g.* [19]) gives that

$$J_u = 1 - \frac{1}{N}$$

while some careful upper and lower bounds (see *e.g.* [20]) show that

$$c_1 N \le J_x(Q_N) \le c_2 N \quad \text{and} \quad c_3 \max\left\{\tfrac{1}{N}, \tfrac{1}{\sqrt{t}}\right\} \le J_{\text{estim}}(Q_N, t) \le c_4 \max\left\{\tfrac{1}{N}, \tfrac{1}{\sqrt{t}}\right\},$$

where c_1, c_2, c_3, c_4 are positive numbers independent of N.

It is interesting to compare the performance of the circle with that of a complete graph, *i.e.*, with the case $Q'_N = \frac{1}{N}\mathbf{1}\mathbf{1}^T$, where in one step the exact average is computed. It is easy to see that the eigenvalues of Q' are 1 with multiplicity 1 and 0 with multiplicity $N-1$, so that $\mathtt{esr}(Q'_N) = 0$, $J_x(Q'_N) = 1 - \frac{1}{N}$, $J_{\text{estim}}(Q'_N, t) = \frac{1}{N}$ for all $t \geq 1$). Intuitively, performance of the circle is much worse, because of the slow flow of information, as opposed to the complete exchange of messages in one single iteration for the complete graph. This intuition is confirmed for most performance indices; however, it is interesting to note that $J_x(Q_N) = J_x(Q'_N)$ is actually the same as for the circle and for the complete graph, thus showing that a different choice of performance metric can lead to significantly different results.

The key point that allows to study the example of the circle is the fact that an expression for the eigenvalues is easily found, thanks to the algebraic structure of Q, which is circulant. The same can be done more in general, for the case of circles with more edges towards neighbors (giving rise to different circulant matrices) and for higher dimension, where the underlying algebraic structure is that of Cayley graphs, Cayley matrices and discrete Fourier transform over Abelian groups (see *e.g.* [17]). The result presented in [20] concerns grids on d-dimensional torus, or grids on d-dimensional cubes with some assumptions of symmetry of the coefficients and suitable border conditions, and in both cases with local neighborhoods (bounded difference among labels of nodes connected by an edge). It states that

$$c_1 f_d(N) \leq J_x \leq c_2 f_d(N) \quad \text{and} \quad c_3 \max\left\{\frac{1}{N}, \frac{1}{(\sqrt{t})^d}\right\} \leq J_{\text{estim}}(t) \leq c_4 \max\left\{\frac{1}{N}, \frac{1}{(\sqrt{t})^d}\right\},$$

where $f_1(N) = N$, $f_2(N) = \log N$ and $f_d(N) = 1$ for all $d \geq 3$, and where c_1, c_2, c_3, c_4 are positive numbers independent of N.

The study of Cayley graphs, although motivated by the algebraic structure that allows to tackle the analysis, is interesting, because they are a simplified and idealized version of communication scenarios of practical interest. In particular, they capture the effects on performance of the strong constraint that communication is local, not only in the sense of a little number of neighbors, but also with a bound on the distance among connected agents. The study of more irregular and realistic scenarios of communication with geometric constraints is the subject of on-going research, where two main directions are being explored. On the one side, there is an interest in the random geometric graph model (points thrown uniformly at random in a portion of space and edges among all pairs of vertices within a given distance r), for which simulations show a behavior very similar to that of a grid (see *e.g.* [20]), but a rigorous theory is still missing: most of known results concern only the essential spectral radius and not all the spectrum. On the other side, there is the idea to study perturbations of known graphs; this is completely different from traditional theory of perturbation of matrices, because here perturbations are not continuous, and are little in the sense that only few edges (with respect to the graph size) are removed or added or receive different weight. In this direction, a useful tool (because of its monotonicity properties with respect to edge insertion) is the analogy between reversible Markov chains and resistive electrical networks, exploited *e.g.* in [5].

We conclude this section by presenting in detail an example that clarifies how comparing two families of graphs by two different performance measures can indeed significantly change the result, leading to a different definition of the 'best' graph. This is a toy example, not very sensible in practice, but easily highlighting which issues can arise.

Example 3.2. Let N be an even number, and consider $\mathscr{G}_N$ a graph consisting of two disconnected complete graphs, each on $N/2$ nodes; Figure 3.1(a) depicts $\mathscr{G}_{10}$ as an example. Associate to each edge a coefficient $2/N$, so that Q_N has the following form:

$$Q_N = \left[\begin{array}{c|c} \frac{2}{N}\mathbf{1}\mathbf{1}^T & 0 \\ \hline 0 & \frac{2}{N}\mathbf{1}\mathbf{1}^T \end{array}\right].$$

We would like to compare performance of this Q_N with the circle presented as Example 3.1, by looking at the essential spectral radius, and then by looking at the estimation error J_{estim}. The eigenvalues of Q_N are simply 1 with multiplicity 2 and the eigenvalue 0 with multiplicity $N-2$, so that $\texttt{esr}(Q_N) = 1$, which is worse than the circle. However, for all $t \geq 1$, $J_{\text{estim}}(Q_N) = \frac{2}{N}$, which is almost as good as the best possible error (the error variance in the case of centralized estimation, $\frac{1}{N}$), as opposed to the circle which, for large N, has a very slow convergence.

Behind computation of the eigenvalues, there is an intuitive explanation of what happens. In the graphs $\mathscr{G}_N$, the essential spectral radius 1 describes the fact that the graph is disconnected, and thus no convergence is possible to the average of all initial values: simply no information can transit from one group to another; nevertheless, the estimation error is very good for large N, because it is the average of $N/2$ measurements, and it is computed very fast, in one iteration, thanks to the complete graph which gives centralized computation within the group of $N/2$ agents. Conversely, in the circle average consensus can be reached asymptotically, as described by the essential spectral radius smaller than one, but convergence is very slow for large N ($\texttt{esr} = 1 - \frac{\pi^2}{N^2} + O(\frac{1}{N^2})$), and a reasonably good estimation error is achieved only after a long time.

The readers concerned with the fact that $\mathscr{G}_N$ is disconnected (and thus violates the assumptions made throughout this chapter) may consider a slightly modified graph $\tilde{\mathscr{G}}_N$, as shown in Figure 3.1(b), still associating a coefficient $2/N$ with each edge; Let us denote by $\tilde{Q}_N$ the matrix so modified. This graph is studied in [10] under the

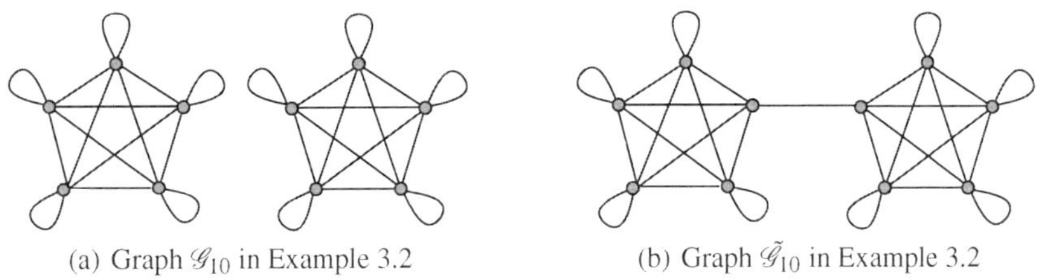

(a) Graph $\mathscr{G}_{10}$ in Example 3.2 (b) Graph $\tilde{\mathscr{G}}_{10}$ in Example 3.2

Fig. 3.1 Communication graphs considered in Example 3.2

name $K_{N/2} - K_{N/2}$ and [10, Prop. 5.1] gives the exact computation of all eigenvalues of $\tilde{Q}_N$: $\Lambda(\tilde{Q}_N)$ has 1 with multiplicity 1, 0 with multiplicity $N-3$ and then $\frac{1}{2} - \frac{2}{N} \pm \frac{1}{2}\sqrt{1 + \frac{8}{N} - \frac{16}{N^2}}$ with multiplicity 1 each. Here the single edge connecting the two subgroups of agents allows only a quite slow convergence ($\mathtt{esr}(\tilde{Q}_N) = 1 - \frac{8}{N^2} + O(\frac{1}{N^2})$, very similar to that of the circle), while the estimation error becomes very good after few iterations ($J_{\text{estim}}(\tilde{Q}_N) \leq \frac{3}{N}$ for all $t \geq 1$).

3.5 Conclusion

In this chapter we have tried to present a comprehensive view of the linear consensus algorithms from a control and estimation perspective, by reviewing the most important results available in the literature, by showing some of the possible applications in control and estimation, and by presenting which are suitable control-based indices of performance for the consensus algorithm design.

We believe that much has still to be done in this area, in particular in two directions. The first direction points to finding which traditional control and estimation problems can be cast as consensus problems. In fact, although not all problems can be cast as averages of local quantities, if they can be approximated as so, we could exploit the effectiveness and strong robustness of consensus algorithms. The second direction addresses the implications of the new control-based performance metrics for the design of the consensus algorithms. In fact, as we illustrated with few toy examples, they give rise to design criteria that can be quite different from the traditional ones.

Acknowledgements. We would like to thank Sandro Zampieri for many useful discussions on the topics of this chapter and for reviewing the original manuscript, and Paolo Frasca for his suggestions. We acknowledge funding from the European Community's Seventh Framework Programme under agreement n. FP7-ICT-223866-FeedNetBack and from the Italian CaRiPaRo Foundation under project WISE-WAI.

References

1. Alighanbari, M., How, J.P.: An unbiased Kalman consensus algorithm. In: Proceedings of IEEE American Control Conference, ACC 2006 (2006)
2. Angeli, D., Bliman, P.-A.: Tight estimates for non-stationary consensus with fixed underlying spanning tree. In: Proceedings of 17th IFAC World Congress, IFAC 2008 (2008)
3. Aysal, T.C., Yildiz, M.E., Sarwate, A.D., Scaglione, A.: Broadcast gossip algorithms for consensus. IEEE Transactions on Signal Processing 57(7), 2748–2761 (2009)
4. Bamieh, B., Jovanovic, M., Mitra, P., Patterson, S.: Coherence in large-scale networks: Dimension dependent limitations of local feedback. In: IEEE Trans. Aut. Cont. (to appear, 2010)
5. Barooah, P.: Estimation and Control with Relative Measurements: Algorithms and Scaling Laws. PhD thesis, University of California, Santa Barbara (2007)

6. Bauso, D., Giarré, L., Pesenti, R.: Nonlinear protocols for optimal distributed consensus in networks of dynamic agents. Systems and Control Letters 55, 918–928 (2006)
7. Bliman, P.A., Ferrari-Trecate, G.: Average consensus problems in networks of agents with delayed communications. In: Proceedings of 44th IEEE Conference on Decision and Control, CDC 2005, vol. 44(8), pp. 1985–1995 (2008)
8. Blondel, V.D., Hendrickx, J.M., Olshevsky, A., Tsitsiklis, J.N.: Convergence in multiagent coordination, consensus, and flocking. In: Proceedings of 44th IEEE Conference on Decision and Control (CDC 2005) and European Control Conference (ECC 2005), pp. 2996–3000 (December 2005)
9. Bolognani, S., Del Favero, S., Schenato, L., Varagnolo, D.: Consensus-based distributed sensor calibration and least-square parameter identification in WSNs. International Journal of Robust and Nonlinear Control 20(2), 176–193 (2010)
10. Boyd, S., Diaconis, P., Parrilo, P., Xiao, L.: Symmetry analysis of reversible Markov chains. Internet Mathematics 2, 31–71 (2005)
11. Boyd, S., Ghosh, A., Prabhakar, B., Shah, D.: Randomized gossip algorithms. IEEE Transactions on Information Theory/ACM Transactions on Networking 52(6), 2508–2530 (2006)
12. Bullo, F., Cortés, J., Martínez, S.: Distributed Control of Robotic Networks. Applied Mathematics Series. Princeton University Press, Princeton (2009)
13. Cao, M., Morse, A.S., Anderson, B.D.O.: Reaching a consensus in a dynamically changing environment– a graphical approach. SIAM Journal on Control and Optimization 47(2), 575–600 (2008)
14. Carli, R., Chiuso, A., Schenato, L., Zampieri, S.: Distributed Kalman filtering based on consensus strategies. IEEE Journal on Selected Areas in Communications 26(4), 622–633 (2008)
15. Carli, R., Chiuso, A., Zampieri, S., Schenato, L.: A PI consensus controller for networked clocks synchronization. In: IFAC World Congress on Automatic Control, IFAC 2008 (2008)
16. Carli, R., Como, G., Frasca, P., Garin, F.: Average consensus on digital noisy networks. In: Proc. of 1st IFAC Workshop on Estimation and Control of Networked Systems (NecSys 2009), September 24-26, 2009, pp. 36–41 (2009)
17. Carli, R., Fagnani, F., Speranzon, A., Zampieri, S.: Communication constraints in the average consensus problem. Automatica 44(3), 671–684 (2008)
18. Carli, R., Frasca, P., Fagnani, F., Zampieri, S.: Gossip consensus algorithms via quantized communication. Automatica 46, 70–80 (2010)
19. Carli, R.: Some issues on the Average Consensus Problem. PhD thesis, Università di Padova, Italy (2008)
20. Carli, R., Garin, F., Zampieri, S.: Quadratic indices for the analysis of consensus algorithms. In: Proc. of the 4th Information Theory and Applications Workshop, La Jolla, CA, USA, pp. 96–104 (Feburary 2009)
21. Cortés, J.: Distributed algorithms for reaching consensus on general functions. Automatica 44, 726–737 (2008)
22. Cortes, J., Martinez, S., Bullo, F.: Robust rendezvous for mobile autonomous agents via proximity graphs in arbitrary dimensions. IEEE Transactions on Automatic Control 51(8), 1289–1298 (2006)
23. Delvenne, J.-C., Carli, R., Zampieri, S.: Optimal strategies in the average consensus problem. In: Proc. of the IEEE Conference on Decision and Control 2007, pp. 2498–2503 (2007)
24. Delvenne, J.-C., Carli, R., Zampieri, S.: Optimal strategies in the average consensus problem. Systems and Control Letters 58, 759–765 (2009)

25. Fagnani, F., Zampieri, S.: Asymmetric randomized gossip algorithms for consensus. In: Proceedings of 17th IFAC World Congress, IFAC 2008 (2008)
26. Fagnani, F., Zampieri, S.: Randomized consensus algorithms over large scale networks. IEEE Journal on Selected Areas in Communications 26(4), 634–649 (2008)
27. Fagnani, F., Zampieri, S.: Average consensus with packet drop communication. SIAM Journal on Control and Optimization 48, 102–133 (2009)
28. Fax, J.F., Murray, R.M.: Information flow and cooperative control of vehicle formations. IEEE Transactions on Automatic Control 49(9), 1465–1476 (2004)
29. Frasca, P., Carli, R., Fagnani, F., Zampieri, S.: Average consensus on networks with quantized communication. International Journal of Robust and Non-Linear Control 19(16), 1787–1816 (2009)
30. Frommer, A., Szyld, D.B.: On asynchronous iterations. Journal of Computation and Applied Mathematics 123, 201–216 (2000)
31. Hatano, Y., Mesbahi, M.: Agreement over random networks. IEEE Transactions on Automatic Control 50(11), 1867–1872 (2005)
32. Huang, M., Manton, J.H.: Coordination and consensus of networked agents with noisy measurements: Stochastic algorithms and asymptotic behavior. SIAM Journal on Control and Optimization 48(1), 134–161 (2009)
33. Jadbabaie, A., Lin, J., Morse, A.S.: Coordination of groups of mobile autonomous agents using nearest neighbor rules. IEEE Transactions on Automatic Control 48(6), 988–1001 (2003)
34. Kar, S., Moura, J.M.F., Ramanan, K.: Distributed parameter estimation in sensor networks: Nonlinear observation models and imperfect communication (submitted, 2008)
35. Kar, S., Moura, J.M.F.: Distributed consensus algorithms in sensor networks with imperfect communication: Link failures and channel noise. IEEE Transactions on Signal Processing 57(5), 355–369 (2009)
36. Kar, S., Moura, J.M.F.: Distributed consensus algorithms in sensor networks: Quantized data and random link failures. IEEE Transactions on Signal Processing 58(3), 1383–1400 (2010)
37. Kashyap, A., Başar, T., Srikant, R.: Quantized consensus. Automatica 43(7), 1192–1203 (2007)
38. Lavaei, J., Murray, R.M.: On quantized consensus by means of gossip algorithm–Part I: Convergence proof. In: Proceedings of the American Control Conference, ACC 2009 (2009)
39. Minc, H.: Nonnegative matrices. John Wiley & Sons, Chichester (1988)
40. Moreau, L.: Stability of continuous-time distributed consensus algorithms. In: Proceedings of the 43rd IEEE Conference on Decision and Control (CDC 2004), vol. 4, pp. 3998–4003 (December 2004)
41. Moreau, L.: Stability of multiagent systems with time-dependent communication links. IEEE Transactions on Automatic Control 50(2), 169–182 (2005)
42. Muthukrihnan, S., Ghosh, B., Schultz, M.H.: First and second order diffusive methods for rapid, coarse, distributed load balancing. Theory of Computing Systems 31, 331–354 (1998)
43. Nedić, A., Olshevsky, A., Ozdaglar, A., Tsitsiklis, J.N.: On distributed averaging algorithms and quantization effects. IEEE Transaction on Automatic Control 54(11), 2506–2517 (2009)
44. Olfati-Saber, R.: Ultrafast consensus in small-world networks. In: Proceedings of the 2005 American Control Conference ACC 2005, vol. 4, pp. 2371–2378 (2005)
45. Olfati-Saber, R.: Algebraic connectivity ratio of ramanujan graphs. In: Proceedings of the 2007 American Control Conference (July 2007)

46. Olfati-Saber, R., Murray, R.M.: Consensus problems in networks of agents with switching topology and time-delays. IEEE Transactions on Automatic Control 49(9), 1520–1533 (2004)
47. Patterson, S., Bamieh, B., El-Abbadi, A.: Convergence rates of distributed average consensus with stochastic link failures. IEEE Transactions on Automatic Control (2009)
48. Penrose, M.: Random Geometric Graphs. Oxford University Press, Oxford (2003)
49. Rajagopal, R., Wainwright, M.J.: Network-based consensus averaging with general noisy channels. Technical Report 751, Dept. of Statistics, UC Berkeley (2008)
50. Ren, W., Beard, R.W.: Consensus seeking in multiagent systems under dynamically changing interaction topologies. IEEE Transactions on Automatic Control 50(5), 655–661 (2005)
51. Ren, W., Beard, R.W., Atkins, E.M.: Information consensus in multivehicle cooperative control. IEEE Control Systems Magazine 27(2), 71–82 (2007)
52. Olfati-Saber, R., Fax, J.A., Murray, R.M.: Consensus and cooperation in multi-agent networked systems. Proceedings of IEEE 95(1), 215–233 (2007)
53. Seneta, E.: Non-negative Matrices and Markov Chains. John Wiley & Sons, Inc., Springer (2006)
54. Seuret, A., Dimarogonas, D.V., Johansson, K.H.: Consensus under communication delays. In: Proceedings of the 47th IEEE Conference on Decision and Control, CDC 2008, pp. 4922–4927 (December 2008)
55. Spanos, D.P., Olfati-Saber, R., Murray, R.M.: Distributed Kalman filtering in sensor networks with quantifiable performance. In: Proccedings of the Information Processing for Sensor Networks IPSN 2005, pp. 133–139 (2005)
56. Strikwerda, J.C.: A probabilistic analysis of asynchronous iteration. Journal of Linear Algebra and its Applications 349, 125–154 (2002)
57. Tahbaz-Salehi, A., Jadbabaie, A.: Small world phenomenon, rapidly mixing Markov chains, and average consensus algorithms. In: Proceedings of IEEE Conference on Decision and Control, CDC 2007, pp. 276–281 (2007)
58. Tahbaz-Salehi, A., Jadbabaie, A.: Consensus over ergodic stationary graph processes. IEEE Transaction on Automatic Control 54(12) (2009)
59. Tanner, H., Jadbabaie, A., Pappas, G.J.: Flocking in fixed and switching networks. IEEE Transaction on Automatic Control 52(5), 863–868 (2007)
60. Tanner, H.G., Christodoulakis, D.K.: The stability of synchronization in local-interaction networks is robust with respect to time delays. In: Proceedings of the 44th IEEE Conference on Decision and Control, CDC 2005, pp. 4945–4950 (December 2005)
61. Tsitsiklis, J.N., Athans, M.: Convergence adn asymptotic agreement in distributed decision problems. IEEE Transactions on Automatic Control 29(1), 42–50 (1984)
62. Xiao, F., Wang, L.: Consensus protocols for discrete-time multi-agent systems with time-varying delays. Automatica 44(10), 2577–2582 (1998)
63. Xiao, L., Boyd, S.: Fast linear iterations for distributed averaging. Systems and Control Letters 53(1), 65–78 (2004)
64. Xiao, L., Boyd, S., Kim, S.-J.: Distributed average consensus with least-mean-square deviation. Journal of Parallel and Distributed Computing 67(1), 33–46 (2007)
65. Xiao, L., Boyd, S., Lall, S.: A scheme for robust distributed sensor fusion based on average consensus. In: Proceedings of the Information Processing for Sensor Networks, IPSN 2005, pp. 63–70 (2005)

Chapter 4
Distributed Optimization and Games: A Tutorial Overview

Bo Yang and Mikael Johansson

Abstract. This chapter provides a tutorial overview of distributed optimization and game theory for decision-making in networked systems. We discuss properties of first-order methods for smooth and non-smooth convex optimization, and review mathematical decomposition techniques. A model of networked decision-making is introduced in which a communication structure is enforced that determines which nodes are allowed to coordinate with each other, and several recent techniques for solving such problems are reviewed. We then continue to study the impact of non-cooperative games, in which no communication and coordination are enforced. Special attention is given to existence and uniqueness of Nash equilibria, as well as the efficiency loss in not coordinating nodes. Finally, we discuss methods for studying the dynamics of distributed optimization algorithms in continuous time.

4.1 Introduction

We are interested in optimization algorithms that can be distributed across many decision-makers. The classical approach to distributed optimization has been decomposition: based on the specific structure of the objective function and constraints, the problem is decomposed into a number of subproblems. These subproblems can be solved independently, but typically require a coordinator to ensure that the local decisions converge to the global optimum; see Figure 4.1. Note how the problem structure imposes a certain computation and communication structure among the individual decision-makers.

In many emerging applications of distributed optimization, however, the situation is the reverse: the communication and computation structure is given and the

Bo Yang
Department of Automation, Shanghai Jiao Tong University, China
e-mail: bo.yang@sjtu.edu.cn, bo.yang@ieee.org

Mikael Johansson
School of Electrical Engineering and ACCESS Linnaeus Center
e-mail: mikaelj@ee.kth.se

A. Bemporad, M. Heemels, and M. Johansson: Networked Control Systems, LNCIS 406, pp. 109–148.
springerlink.com

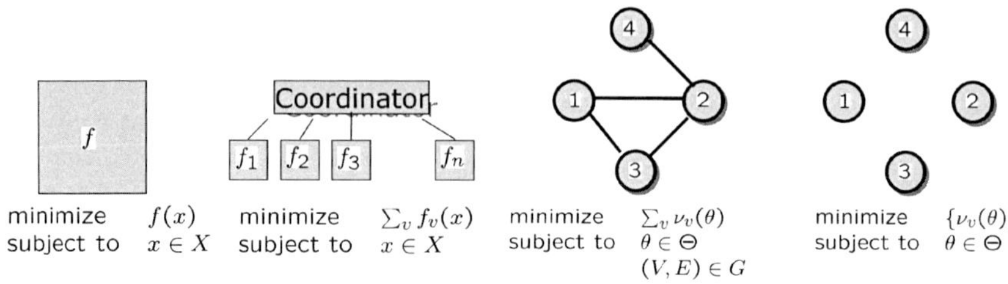

Fig. 4.1 Schematic illustration of centralized optimization, decomposition, networked optimization and non-cooperative optimization

implementation of a centralized coordinator is undesirable or infeasible. One example is systems where nodes can only coordinate their decisions with their immediate neighbors. In this case, we are restricted to use optimization algorithms that respect the communication structure; see Figure 4.1. In some applications, it is desirable to avoid coordination altogether, either because it is unlikely that nodes would actually cooperate, or as a way to eliminate complex coordination protocols and associated traffic overhead, see Figure 4.1. We will consider all the three classes of distributed optimization techniques in this chapter: decomposition, networked-optimization, and non-cooperative games. Although we do not make any restrictions on the computational model, it is good to note that often we will not have access to a closed-form mathematical expression for the per-node objective functions. Rather, these can only be evaluated by applying our best current decision to an underlying engineering system and observe its performance. We will hence also investigate the consequences of having a real system acting as oracle for our optimization algorithm.

This chapter is organized as follows: we first discuss gradient and subgradient method for convex optimization. Although these algorithms do not have the best theoretical properties, they are easy to implement and surprisingly robust to noise and errors. We then proceed to study mathematical decomposition techniques, focusing primarily on dual and primal decomposition. Next, we focus on networked optimization problems where an underlying communication graph dictates which nodes can coordinate with each other. Our final methodological discussion concerns game theory. Finally, the chapter is concluded by methods for studying the dynamics of optimization algorithms.

4.2 Convex Optimization Using First-Order Methods

We consider optimization problem on the form

$$\begin{array}{ll} \text{minimize} & f_0(x) \\ \text{subject to} & x \in X \end{array} \tag{4.1}$$

Here $x \in \mathbb{R}^n$ is the vector of decision variables that must belong to the feasible set X and f_0 is the convex objective function that we would like to minimize. We will assume that X is closed, convex and non-empty and that $\mathbf{dom} f_0 \subseteq X$. It is sometimes useful to characterize the constraint set explicitly. We then use the notation

$$\begin{array}{lll} \text{minimize} & f_0(x) & \\ \text{subject to} & f_i(x) \leq 0 & i=1,\dots,m \\ & h_i(x) = 0 & i=1,\dots,p \end{array} \tag{4.2}$$

Hence, the feasible set is the set of x that satisfies the constraint equations, $X = \{x \mid f_i(x) \leq 0,\ h_i(x) = 0\}$. Note that for X to be convex, $f_i(x)$ have to be convex and $h_i(x)$ must be affine functions of x. There is a vast literature on methods for solving convex optimization problem; some good starting points include the text books [10, 5, 49, 53] and slightly more advanced text are [57, 24]. In recent years, interior-point methods have become the method of choice for solving large-scale convex programming, due to their attractive theoretical properties and strong practical performance. However, our focus is different: we look for methods that can be distributed across many nodes. In this case, first order (gradient) methods are easier to apply and will be the central workhorse throughout this chapter.

4.2.1 Gradient Methods for Smooth Problems

Let us start with the most basic problem of minimizing a smooth convex function without constraints, *i.e.*

$$\text{minimize } f_0(x)$$

The gradient method then takes the form

$$x^{(t+1)} = x^{(t)} - \alpha^{(t)} \nabla f_0(x^{(t)}) \tag{4.3}$$

i.e. new iterates are computed by adjusting the current iterate in the direction of the negative gradient. In its most basic form the step size is constant, *i.e.* $\alpha^{(t)} = \alpha$, and convergence is guaranteed using the following typical result.

Proposition 4.1. *Assume that $f_0(x)$ is a convex function with* $\mathbf{dom} f_0 = \mathbb{R}^n$ *whose gradient is Lipschitz continuous with constant $L > 0$, i.e.*

$$\|\nabla f_0(x) - \nabla f_0(y)\|_2 \leq L\|x-y\|_2 \qquad \forall x, y$$

Assume that f_0 has finite optimal value $f_0^\star$ with minimizer $x^\star$. Let $\{x^{(t)}\}$ be a sequence generated by the gradient iteration (4.3) with a constant stepsize $\alpha^{(t)} = \alpha$ satisfying $0 < \alpha \leq 1/L$. Then

$$f_0(x^{(t)}) - f_0^\star \leq \frac{1}{2\alpha t}\|x^\star - x^{(0)}\|_2^2$$

This result tells us several things. First, the gradient iteration improves the objective function value in each step. Second, the iterates converge asymptotically to the optimum. Third, ε accuracy can be achieved in $O(1/\varepsilon)$ iterations for any continuously differentiable function with Lipschitz continuous gradient.

Practical performance can be improved by time-varying step sizes, either found via line search in each iteration or predetermined ("open-loop"). Common open-loop step-size sequences include the square summable but not summable,

$$\sum_{t=1}^{\infty} \alpha^{(t)} = \infty, \quad \sum_{t=1}^{\infty} (\alpha^{(t)})^2 < \infty \tag{4.4}$$

such as $\alpha^{(t)} = a/(b+t)$ for $a > 0, b \geq 0$, or diminishing stepsizes

$$\sum_{t=1}^{\infty} \alpha^{(t)} = \infty, \quad \lim_{t \to \infty} \alpha^{(t)} = 0 \tag{4.5}$$

such as $\alpha^{(t)} = a/\sqrt{t}$ for $a > 0$. However, the complexity of the method is not improved but $O(1/\varepsilon)$ iterations are still needed to achieve ε accuracy.

An important performance measure of optimization algorithms is their convergence rate. If we assume that $f_0(x)$ is strongly convex, *i.e.* that there exists a positive constant l such that

$$l\|x - x^\star\|_2^2 \leq f_0(x - x^\star)$$

and that f_0 is twice continuously differentiable with Lipschitz-continuous gradient. Then, in a neighborhood of $x^\star$ where f can be approximated as

$$f_0(x) \approx f_0(x^\star) + f_0'(x)(x - x^\star) + f_0''\|x - x^\star\|_2^2$$

one can show that the distance between optimal point and the iterates produced by the gradient method decreases geometrically

$$\|x^{(t)} - x^\star\| \leq cq^t$$

for some positive constants c and q with $q < 1$. In the optimization literature, one then says that the method converges linearly, since the distance to the optimal set in iteration t is a linear function of the distance to the optimal set in iteration $t-1$. The convergence rate q of the basic gradient method can be improved by multi-step methods. One of the first such method was the heavy-ball method due to Polyak:

$$x^{(t+1)} = x^{(t)} - \alpha^{(t)} \nabla f_0(x^{(t)}) + \beta^{(t)}(x^{(t)} - x^{(t-1)}) \tag{4.6}$$

The following result reveals the optimal constant step-size parameters α and β and quantifies the speed-up compared to the classical gradient iteration [53].

Proposition 4.2. *Let $x^\star$ be a nonsingular minimum point of $f_0(x) : \mathbb{R}^n \mapsto \mathbb{R}$ and assume that f_0 is twice continuously differentiable and satisfies*

$$lI_n \preceq f_0''(x^\star) \preceq LI_n$$

Then, we can find an $\varepsilon > 0$ such that for any $x(0), x(1)$ with $\|x(0) - x^\star\|_2 \leq \varepsilon$ and $\|x(1) - x^\star\|_2 \leq \varepsilon$, both the gradient and the heavy-ball method produce iterates that converge to $x^\star$ with geometric progression

$$\|x^{(t)} - x^\star\| \leq c(\delta)(q+\delta)^t, \qquad 0 \leq q < 1, \, 0 < \delta < 1-q$$

For the gradient method, the smallest achievable value of q is $(L-l)/(L+l)$ obtained for $\alpha^{(t)} = 2/(L+l)$. For the heavy-ball method, the smallest achievable value of q is $(\sqrt{L} - \sqrt{l})/(\sqrt{L} + \sqrt{l})$ obtained for the step size parameters $\alpha^{(t)} = 4/(\sqrt{L} + \sqrt{l})^2$, $\beta^{(t)} = (\sqrt{L} - \sqrt{l})^2/(\sqrt{L} + \sqrt{l})^2$.

If the problem is ill-conditioned, *i.e.* if L/l is large, then heavy-ball improves the value of q by roughly a factor of $\sqrt{L/l}$. Since multi-step methods can improve the convergence rate of the gradient iteration, it is interesting to see if also the complexity can be improved from $O(1/\varepsilon)$ to the theoretical lower bound $O(1/\sqrt{\varepsilon})$. The answer to this question was given by Nesterov [47], who constructed a family of multi-step methods on the form

$$\begin{aligned} x^{(t)} &= \hat{x}^{(t-1)} - \alpha^{(t)} \nabla f_0(\hat{x}^{(t-1)}) \\ \hat{x}^{(t)} &= x^{(t)} + \beta^{(t)}(x^{(t)} - x^{(t-1)}) \end{aligned} \tag{4.7}$$

For appropriate choices of $\alpha^{(t)}$ and $\beta^{(t)}$, these methods have complexity $O(1/\sqrt{\varepsilon})$ and are thus order optimal. In most cases, $\alpha^{(t)}$ and $\beta^{(t)}$ are time-varying, which sometimes makes it inconvenient to use them in distributed optimization, but when f_0 is strongly convex, the constant step-sizes $\alpha^{(t)} = 1/L$ and $\beta^{(t)} = (\sqrt{L} - \sqrt{l})/(\sqrt{L} + \sqrt{l})$ also yield an optimal scheme.

The gradient methods can also be extended to deal with problems with constraints. In this case, one typically uses *projected gradient* methods,

$$x^{(t+1)} = P_X\{x^{(t)} - \alpha^{(t)} \nabla f_0(x^{(t)})\} \tag{4.8}$$

where $P_X\{x\}$ denotes the (Euclidean) projection of x onto the constraint set X. Similar convergence results hold as for the unprojected gradient method (see, *e.g.*, [5]).

Example 4.1. To get a feel for the different gradient methods introduced above, consider the problem of minimizing

$$f_0(x) = \sum_{k=1}^{n} k^2 \frac{x_k^2}{2}$$

Since the Lipschitz constant equals n^2, the gradient method uses $\alpha^{(t)} = 1/(n+1)^2$; the heavy-ball method uses $\alpha^{(t)} = 4/(n+1)^2, \beta^{(t)} = (n-1)^2/(n+1)^2$, and Nesterov's method uses the updates above with $\alpha^{(t)} = 1/n$. The figures below show the evolution objective function for $n = 1$ (left) and $n = 7$ (right). We can notice several things: first, the heavy-ball method is perfectly tuned towards a

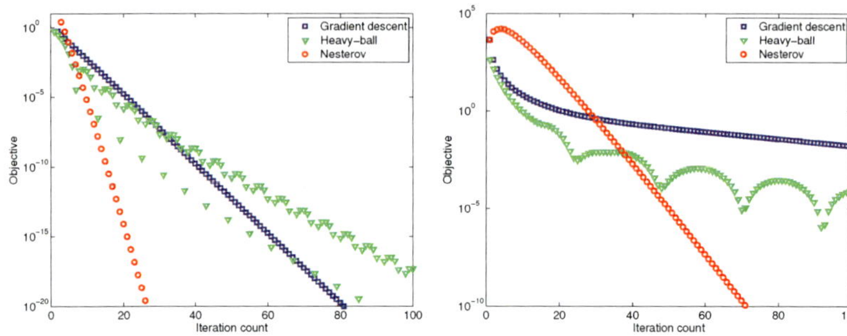

Fig. 4.2 Comparison of ordinary gradient descent, the heavy-ball method, and Nesterov's scheme

second-order quadratic function and has the best convergence rate for $n = 2$ and Nesterov's method does not show any advantage over the gradient scheme. Note how Nesterov's method is not a descent method as the objective function sometimes increases. Moving to $n = 7$ we see that the asymptotic convergence rate of the heavy-ball method is still superior, but that is too aggressive initially. The improvement of Nesterov's scheme compared to the ordinary gradient method is now significantly improved.

4.2.2 *Subgradient Methods for Non-smooth Problems*

In many situations when we do distributed optimization, we will need to work with objective functions that are not smooth. One natural extension of the gradient of a convex function is the concept of a *subgradient*. Like the gradient, the subgradient of a convex function is a global underestimator, *i.e.*

Definition 4.1. *A vector $g \in \mathbb{R}^n$ is a* subgradient *of f at x if*

$$f(y) \geq f(x) + g^T(y - x) \text{ for all } y \in \mathbf{dom} f$$

The set of all g satisfying the inequality is called the subdifferential *of f at x and denoted $\partial f(x)$.*

The *subgradient method* for unconstrained minimization of a non-smooth convex function now takes the form

$$x^{(t+1)} = x^{(t)} - \alpha^{(t)} g^{(t)} \tag{4.9}$$

where $g^{(t)} \in \partial f(x^{(t)})$, and the result corresponding to Proposition 4.1 reads

Proposition 4.3. *Assume that $f_0(x)$ is a convex function with $\mathbf{dom} f_0 = \mathbb{R}^n$ and that f_0 is Lipschitz continuous with constant $L > 0$, i.e.*

$$\|f_0(x) - f_0(y)\|_2 \leq L\|x - y\|_2 \qquad \forall x, y$$

Assume that f_0 has finite optimal value $f_0^\star$ with minimizer $x^\star$. Let $\{x^{(t)}\}$ be a sequence generated by the subgradient iteration (4.9) with a constant stepsize $\alpha^{(t)} = \alpha$. Then

$$\min_{0\leq k\leq t} f_0(x^{(k)}) - f_0^\star \leq \frac{\|x^\star - x^{(0)}\|_2^2 + t\alpha^2 L^2}{2\alpha t}$$

There are several important differences between this result and the corresponding result for the gradient method: the analysis only establishes properties of the *best* iterate found so far and says nothing about the most recent iterate. In fact, the subgradient method is not a descent method and the objective function typically does not improve in each iteration. Moreover, under fixed stepsize even the best iterate does not converge to the optimum when $t \to \infty$: all we can say is that $\liminf_{t\to\infty} f_0(x^{(t)})$ is approximately $L^2\alpha/2$ suboptimal. If we change from constant to divergent stepsize sequences (4.4) or (4.5), then $\liminf_{t\to\infty} f_0(x^{(t)}) \to f_0^\star$. It is possible to show that the subgradient method can achieve ε-convergence in $O(1/\varepsilon^2)$ iterations, which is considerably slower than the gradient methods (see, *e.g.* [62]). Only recently, Nesterov has demonstrated how smoothing techniques can be applied to non-smooth optimization to recover the $O(1/\varepsilon)$ complexity of the basic gradient method [48].

It is possible to make slightly stronger statements about the subgradient method by considering the *Cesàro* averages

$$\tilde{x}^{(t)} = \frac{\sum_{k=0}^{t} \alpha^{(k)} x^{(k)}}{\sum_{k=0}^{t} \alpha^{(k)}}$$

Under the same conditions as Proposition 4.3, if $x(t)$ are generated by the subgradient iteration (4.9) with a constant stepsize $\alpha^{(k)} = \alpha$, then

$$f_0(\tilde{x}^{(t)}) - f_0^\star \leq \frac{\|x(0) - x^\star\|_2^2 + t\alpha^2 L^2}{2\alpha t}$$

This inequality establishes that $\lim_{t\to\infty} f_0(\tilde{x}^{(t)}) \leq f^\star + \alpha L^2/2$. Similarly, the corresponding analysis for divergent step-size sequences (4.4) or (4.5) establishes that the Cesàro averages converge asymptotically to the optimal set.

As for the gradient method, the subgradient method can be extended to handle constrained minimization by projecting the iterates onto the constraint set. The *projected subgradient method* takes the form

$$x^{(t+1)} = P_X\left\{x^{(t)} - \alpha^{(t)} g^{(t)}\right\}$$

Again, very convergence results similar to those of the (unprojected) subgradient method can be established also for the projected subgradient method [62].

4.2.3 Incremental Subgradient Methods

In many applications, we will encounter objective functions on the form

$$f_0(x) = \sum_{i=1}^{N} f_{0i}(x)$$

where each component f_{0i} can be evaluated in parallel (on different machines, or by different decision-makers). The projected subgradient method

$$x^{(t+1)} = P_X \left\{ x^{(t)} - \alpha^{(t)} \sum_{i=1}^{N} g_{0i}^{(t)} \right\}$$

would need to collect all the subgradients g_{0i} of the individual objective function components f_{0i} at the current iterate $x^{(t)}$ to perform an iteration. In contrast, *incremental subgradient methods* cycle through the components and make incremental changes of the iterate in the direction of the negative subgradient of each component,

$$\begin{aligned} x_i^{(t)} &= P_X \left\{ x_{i-1}^{(t)} - \alpha^{(t)} g_{0i}^{(t)} \right\}, \; i = 1, \ldots, N \\ x_0^{(t+1)} &= x_N^{(t)} \end{aligned} \tag{4.10}$$

Here $x_i^{(t)}$ is the iterate computed by accounting for component i, and the subgradient is evaluated at the iterate produced by component $i-1$ at the same iteration, *i.e.*

$$g_{0i}^{(t)} \in \partial f_{0i}(x_{i-1}^{(t)})$$

The first line of (4.10) describes how the inner iterations cycle through components, one-by-one in a given order, and makes incremental updates in the iterate. The second line describes how a new round of iterations are started by passing the iterate produced by the last component to the first. By analyzing how the iterates behave at the beginning of each iteration round, very similar convergence results as for the standard subgradient method can be obtained

Proposition 4.4. *Assume that $f_{0i}(x)$ are convex functions with $X \subseteq \mathbf{dom} f_{0i}$ and that f_{0i} are Lipschitz continuous with constants $L_i > 0$, i.e.*

$$\|f_{0i}(x) - f_{0i}(y)\|_2 \leq L_i \|x - y\|_2 \qquad \forall x, y$$

Assume that $f_0 = \sum_i f_{0i}$ has finite optimal value $f_0^\star$ with minimizer $x^\star$. Let $\{x^{(t)}\}$ be a sequence generated by the incremental subgradient iteration (4.10) with a constant stepsize $\alpha^{(t)} = \alpha$. Then

$$\min_{0 \leq k \leq t} f_0(x_0^{(k)}) - f_0^\star \leq \frac{\|x^\star - x_0^{(0)}\|_2^2 + t\alpha^2 L^2}{2\alpha t}$$

where $L = \sum_i L_i$.

Analysis for diminishing step-size rules can be found in, *e.g.* [45]. Several variations, including methods where a new ordering of the components are chosen at random at the beginning of each outer iterations, have also been proposed [45]. It turns out that the expected convergence rate is better for the randomized method than the deterministic ones that uses the same fixed update order throughout.

Example 4.2. To illustrate the subgradient and incremental subgradient methods, consider the problem of minimizing

$$f_0(x) = \sum_{k=1}^{n} k|x_k|$$

Since the f_0 has Lipschitz constant n, targeting an accuracy of ε gives a constant step-size of $\alpha = 2\varepsilon/n^2$. We consider $n = 4$ and $\varepsilon = 0.1$ which gives $\alpha = 1/80$. Figure 4.3 shows how the objective function evolves for fixed (dark color) and diminishing (light color) stepsize $\alpha^{(t)} = 1/t$. In both cases, the iterates oscillate around the optimum: for fixed stepsize, there is a sustained oscillation with fixed amplitude (below the target accuracy, as predicted by theory); for diminishing step-size, the magnitude of the oscillations decreases (due to the decreasing stepsize of the divergent step-size rules) and the best function value found during the iterations will asymptotically converge to the optimum.

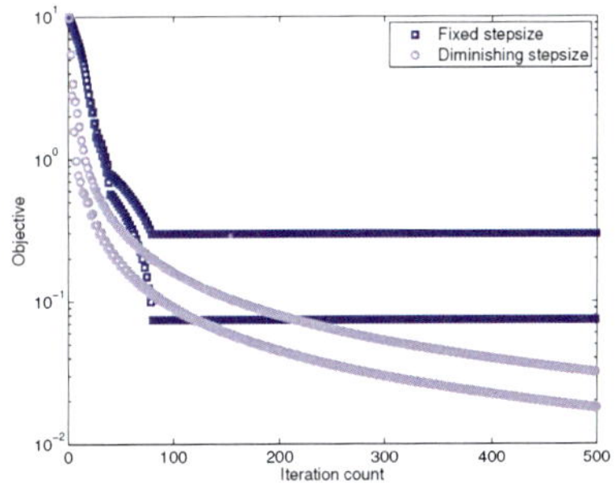

Fig. 4.3 Comparison of subgradient method for fixed and diminishing stepsizes

4.3 Decomposition Techniques

The basic idea of decomposition techniques is to exploit problem structure to divide a complex optimization problem into subproblems that are easier to solve. The subproblems are then coordinated towards the globally optimal solution by the (repeated) solution of a master problem, see Figure 4.1.

One can essentially trace two different motivations for using decomposition in the mathematical programming literature. One line of work is motivated by the need to solve very large-scale optimization problems, *e.g.* [15, 4, 37, 21, 6]. Another line of work is motivated by decentralization of decision-making in organizations or engineering systems, *e.g.* [3, 65, 14, 25]. In the first line of work, subproblems are typically easier to solve simply because they are smaller (in terms of decision variables or constraints) or because they have a special structure that can be exploited. In the other line of work, it is not the computations that complicate the original problem but the fact that the overall problem combines variables (decisions) that belong to separate components in an underlying system. Implementation advantages are then obtained if we can find a problem decomposition that places the main computations within individual components and restricts the coordination overhead among these.

In this section, we will review the basic primal and dual decomposition techniques and some variations.

4.3.1 Dual Decomposition

From a large-scale optimization perspective, the basic idea with dual decomposition is to exploit the structure of the dual optimization problem to improve computational efficiency. The theoretical foundation for this is one of Lagrangean duality . Associated to the optimization problem (4.2) is the *Lagrangean*

$$L(x,\lambda,\mu) = f_0(x) + \sum_i \lambda_i f_i(x) + \sum_i \mu_i h_i(x)$$

Here, the constraints of the original problem have been relaxed and are accounted for by adding a weighted sum of the constraint functions to the objective. Note that if $\lambda \succeq 0$ and $x \in X$, then $L(x,\lambda,\mu) \leq f_0(x)$, and that the *dual function*

$$q(\lambda,\mu) = \inf_x L(x,\lambda,\mu) \tag{4.11}$$

is a lower bound to the optimal value of the original problem. It is hence natural to try to maximize this lower bound, which leads to the dual problem

$$\begin{array}{l} \text{maximize } q(\lambda,\mu) \\ \text{subject to } \lambda \succeq 0 \end{array} \tag{4.12}$$

The fact that the optimal value of this program is always a lower bound to the original problem is called *weak duality* . For convex problems, which is the focus of this chapter, relatively mild conditions guarantee that the optimal value of the dual problem coincides with the optimal value of the (original) primal problem. We then say that *strong duality* holds. Conditions for strong duality are known as constraint qualification, and the most well-known result is due to Slater.

Proposition 4.5. *Consider the convex optimization problem (4.2). If there exists a strictly feasible point $\check{x}$ satisfying*

$$\begin{aligned} f_i(\check{x}) &< 0 \qquad i = 1,\ldots,m \\ h_i(\check{x}) &= 0 \qquad i = 1,\ldots,p \end{aligned}$$

then strong duality holds.

For alternative constraint qualification theorems and stronger version of Slater's conditions, see *e.g.* [5].

To solve the dual problem using a first-order method, we need a gradient or subgradient of the dual function at the current dual variables (λ, μ). To this end, the following result is useful.

Proposition 4.6. *The dual function (4.11) is concave in λ and μ, and a subgradient of $-q(\lambda, \mu)$ at (λ, μ) is given by*

$$-\left(f_1(x^\star(\lambda,\mu)),\ \cdots,\ f_m(x^\star(\lambda,\mu)),\ h_1(x^\star(\lambda,\mu)),\ \cdots,\ h_p(x^\star(\lambda,\mu))\right)$$

where $x^\star(\lambda,\mu) = \arg\inf_x L(x,\lambda,\mu)$. If $f_0(x)$ is strictly convex in all variables, then the dual function is continuously differentiable.

This result tells us that the dual problem is always convex (in fact, it is convex even if the original problem is not), and hence can be solved using gradient or subgradient techniques depending on if the original problem is strictly convex or not.

As we already mentioned, the basic idea of dual decomposition is to explore structure in the dual function to improve computational efficiency and/or ensure decentralization of decisions. The following two examples, taken from [9], illustrate the ideas.

Example 4.3 (Dual decomposition of coupling constraint). To illustrate the dual decomposition technique, consider the following problem

$$\begin{aligned} &\text{minimize } \varphi_1(x_1) + \varphi_2(x_2) \\ &\text{subject to } x_1 + x_2 \leq x_{\text{tot}} \end{aligned}$$

If it were not for the coupling constraint on the total resource, the problem would be separable and the optimal allocations $x_1^\star$ and $x_2^\star$ could easily be found. Introducing a dual variable λ for the total resource constraint, we find the Lagrangian

$$\begin{aligned} L(x,\lambda) &= \varphi_1(x_1) + \varphi_2(x_2) + \lambda(x_1 + x_2 -_{\text{tot}}) = \\ &= \varphi_1(x_1) + \lambda x_1 + \varphi_2(x_2) + \lambda x_2 - \lambda x_{\text{tot}} \end{aligned}$$

which is separable in x_1 and x_2. Hence, so is the dual

$$\begin{aligned} q(\lambda) = \inf_x L(x,\lambda) &= \underbrace{\inf_{x_1}\{\varphi_1(x_1) + \lambda x_1\}}_{q_1(\lambda)} + \underbrace{\inf_{x_2}\{\varphi_2(x_2) + \lambda x_2\}}_{q_2(\lambda)} - \lambda x_{\text{tot}} \\ &= q_1(\lambda) + q_2(\lambda) - \lambda x_{\text{tot}} \end{aligned}$$

The dual function can thus be evaluated in parallel, by letting one decision-maker find the x_1 that minimizes $\varphi_1(x_1) + \lambda x_1$ and another decision-maker find the x_2 that minimizes $\varphi_2(x_2) + \lambda x_2$, and then compute the sum above. Coordination of the two decision-makers is done by adjusting λ to maximize $g(\lambda)$ (*i.e.* to solve the dual problem), *e.g.* using the subgradient iteration

$$\lambda^{(t+1)} = \max\left\{\lambda^{(t)} + \alpha^{(t)}\left(x_{\text{tot}} - x_1^\star(\lambda^{(t)}) - x_2^\star(\lambda^{(t)})\right), 0\right\}$$

Convergence of the Lagrange multipliers (and hence of the dual function) is guaranteed using the classical results on projected gradient and subgradient iterations.

The above example also illustrates why dual decomposition is sometimes referred to as *price-directive decomposition*. Interpreting the dual variable as the unit price for the common resource, the system is directed towards its optimal operation by appropriate pricing of the common resource. Constraints on the common resource are not explicitly enforced, but the demand is asymptotically aligned with the supply using a simple pricing strategy: increase the prices if the resource is in shortage and decrease the price if the resource is in excess. To exercise this pricing interpretation further, and to be able to make a link to the game-theoretic methods descried later in this chapter, we consider the following application to rate allocation in communication networks [30].

Example 4.4. Consider a communication system where N flows are routed through the same bottleneck link with capacity c. To find how to optimally allocate the available bandwidth to the flows, consider the following optimization problem

$$\begin{array}{ll} \text{maximize}_x & \sum_i u_i(x_i) \\ \text{subject to} & \sum_i x_i \le c \\ & x_i \ge 0,\ i = 1, \cdots, N \end{array}, \tag{4.13}$$

Here, x_i is the rate allocated to flow i and the constraints encode that rates are positive and the total communication rate cannot exceed capacity. Associated to each flow i is a concave, strictly increasing, and continuously differentiable utility function with domain $x_i \ge 0$. The utility function $u_i(\cdot)$ is used to represent the degree of satisfaction when user i is allocated rate x_i. Concavity here corresponds to the assumption of *elastic* traffic. Note that (4.13) is a convex optimization problem that can be put into our standard form by replacing maximize $\sum_i u_i(x_i)$ with minimize $\sum_i -u_i(x_i)$ and introducing $f_i(x_i) = -u_i(x_i)$. Hence, the problem (4.13) can be solved by dual decomposition with Lagrange multiplier λ. The associated partial Lagrangian is

$$L(x, \lambda) = \left\{ \sum_i -u_i(x_i) + \lambda \left(\sum_i x_i - c \right) \;\middle|\; x_i \ge 0,\ i = 1, \ldots, N \right\}.$$

In the dual decomposition framework, each user is charged a common price λ per unit flow and sets its rate to maximize $u_i(x_i) - \lambda x_i$.

One could also imagine alternative mechanisms for allocating the capacity. One such mechanism is biding: each user i submits a bid $b_i \ge 0$ of the amount of money that the user is willing to pay to get a share of the bandwidth. Each user is charged the same price μ per unit flow, leading to a rate allocation of $x_i = b_i/\mu$. In this case, given a price $\mu \ge 0$, user i acts to maximize the following payoff function over $b_i \ge 0$:

$$w_i(b_i, \mu) = u_i\left(\frac{b_i}{\mu}\right) - b_i. \tag{4.14}$$

It was shown in [30] that if the link manager sets price to "clear the market", *i.e.*

$$\mu = \frac{\sum_i b_i}{c}. \tag{4.15}$$

there is a pair (b, μ) solving (4.14) – (4.15), which also solves (4.13). In the above situation, user is assumed to be a price taker, *i.e.* the payoff function w_i (4.14) takes the price μ as a fixed parameter. Price-anticipating users will realize that μ is set according to (4.15). This makes the model a game between N users, which will be discussed more detailed in Section 4.5.2.3.

The next example illustrates how dual decomposition can be used when variables couple the objective functions in an otherwise decoupled optimization problem.

Example 4.5 (Dual decomposition of coupling variable). The dual decomposition technique can also be used to decouple problems in which the same decision variable appears in several objective functions. One of the simplest instances is

$$\text{minimize } \varphi_1(x) + \varphi_2(x)$$

To decouple the problem, introduce x_1 and x_2 to reflect the two decision-makers' respective view of the optimal variable x, *i.e.* consider the equivalent problem

$$\begin{aligned} &\text{minimize } \varphi_1(x_1) + \varphi_2(x_1) \\ &\text{subject to } x_1 = x_2 \end{aligned}$$

The problem now has the same form as the previous example, and the same simple steps yield an algorithm where decision-makers optimize there individual decision variables, and the optimal value is found by adjusting the "consistency price" μ.

A drawback with dual decomposition is that a solution to the dual problem (4.12), even under strong duality, only provides the optimal value of the primal problem (4.2) but not necessarily the optimal primal decision variables. For non-optimal values of λ and μ, the primal iterates $x^\star(\lambda, \mu) = \arg\inf_x L(x, \lambda, \mu)$ are typically not even feasible to the original problem. Hence, if constraint violations cannot be tolerated, the dual decomposition method needs to be complemented by a method for recovering primal feasible solutions. When the dual function is differentiable, the primal iterates will converge to their optimal values as the dual variables approach optimality. However, as illustrated next, the primal iterates typically do not converge when the dual function is non-smooth.

Example 4.6. Consider the following specific instance of the problem in Example 4.3

$$\begin{aligned} &\text{minimize } 2|x_1 - 2| + 4|x_2 - 4| \\ &\text{subject to } x_1 + x_2 = 5 \\ &\qquad\qquad 0 \le x_1 \le 10,\ 0 \le x_2 \le 10 \end{aligned}$$

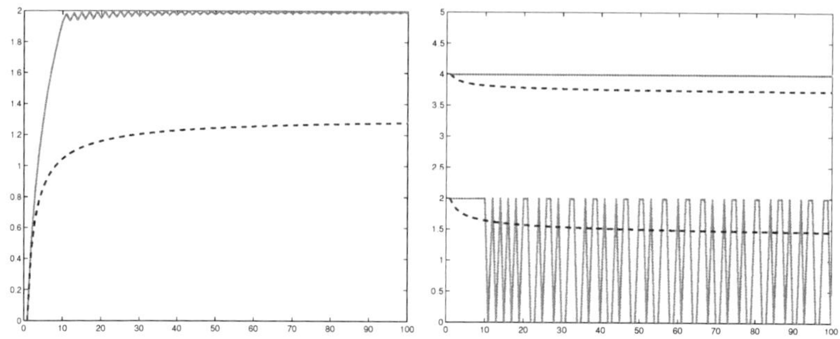

Fig. 4.4 When the dual function is non-smooth, the dual objective (left, full lines) converges while the primal iterates (right, full lines) do not. When the dual function is differentiable, both objective and iterates converge asymptotically (dashed lines)

Clearly, the optimal solution is $x_1^\star = 1, x_2^\star = 4$ with optimal value $f^\star = 2$. The full light blue lines in Figure 4.2 (left) show the evolution of the dual function for the step-size rule $\alpha(t) = 1/(1+t)$. Note that the dual function is a lower bound to the optimal value and that the dual problem converges to the optimal value. However, as revealed in Figure 4.2(right), the primal variables do not converge but oscillate, in this case, between their minimal value and unconstrained optimum.

For comparison, we change the objective function to

$$2(x_1-2)^2+4(x_2-4)^2$$

which is smooth and strictly convex. The optimal solution is now $x_1^\star = 4/3, x_2^\star = 11/3$ and the optimal value is $f^\star = 4/3$. The dashed lines in Figure 4.2(left) show how the dual objective converges. In contrast to the non-smooth example, the iterates now also converge to their optimal values, see Figure 4.2(right).

Techniques for recovering primal optimal solutions are developed and reviewed in [35, 46]. Since differentiability of the dual function gives several advantages, many techniques have been proposed to ensure dual differentiability. A simple approach is to regularize the objective function, *e.g.* to replace $f_0(x)$ by $f_0(x)+\varepsilon\|x\|_2^2$ for some small positive constant. A more elegant approach is to use proximal optimization or augmented Lagrangian techniques described next.

4.3.2 Augmented Lagrangian and Proximal Point Methods

The basic idea of the proximal point method is to ensure strict convexity by introducing an artificial variable y and re-write the original problem (4.1) as

$$\begin{array}{ll} \text{minimize} & f_0(x)+\frac{1}{2c}\|x-y\|_2^2 \\ \text{subject to} & x \in X,\ y \in \mathbb{R}^n \end{array} \tag{4.16}$$

for some positive constant $c>0$. This problem can be considered as one in y only:

$$\begin{array}{ll} \text{minimize } f_c(y) \\ \text{where} \quad f_c(y) = \inf_{x \in X} f_0(x) + \frac{1}{2c}\|y-x\|_2^2 \end{array} \tag{4.17}$$

Now, $f_c(y)$ is continuously differentiable with

$$\nabla f_c(y) = \frac{1}{c}\left(y - x^\star(y)\right)$$

where

$$x^\star(y) = \arg\inf_{x \in X} f_0(x) + \frac{1}{2c}\|y-x\|_2^2$$

so the problem is readily solved using the gradient iteration

$$y^{(t+1)} = y^{(t)} - c\nabla f_c(y^{(t)}) = x^\star(y^{(t)})$$

A related technique is the method of augmented Lagrangians, which is most readily explained on convex problems with equality constraints

$$\begin{array}{l} \text{minimize } f(x) \\ \text{subject to } Ax = b \end{array} \tag{4.18}$$

We re-write the problem as

$$\begin{array}{l} \text{minimize } f(x) + \frac{1}{2c}\|Ax-b\|_2^2 \\ \text{subject to } Ax = b \end{array} \tag{4.19}$$

Introduce multipliers μ for the constraints and form the *augmented Lagrangian*

$$L_a(x,\mu) = f(x) + \mu^T(Ax-b) + \frac{1}{2c}\|Ax-b\|_2^2$$

Now, the method of multipliers is essentially dual decomposition applied to (4.19), *i.e.* we run the iterations

$$\begin{aligned} x^{(t+1)} &= \arg\min_x L_a(x, \mu^{(t)}) \\ \mu^{(t+1)} &= \mu^{(t)} + c(Ax^{(t+1)} - b) \end{aligned}$$

It is possible to show that these iterations are equivalent to what would have been derived using proximal minimization of the dual of (4.18), see *e.g.* [6, Ch. 3]. Next, we illustrate how the techniques work on simple problem from Example 4.5.

Example 4.7. Consider the problem from Example 4.5, introduce an auxillary variable y, and re-write the problem as

$$\begin{array}{ll} \text{minimize } \varphi_1(x_1) + \varphi_2(x_2) \\ \text{subject to } x_1 = y \\ \qquad\qquad\quad x_2 = y \end{array}$$

The augmented Lagrangian is thus

$$L_a(x,y,\mu) = \varphi_1(x_1) + \varphi_2(x_2) + \mu_1(x_1 - y) + \mu_2(x_2 - y) + \frac{1}{2c}\left((x_1 - y)^2 + (x_2 - y)^2\right)$$

and performing alternating minimization of the primal variables, we find

$$\begin{aligned} x_i^{(t+1)} &= \arg\min_z \varphi_i(z) + \mu_i^{(t)} z + \frac{1}{2c}(z - y)^2, \quad i = 1,2 \\ y^{(t+1)} &= \frac{x_1^{(t+1)} + x_2^{(t+1)}}{2} - c\frac{\mu_1^{(t)} + \mu_2^{(t)}}{2} \\ \mu_i^{(t+1)} &= \mu_i^{(t)} + (x_i^{(t+1)} - y^{(t+1)}), \quad i = 1,2 \end{aligned}$$

In fact, the iterations can be simplified by noting that at optimality, $\mu_1^\star + \mu_2^\star = 0$ and that if we initialize the multipliers such that $\mu_1^{(0)} + \mu_2^{(0)} = 0$, it will hold that $\mu_1^{(t)} + \mu_2^{(t)} = 0$ for all t. Thus, the simplified iterations can be written as

$$\begin{aligned} x_i^{(t+1)} &= \arg\min_z \left\{ \varphi_i(z) + \mu_i^{(t)} z + \frac{1}{2c}(z - \overline{x}^{(t)})^2 \right\}, \quad i = 1,2 \\ \mu_i^{(t+1)} &= \mu_i^{(t)} + (x_i^{(t+1)} - \overline{x}^{(t+1)}), \quad i = 1,2 \end{aligned}$$

where $\overline{x}^{(t)} = (x_1^{(t)} + x_2^{(t)})/2$.

Although the proximal and augmented Lagrangian techniques are guaranteed to converge for all values of c, the choice of c influences both the numerical conditioning of the subproblems and the converge speed of the method, see [6].

4.3.3 Primal Decomposition

Primal decomposition is also called *resource-directive* decomposition. Rather than introducing a pricing scheme for the common resources, the primal decomposition approach sequentially updates the resource allocation to minimize the global system objective. Contrary to dual decomposition, the iterates generated by primal decomposition techniques are always feasible (by construction) and converge asymptotically to their optimal values.

The theory for primal decomposition is built around the concept of primal function of an optimization problem. The *primal function* $p(u)$ of the problem (4.2) is the optimal value of the perturbed problem

$$\begin{aligned} &\text{minimize } f_0(x) \\ &\text{subject to } f_i(x) \le u_i, \ i = 1,\dots,m \\ &\qquad\qquad h_i(x) = 0, \ i = 1,\dots,p \end{aligned} \tag{4.20}$$

The domain of p is the set of perturbations u for which there is a feasible primal solution x and, hence, $p(u) < \infty$. We denote the domain of p by P:

$$P = \{u \mid p(u) < \infty\}$$

Contrary to the dual function (which is always concave), convexity of the primal function and its domain requires convexity of the original problem:

Proposition 4.7. *Consider the convex optimization problem (4.2) and assume that $p(u) > -\infty$ for all $u \in P$. Then P is a convex set and $p(u)$ is convex over P.*

In order to minimize $p(u)$ we also need to be able to compute (at least) a subgradient. The following characterization is then useful.

Proposition 4.8. *Consider the convex optimization problem (4.2) with optimal dual variables $(\lambda^\star, \mu^\star)$. Then*

$$p(0) \le p(u) + u^T \lambda^\star$$

i.e., $-\lambda^\star$ is a subgradient of p at $u = 0$.

Next, we demonstrate how primal decomposition can be applied to Example 4.3.

Example 4.8. The key trick in applying primal decomposition to a problem with a complicating constraint such as Example 4.3 is to introduce variables r_i that represent the amount of the common resource allocated to subproblem i and re-write it as

$$\begin{array}{ll} \text{minimize} & \varphi_1(x_1) + \varphi_2(x_2) \\ \text{subject to} & x_1 \le r_1, \quad x_2 \le r_2 \\ & r_1 + r_2 \le x_{\text{tot}} \end{array}$$

For notational simplicity, introduce the constant $r_{\text{tot}} = x_{\text{tot}}$ and define functions $v_i(r_i) = \inf\{\varphi_i(x_i) \mid x_i \le r_i\}$. The problem can then be equivalently written as

$$\begin{array}{ll} \text{minimize} & \sum_i v_i(r_i) \\ \text{subject to} & \sum_i r_i \le r_{\text{tot}} \end{array} \tag{4.21}$$

Now, from the results on the primal function, v_i is a convex, possibly non-smooth function and a subgradient of v_i at r_i is given by the optimal dual variable $\lambda_i^\star(r_i)$ associated with the inequality constraint of the primal minimization problem

$$\begin{array}{ll} \text{minimize} & \varphi_i(x_i) \\ \text{subject to} & x_i \le r_i \end{array}$$

Hence, (4.21) is a convex optimization problem that can be solved using a projected subgradient method

$$r^{(t+1)} = P_R\{r^{(t)} + \alpha^{(t)} \lambda^\star(r^{(t)})\}$$

where $r^{(t)} = (r_1^{(t)}, r_2^{(t)})$, $\lambda^\star(r^{(t)}) = (\lambda_1^\star(r_1^{(t)}), \lambda_2^\star(r_2^{(t)}))$ and $P_R\{\cdot\}$ denotes projection onto the total budget constraint $r_1 + r_2 \le r_{\text{tot}}$.

It is useful to compare the solution mechanisms suggested by primal and dual decomposition. In primal decomposition, the master problem assigns resources to the subsystems. The subsystems then optimize their operation to "perform their best" with the amount of resources they have been assigned (compute the optimal x_i and the associated value of $v_i(r_i)$ above), and return a Lagrange multiplier vector back to the coordinator. It is well-known from sensitivity theory that the Lagrange multipliers indicate the potential cost reduction that can be achieved by an additional amount of resource. In our case, if we assume that φ_i are (in addition to convex) smooth and decreasing, and r_i are scalar variables, then the first-order optimality conditions yield that $\lambda_i(r_i) = -\varphi_i'(r_i)$. In other words, the Lagrange multipliers signal exactly the amount of cost reduction that a subsystem could generate if it would obtain a small additional amount of resource. When the master problem updates the resource allocation via the projected (sub)gradient iteration, it effectively shifts resources from subsystems with small predicted cost reduction to subsystems with large predicted cost reduction. The local decisions (iterates x_i) taken by nodes are, by construction, feasible to the original problem.

In the dual decomposition method, on the other hand, the coordinator announces a dual variable λ which, by similar reasoning as above, can be interpreted as the unit cost of the common resource. Subsystems then optimize their operation accounting both for their operational cost ($\varphi_i(x_i)$) and the resource cost (λx_i). The coordinator then updates the price to align supply and demand: increase the price if the resource is overbooked and decrease the price otherwise. The iterates are, in general, not guaranteed to be feasible to the original problem.

4.4 Networked Optimization

With a basic understanding of first-order methods and decomposition techniques, we are ready to consider *networked optimization* problems. Contrary to decomposition techniques, in which the problem structure determines how the original problem is divided into subproblems, networked optimization problems arise when the communication structure (which decision-makers are allowed to coordinate with each other) is fixed and the computation structure has to be tailored to match. This class of optimization algorithms have also been termed *multi-user* optimization, since the set-up can be used to model a multi-agent system in which agents cooperate and exchange information with neighbors to find a globally optimal decision [33]. Although decomposition techniques will turn out to be useful also in this context, decomposition does not necessarily yield distributed optimization algorithms unless the master- and sub-problems can be distributed.

To study networked optimization, we consider problems on the form

$$\begin{aligned} &\text{minimize } \textstyle\sum_{v\in\mathscr{V}} f_v(x_v,\theta) \\ &\text{subject to } x_v \in X_v,\ \theta\in\Theta \end{aligned} \tag{4.22}$$

with an associated *communication graph* $\mathscr{G} = (\mathscr{V}, \mathscr{E})$; see Figure 4.1. The vertices of the graph represent decision-makers, and the edges encode which agents can exchange information to coordinate their decisions. Each node has a set of local (private) decision variables x_v. These variables are under exclusive control of node v and only influences the local loss function $f_v(x_v, \theta)$. The global variables θ, on the other hand, impact the loss functions of multiple nodes. We focus on convex problems and assume that the constraint sets X_v and Θ are non-empty, closed and convex and that the local loss functions $f_v(x_v, \theta)$ are convex in (x_v, θ).

We can eliminate the local variables by introducing new loss functions

$$v_v(\theta) = \inf_{x_v \in X_v} f_v(x_v, \theta)$$

and reformulate (4.22) as

$$\begin{array}{l} \text{minimize } \sum_{v \in \mathscr{V}} v_v(\theta) \\ \text{subject to } \theta \in \Theta \end{array} \tag{4.23}$$

Clearly, under our assumptions $v_v : \mathbb{R}^n \mapsto \mathbb{R}$ are convex but possibly non-smooth.

4.4.1 Networked Optimization via Dual Decomposition

The most direct approach to networked optimization is dual decomposition. First, introduce local decision variables θ_v at each node $v \in \mathscr{V}$ and rewrite (4.23) as

$$\begin{array}{lll} \text{minimize } \sum_{v \in \mathscr{V}} v_v(\theta_v) & & \\ \text{subject to } \theta_v = \theta_w & & \forall (v, w) \in \mathscr{E} \\ \qquad\qquad\quad \theta_v \in \Theta & & \forall v \in V \end{array}$$

The problem would be separable if it were not for the edge-wise coupling constraints. It is thus natural to relax these constraints by dual decomposition. To this end, introduce multipliers $\mu_{(v,w)}$ for all $(v, w) \in \mathscr{E}$ and form the partial Lagrangean

$$L(\theta, \mu) = \{\sum_v v_v(\theta_v) + \mu_{(v,w)}(\theta_v - \theta_w) \mid \theta_v \in \Theta\}$$

with associated dual function

$$g(\mu) = \inf_{\theta_v \in \Theta_v} \sum_v v_v(\theta_v) + \theta_v \sum_w (\mu_{(v,w)} - \mu_{(w,v)})$$

Since the problem is equality-constrained, the dual problem is unconstrained and can be solved using a standard subgradient optimization

$$\mu_{(v,w)}^{(t)} = \mu_{(v,w)}^{(t)} + \alpha^{(t)}(\theta_v^\star(\mu^{(t)}) - \theta_w^\star(\mu^{(t)})) \tag{4.24}$$

where $\theta^\star(\mu^{(t)}) = \arg\inf_\theta L(\theta, \mu^{(t)})$. The following example illustrates how the technique applies to a networked least-squares problem.

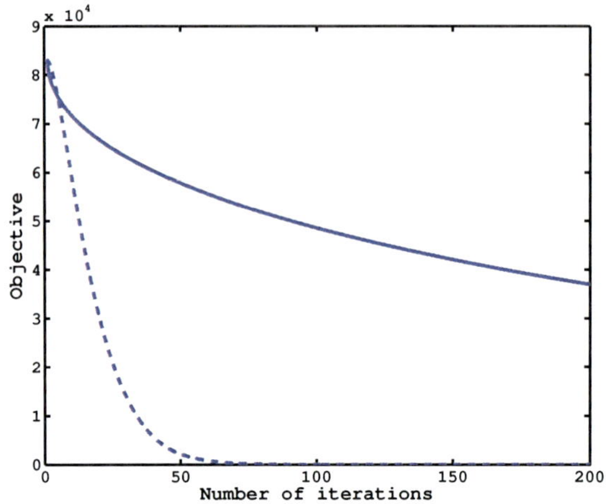

Fig. 4.5 Distributed least-squares: gradient (full) and accelerated gradient (dashed)

Example 4.9. Consider a networked least-squares problem, where

$$f_v(x_v, \theta) = \frac{1}{2}(\theta - z_v)^2$$

Introducing local variables θ_v and following the procedure outlined above, we find

$$v_v(\theta_v) = \frac{1}{2}(\theta_v - z_v)^2$$

In this case, it is also possible to find an explicit expression for $\theta_v^\star(\mu^{(t)})$,

$$\theta_v^\star(\mu^{(t)}) = z_v - \sum_w (\mu_{(v,w)}^{(t)} - \mu_{(w,v)}^{(t)})$$

This expression, together with the iteration (4.24), defines a networked optimization algorithm that is guaranteed to find the optimal solution, provided that the step-length sequence $\{\alpha^{(t)}\}$ is chosen appropriately.

To implement the algorithm, note that each link needs information about $\theta_v^\star$ and $\theta_w^\star$, *i.e.* the current version of the local decision variables of the two nodes connected to the link. To compute these decisions, on the other hand, nodes need to know the Lagrange multipliers of all links that they are connected to.

Since the optimization problem is strictly convex, the dual function is smooth we can use a higher-order method to accelerate convergence. We omit the details here but evaluate both the direct and the accelerated method to a ring network of $N = 100$ nodes. Figure 4.5 shows how the methods converge, with a significant speed-up advantage of the accelerated method.

While dual decomposition is classical, the first applications to networked optimization (and estimation) in the spirit above the authors are aware of are the work by

Fawal, Georges and Bornard [18] and the one by Rabbat and Nowak [54] (who also analyze convergence of the method for time-varying graphs). For ease of exposition, we have introduced one Lagrange multiplier for every edge in the graph, but it is possible to only introduce dual variables for a subset of links (as long as the underlying graph is connected) [60]. Finally, it is also possible to apply alternative decomposition techniques, such as the augmented Lagrangian method, to networked optimization [18, 60].

4.4.2 Consensus-Subgradient Schemes

In the dual decomposition approach, asymptotic agreement on the global decision variable is enforced by adjusting Lagrange multipliers. However, agreement among nodes in a network can be achieved by many different techniques [50, 52, 68], indicating that there could be a rich family of algorithms for networked optimization. We will consider one such class of algorithms which combines subgradient optimization and consensus algorithms. Although consensus algorithms are described in depth in other chapters in this book, we will give a brief overview for completeness.

The consensus problem considers the design of protocols that ensure that all nodes in a network agree on a common quantity. In the area of multi-agent system, a lot of attention has been focused on conditions where linear iteration on the form

$$x_v(t+1) = x_v(t) - \sum_{w:(v,w)\in\mathscr{E}} W_{(vw)}(x_v - x_w) \tag{4.25}$$

converges so that all node variables equal the average of the initial values, *i.e.*

$$\lim_{t\to\infty} x_v(t) \to \frac{1}{|\mathscr{V}|}\sum_v x_v(0)$$

It is convenient to write the iterations in matrix form:

$$x(t+1) = Wx(t), \tag{4.26}$$

where the i-th element of the vector $x(t)$ corresponds to the local value at node i, $x_i(t)$. We make the following assumptions on W.

Assumption 4.1 (Consensus Matrix Properties). *The weight matrix W satisfies*

$$\begin{aligned} &i) \quad [W]_{ij} = 0, \ \textit{if } (i,j) \notin \mathscr{E} \textit{ and } i \neq j, \\ &ii) \quad W\mathbf{1}_N = \mathbf{1}_N, \ \rho\left(W - \frac{\mathbf{1}_N\mathbf{1}_N^T}{N}\right) < 1, \\ &iii) \quad W = W^T, \end{aligned}$$

where $\rho(\cdot)$ is the spectral radius and $\mathbf{1}_N \in \mathbb{R}^N$ is the column vector of all ones.

The first assumption restricts nodes to only communicate with their immediate neighbors, while *iii)* encodes the assumption that the underlying graph is undirected (or that all links are bidirectional) and that the same weight is used in the consensus iterations for both directions of the same link. Finally, the second assumption ensures asymptotic average consensus:

Lemma 4.1. *If the weight matrix $W \in \mathbb{R}^{N\times N}$ is symmetric, then the limit*

$$\lim_{k\to\infty} W^k = \frac{\mathbf{1}_N\mathbf{1}_N^T}{N} \tag{4.27}$$

holds if and only if Assumption 4.1 ii) holds for W.

Proof. See, *e.g.*, [68, Theorem 1]. ∎

There are many ways of selecting W to satisfy the above conditions using either centralized or decentralized information, see *e.g.* the chapter by Garin and Schenato in this book or references [50, 8]. We will only present one example, namely the Metropolis-Hastings scheme [23, 8]:

Lemma 4.2 (Metropolis-Hastings). *If the graph $\mathscr{G}$ is connected, then W fulfills Assumption 4.1 if the elements of W are set to*

$$[W]_{vw} = \begin{cases} \min\{d_v^{-1}, d_w^{-1}\} & \text{if } (v,w)\in\mathscr{E} \text{ and } v\neq w \\ \sum_{(v,w)\in\mathscr{E}} \max\{0, d_v^{-1}-d_w^{-1}\} & \text{if } v=w \\ 0 & \text{otherwise.} \end{cases} \tag{4.28}$$

where d_v denotes the degree (number of neighbors) of node v.

Note that the iteration (4.25) is simply a method for distributed computation of a network-wide average. To make use of the algorithm for networked optimization, we must first compute the local quantity that should be averaged across nodes and then execute one or more iterations of the consensus algorithm. As an example, consider the consensus-subgradient method from [44] and further developed in [27]. The basic idea of this algorithm is to interpret the gradient of the objective function

$$\frac{\partial}{\partial\theta}\nu(\theta) = N\frac{1}{N}\sum_v \nu_v'(\theta)$$

as (N times) the average of the gradients of the individual node objective functions. Hence, for smooth objectives with $\Theta = \mathbb{R}^n$ it is natural to consider the algorithm

$$\theta_v^{(t+1)} = \sum_{w\in\mathscr{V}} [W^\varphi]_{vw}\left(\theta_w^{(t)} - \alpha^{(t)}\nu_w'(\theta_w^{(t)})\right)$$

Here, W^φ denotes that the consensus iteration is performed φ times. Hence, every node v maintains a vector of θ_v of local variables, and computes a desired next iterate by accounting only for the local gradient. Nodes then perform consensus iterations to agree on the next iterate and execute this one. Clearly, if all nodes start with the equal initial values, $\theta_v(0) = \theta_w(0)$ for all $(v,w) \in \mathscr{E}$ and we let $\varphi \to \infty$ we recover the classical gradient iteration. More surprisingly, the method still works when nodes start with different initial values and perform a finite number of iterations, even when the objective functions are not differentiable and θ is subject to (convex) constraints. The constrained non-smooth version of algorithm reads

$$\theta_v^{(t+1)} = P_\Theta\left[\sum_w [W^\varphi]_{vw}\left(\theta_w^{(t)} - \alpha^{(t)} g_w(\theta_w^{(t)})\right)\right], \tag{4.29}$$

where $\theta_v^{(0)} \in \Theta$ for all $v \in \mathscr{V}$, $g_v(\theta_v^{(t)}) \in \partial f_v(\theta_v^{(t)})$, $\varphi \in \mathbb{N}$, and $[W^\varphi]_{vw}$ denotes the element of W^φ in the v-th row and w-th column. Hence, using this scheme, agents maintain their local variable vector $\theta_v^{(t)}$ and compute a desired next iterate $\theta_v^{(t)} - \alpha^{(t)} g_v(\theta_w^{(t)})$. They then announce this desired iterate to their neighbors and update their own decision taking the neighbors desires into account. More specifically, (4.29) implies that each agent runs φ number of consensus iterations with its neighbors defined by (4.26) to determine its next iterate. The total number of iterations in the algorithm up to step t is therefore $t\varphi$. To give the flavor of the method, we give the following result

Theorem 4.2. *Consider Problem (4.23) with $\Theta = \mathbb{R}^n$ and assume that $\|g_v\|_2 \le L$ for all $g_v \in \partial v_v$ and all $v \in \mathscr{V}$. Let $\{\bar{\theta}_v(t)\}_{t=0}^\infty$ be the Cesàro averages*

$$\tilde{\theta}_v^{(t)} = \frac{\sum_{k=0}^t \alpha^{(k)} \theta_v^{(k)}}{\sum_{k=0}^t \alpha^{(k)}}$$

of the iterates $\theta_v^{(t)}$ produced by (4.29) with a consensus matrix satisfying Assumption 4.1. Define

$$\gamma = \rho\left(W - \frac{\mathbf{1}\mathbf{1}^T}{N}\right)$$

and $\beta^{(0)}$ such that $\|\theta_v^{(0)} - |\mathscr{V}|^{-1}\sum_{w\in\mathscr{V}} \theta_w^{(0)}\| \le \beta^{(0)}$. Then, under diminishing stepsizes satisfying $\alpha^{(t+1)}/\alpha^{(t)} > \mu$, if

$$\varphi \ge \frac{\log(\mu\beta^{(0)}) - \log(4\sqrt{|\mathscr{V}|}n(\beta^{(0)} + \alpha^{(0)}L^2)}{\log(\gamma)}$$

we have that

$$\lim_{t\to\infty} v(\tilde{\theta}_v^{(t)}) = v^\star, \quad \forall v \in \mathscr{V}.$$

More extensive analysis results for several variants of the method can be found in [27].

4.4.3 Networked Incremental Subgradient Methods

The use of classical incremental subgradient methods in a networked setting would suggest a message-passing solution where in each iteration, one node receive a token that contains the most recent iterate from another node in the network, computes the next iterate by taking a step in the negative subgradient of its own local objective function, and then passes the token to another node. A drawback with the incremental subgradient method is that the analysis assumes that the token is passed around in a ring (*i.e.* it performs cyclic rounds in which each component is updated once).

An alternative approach, tailored to the networked setting, was proposed and analyzed in [28]. Here, nodes pass their updated iterate to a *random neighbor* and the need for organizing the nodes into an underlying logical ring is removed. The algorithm is shown to converge as long as each component of the objective function is updated with equal probability. In other words, the token should perform an unbiased random walk on the graph. To design such a random walk, consider a Markov chain in which each state is associated to a node in the underlying network. To ensure that the token is passed only between neighboring nodes, the transition matrix W must fulfill the sparsity constraint that $W_{vw} = 0$ if $(v,w) \notin \mathscr{E}$. Furthermore, the analysis requires that the Markov chain is irreducible, aperiodic, and its stationary distribution is uniform. It turns out that matrices W constructed using Lemma 4.2 satisfy these assumptions. Now, the algorithm takes the form

$$\theta^{(t+1)} = P_{\Theta}\left\{\theta^{(t)} - \alpha g_w\right\}, \tag{4.30}$$

where $w^{(t)}$ is the state of Markov Chain in iteration t, α is a fixed step-length parameter and $g_w \in \partial f_w(\theta_w^{(t)})$. The following result establishes its convergence

Proposition 4.9. *Let* $\{\theta^{(t)}\}_{t=0}^{\infty}$ *be generated by* (4.30). *With probability 1*

$$\begin{cases} \liminf_{t\to\infty} v\left(\theta^{(t)}\right) = v^\star, & \text{if } f^\star = -\infty \\ \liminf_{t\to\infty} v\left(\theta^{(t)}\right) \leq v^\star + \frac{\alpha L^2 K}{2}, & \text{if } v^\star > -\infty. \end{cases}$$

where K is an upper bound on the second moment of the recurrence time for all states in the Markov Chain.

Several variations and extensions of the basic method are given in [28], including stronger results (lim sup rather than lim inf) for the running average of the iterate that a specific node sees during the course of the algorithm. The same paper also compares the predicted convergence rates for the networked incremental subgradient method with the provable convergence for the classical (deterministic or randomized) incremental subgradient method on several classes of graphs. Extensions to time-varying graph topologies and diminishing step-sizes can be found in [55].

4.5 Game Theory in Distributed Optimization

Game theory is typically used to model and counteract selfish behaviors in distributed systems. However, there are several reasons for us to introduce game theory in this chapter. First, networked optimization usually consists of multiple agents who can observe and react to their environment. Game theory offers a powerful tool set to analyze interactions between such intelligent entities. Second, although network components or agents would like to cooperate, it might be impractical or impossible to exchange the information required to implement any of the distributed optimization techniques described so far. It might then be better for agents to optimize their local or private objective and react to limited network information. In these cases, we use non-cooperation to capture limited information. The third reason is that game theory provides a way to predict, analyze or even to improve the outcome of a non-cooperative interaction, *e.g.* the notation of equilibrium.

4.5.1 Basics of Game Theory

We will mainly focus on the discussion of non-cooperative game model with complete information[1] and finite number of players. Formally, a normal or strategic form [2] of a game Ξ is given by $\Xi = \left\{\mathscr{V}, \{\mathscr{S}_v\}_{v\in\mathscr{V}}, \{w_v\}_{v\in\mathscr{V}}\right\}$, where $\mathscr{V}$ is the set of agents or players; $\mathscr{S}_v$ is the set of strategiesof player v and w_v is its payoff function[3], which is a function of the strategy x_v chosen by player v and the strategies chosen by other players, denoted as x_{-v}. We will use $\mathscr{S} = \Pi_{v=1}^{|\mathscr{V}|}\mathscr{S}_v$ to denote the Cartesian product of sets and use $x = [x_v, x_{-v}] = \left[x_1, x_2, \cdots, x_{|\mathscr{V}|}\right]$, $\forall v \in \mathscr{V}$ to refer to a vector. Based on this model, the Nash equilibrium steady-state of a game can be defined. The definition is based on the concept of a best response correspondence.

Definition 4.2. *Let* $\Xi = \left\{\mathscr{V}, \{\mathscr{S}_v\}_{v\in\mathscr{V}}, \{w_v\}_{v\in\mathscr{V}}\right\}$ *be a strategic game. For any* $x_{-v} \in \mathscr{S}_{-v}$ *we define the best response correspondence* $BR_v(x_{-v})$ *as,*

$$BR_v(x_{-v}) = \left\{x_v \in \mathscr{S}_v | w_v(x_v, x_{-v}) \geq w_v\left(x_v', x_{-v}\right) \text{ for all } x_v' \in \mathscr{S}_v\right\}.$$

A strategy profile x^* *is a* Nash equilibrium *if and only if* $x_v^* \in BR_v\left(x_{-v}^*\right)$ *for all* $v \in \mathscr{V}$.

The above definition is referred to pure Nash equilibrium. As seen in the above definition, the pure Nash equilibrium consists of selecting one element from each player's possible action sets and is also stable for each player to deviate from the equilibrium. A mixed strategy consists of selecting multiple elements from each

[1] The complete information means that every player knows the payoff of the others.

[2] When a game is presented in normal form or strategic form, it is presumed that each player acts simultaneously or, at least, without knowing the actions of the other. If players have some information about the choices of other players, the game is usually presented in extensive form.

[3] Hereafter, we consider each player chooses strategy to maximize its payoff. The game can be defined accordingly if each play chooses strategies to minimize its own cost function.

player's action set and run the series of actions with some probability. The associated equilibrium is then called a mixed Nash equilibrium. In the following we will assume the strategic game Ξ is static, *i.e.* it is played in one-shot with complete information of payoffs, with a finite number of players and finite strategy sets. As shown later, although best response and gradient methods are usually adopted to update players' strategies to optimize the *current* payoff in reaction to other players' strategies, the game is still called a static game. If players choose their strategies to maximize their payoff functions averaged over the whole game duration, the game is called a repeated game.

4.5.2 Properties of Nash Equilibria

4.5.2.1 Existence of Nash Equilibria

The strategy tuples corresponding to Nash equilibria are consistent predictions of the outcome of a game. The first question after defining a game is whether there exists an equilibrium. Generally, proving the existence of an equilibrium involves proving the existence of a solution to a fixed-point problem [7]. However, a number of sufficient conditions for the existence of Nash equilibria have been developed for games with particular structure of strategy sets and payoff functions.

Theorem 4.3 (Debreu, Glicksberg, Fan [20]). *Let* $\Xi = \left\{\mathscr{V}, \{\mathscr{S}_v\}_{v\in\mathscr{V}}, \{w_v\}_{v\in\mathscr{V}}\right\}$ *be a strategic game, where* $\mathscr{V}$ *is a finite set. If* $\forall v \in \mathscr{V}$, $\mathscr{S}_v$ *is a non-empty compact and convex subset of a finite-dimensional Euclidean space;* $w_v(x)$ *is a continuous function in the profile of strategies* x *and quasi-concave in* x_v*; then the game has at least one pure Nash equilibrium.*

The power control game in [22] has been shown to be a quasi-concave game with a compact convex strategy set and hence has at least one pure Nash equilibrium. The multiple access game in [71] has a concave payoff function, which is a special case of the above theorem and thus it has a pure Nash equilibrium.

If a game does not have quasi-concave payoffs, one may turn to supermodular games [66] and potential games [43] to argue about the existence of Nash equilibria. Let us first look at the definition of supermodular game.

Definition 4.3 ([66]). *The strategic form game* Ξ *is called supermodular if:* $\forall v \in \mathscr{V}$, $\mathscr{S}_v$ *is a compact subset of* $\mathbb{R}$; w_v *is upper semi-continuous in* x; $\forall v \in \mathscr{V}$, $\forall x_{-v} \succeq x'_{-v}$ *the quantity* $w_v(x) - w_v\left(x_v, x'_{-v}\right)$ *is non-decreasing in* x_v.

Intuitively, the definition means that the marginal payoff of increasing a player's strategy rises with increases in the other players' strategies. This implies that the best response of a player is a non-decreasing function of other players' strategies. Supermodular games are interesting for several reasons. First, many applied models satisfy the assumptions of supermodular games. Second, they have the remarkable property that many solution concepts yield the same predictions. Finally, they tend to be analytically appealing – they have nice comparative statical properties and behave well under various learning rules.

Theorem 4.4 ([66]). *If Ξ is an supermodular game, it has at least one pure Nash equilibrium.*

Applications of supermodular games include the pricing-based power control algorithm designed in [26] for solving a sum utility maximization problem. In [26], each wireless transmitter adapts transmission power and charges interference price to interfering transmitters based on best response update in an implicit supermodular game. Sum rate maximization, which is a special case in [26], has been solved by [13] using gradient algorithm based on dual decomposition. As opposed to gradient methods that might need a small stepsize to converge to the optimum at the price of slow convergece, the convergence of the best response algorithm in the fictitious game [26] is ensured by supermodular game theory without appealing to stepsize.

Another particular game possessing equilibrium is the potential game.

Definition 4.4 ([43]). *A strategic game Ξ is called*

1. *an exact potential game if there exists a function $\Phi : \mathscr{S} \to \mathbb{R}$ such that for all $v \in \mathscr{V}$, and $(x_v, x_{-v}) \in \mathscr{S}$, $x'_v \in \mathscr{S}_v$:*

$$w_v(x_v, x_{-v}) - w_v(x'_v, x_{-v}) = \Phi(x_v, x_{-v}) - \Phi(x'_v, x_{-v}). \tag{4.31}$$

   If the payoff function w_v is continuously differentiable, then (4.31) is equivalent to

$$\frac{\partial w_v(x_v, x_{-v})}{\partial x_v} = \frac{\partial \Phi(x_v, x_{-v})}{\partial x_v}, \ \forall v \in \mathscr{V}. \tag{4.32}$$

2. *an ordinal potential game if there exists a function $\Phi : \mathscr{S} \to \mathbb{R}$ such that for all $v \in \mathscr{V}$, and $(x_v, x_{-v}) \in \mathscr{S}$, $x'_v \in \mathscr{S}_v$: $sgn(w_v(x_v, x_{-v}) - w_v(x'_v, x_{-v})) = sgn(\Phi(x_v, x_{-v}) - \Phi(x'_v, x_{-v}))$.*

For definitions of min-max potential games and state-based potential games, please refer to [56] and [41], respectively.

Theorem 4.5. *If Ξ is a potential game with a finite number of players, compact strategy sets, and continuous payoffs, then it has at least one pure Nash equilibrium.*

One good example of potential game is the application to cooperative control, where multiple agents interact with each other to achieve a common target, *e.g.* the consensus problem [51] and vehicle formation problem [56]. There are two advantages when the cooperative control is modelled as a potential game. First, for cooperative control each agent updates its strategy by evaluating its effects on the common objective function. This evaluation will require to observe the decisions of all agents [40]. In the potential game formulation, each player has local payoff function that captures the player's marginal contribution to the potential function, which is the common objective of cooperative control. The implementation overhead is reduced in the game setting. Thus, one critical point in applying potential game to cooperative control is to assign each player a reasonable payoff function that is aligned with the potential function. Secondly, in cooperative control problems formulated as a potential game, there are many learning algorithms adaptive to time-varying environment with guaranteed convergence to Nash equilibria.

4.5.2.2 Uniqueness of Nash Equilibrium

After establishing the existence of a Nash equilibrium the next issue is to study its uniqueness. Uniqueness of Nash equilibria is both critical for predicting outcome of a game and important for convergence issues. A general result is given by Rosen [58] to ensure that a game has a unique Nash equilibrium. Another value of [58] is that it points out a way to select one equilibrium if there are multiple equilibria. See [36] and the references therein for more discussions. The sufficient conditions in [58] change the structure of a game and restrict the payoff and strategy sets. When the best response of a game can be explicitly expressed, there are some weaker conditions that guarantee uniqueness of Nash equilibrium. The first result is based on properties of contraction mapping.

Definition 4.5 ([67]). *$M(\cdot): X \to X$, where X is a subset of $\mathbb{R}^{|\mathcal{V}|}$, is a contraction mapping if there exists $\varepsilon \in (0,1)$ such that $\|M(x) - M(y)\| \leq \varepsilon \|x - y\|$*[4]*, $\forall x, y \in X$.*

Theorem 4.6. *Suppose that $M(\cdot): X \to X$ is a contraction mapping and X be a closed subset of $\mathbb{R}^{|\mathcal{V}|}$. Then M has a unique fixed point x^* that is globally asymptotically stable.*

When the contraction mapping theorem is applied, the best response correspondence is presumed to be a mapping. In the context of game theory, if the best response mapping $BR(\cdot): \mathscr{S} \to \mathscr{S}$ is a contraction mapping, the sequence $\left\{x^{(k)}\right\}$ generated by $x^{(k+1)} = BR\left(x^{(k)}\right)$ converges to a unique fixed point x^* from any initial strategy profiles $x(0) \in \mathscr{S}$. According to the definition of Nash equilibrium, this fixed point x^* is the unique Nash equilibrium. The application of contraction mapping theorem to study uniqueness of Nash equilibrium can be found in [69] for scalar strategies and in [63] for vector strategies.

In addition to contraction mapping, if the best response correspondences satisfy some nice properties there can be a unique intersection in the strategy space. One of such functions or correspondences is called standard function [72], which was generalized by [64] to include type-II standard function. Together this type of function is called two-sided scalable function.

Definition 4.6 ([64]). *A vector function $I(\cdot): X \to X$ with X a subset of $\mathbb{R}$, is said to be two-sided scalable, if for all $\beta > 1$, $\forall x \in X$, $\forall x' \in X$, $(1/\beta)x \leq x' \leq \beta x$ implies*

$$\frac{1}{\beta} I(x) < I(x') < \beta I(x).$$

Theorem 4.7 ([64]). *If $I(\cdot): X \to X$ is two-sided scalable and a fixed point exists, then $I(\cdot)$ has a unique fixed point, which is globally asymptotically stable.*

Once again, if the best response mapping $BR(\cdot): \mathscr{S} \to \mathscr{S}$ of a strategic game is two sided scalable it converges to the unique Nash equilibrium when there is a Nash

[4] Here $\|\cdot\|$ is some norm.

equilibrium of the game. Note that the above theorem can not guarantee the existence of Nash equilibrium. To study existence of Nash equilibrium one has to invoke Theorem 2 in [20] or turn to Brouwer's Fixed-Point Theorem in [7].

4.5.2.3 Efficiency of Nash Equilibrium

The Nash equilibrium discussed above provides a solution to multi-objective optimization problem where no agent can increase its performance through individual effort. Thus, it is an outcome of distributed decision making which could be less efficient than a possible scheme through cooperation between agents and/or as a result of centralized optimization. Equilibrium efficiency is also a criterion to select one from multiple equilibria, if there are more than one equilibrium.

A well known efficiency criterion is *Pareto optimality*. An outcome of a game is Pareto optimal if there is no other outcome that makes every player at least as well off and at least one player strictly better off. In other words, a Pareto-optimal outcome cannot be improved upon without hurting at least one player. It should be noted that a Nash equilibrium solving a social optimal problem is Pareto-optimal. However, in reality the competitive solution is far from social optimal or Pareto-optimal. One may expect to quantify the performance gap between the social optimal and Nash equilibria. The performance gap is known as the *price of anarchy [34]*, defined next.

Let us first represent the social performance achieved by all players at a given Nash equilibrium x as

$$SUM(x) = \sum_{v\in\mathscr{V}} \upsilon_v(x_v, x_{-v}),$$

where $\upsilon_v(x)$ denotes the utility (or payoff) of player v at equilibrium x. The social optimum OPT is defined to be the maximum $SUM(x)$ achieved by all players.

Definition 4.7. *The price of anarchy of a game is the worst-case efficiency ratio among all pure strategy Nash equilibria,*

$$PoA = \min_{x\in\mathscr{X}} \frac{SUM(x)}{OPT}$$

Example 4.10. Given the example considered in (4.14), in the case of price-anticipating users the problem turns into a game. Each user chooses b_i to maximize its payoff

$$w_i(b_i, b_{-i}) = \begin{cases} u_i\left(\frac{b_i}{\sum_{v=1}^{N} b_v} c\right) - b_i & \text{if } b_i > 0 \\ u_i(0) & \text{if } b_i = 0 \end{cases} \tag{4.33}$$

over non-negative b_i. Note that the payoff function in (4.33) maybe discontinuous at $b_i = 0$, if $\sum_{v\neq i} b_v = 0$. This discontinuity may preclude the existence of a Nash equilibrium [29]. The authors in [29] explore the effects of price-anticipation and prove that the price of anarchy is a 25% efficiency loss compared with the maximum possible aggregate utility in (4.13) or (4.14).

The reason for low efficiency of Nash equilibrium in a non-cooperative game is that each player aims to optimize its own performance without regarding the cost it imposes on others. Pricing (or taxation) has been proven to be an efficient way to improve efficiency. Here, we do not use price to generate revenue for the system but use price to encourage players to use system resources more efficiently rather than the aggressive competition of the purely non-cooperative game. A pricing scheme is called incentive compatible if pricing enforces a Nash equilibrium that improves social welfare. An efficient price should reflect accurately the costs of usage of a resource and must take into account the individual player's effects on system. In [59], it has been shown that the utility in the energy-efficient power control game can be improved when some players deviate somewhat from the Nash equilibrium, *i.e.* the resulting equilibrium is not Pareto-optimal. To restrict interference (negative results of selfish behaviors), the authors use a usage-based pricing to improve the equilibrium utilities. However, this linear usage-based pricing in [59] is far from social optimum since it does not take into account individual player's effects on system performance. Maximizing the sum of coupled utilities in a non-cooperative environment usually involves some control message passing. To illustrate this, let us consider a social optimal problem,

$$\max_{x \in \mathscr{S}} \sum_{v \in \mathscr{V}} u_v(x), \tag{4.34}$$

where $u_v(x)$ is assumed to be differentiable. One of the necessary conditions for $x^* \in \mathscr{S}$ to be optimal for (4.34) is

$$\frac{\partial u_v(x^*)}{\partial x_v} + \sum_{m:m \neq v} \frac{\partial u_m(x^*)}{\partial x_v} = \lambda_v^*, \forall v \in \mathscr{V} \tag{4.35}$$

where λ_v^* is a Lagrange multiplier used to regulate the strategy selection within the strategy set of player v. In a non-cooperative game, where each player selfishly tries to optimize its own payoff rather than the common objective in (4.34), the condition (4.35) can serve as a guideline to design an incentive compatible pricing scheme. Since each player solves

$$\max_{x_v \in \mathscr{S}_v} w_v(x), \ \forall v \in \mathscr{V}, \tag{4.36}$$

where $w_v(x) = u_v(x) - \kappa_v(x)$, the unit price $\kappa_v(x)$ that user v is charged for the common resource should be based on $\sum_{m:m \neq v} \frac{\partial u_m(x)}{\partial x_v}$. Specifically, $\kappa_v(x)$, $\forall v \in \mathscr{V}$ should be designed to guarantee that the Nash equilibrium for (4.36) also satisfies the first order necessary conditions of (4.34).

Applications of these techniques to communication systems can be found in, for example, [26, 39]. To bring the competitive players to solve a social utility maximization problem in (4.34), the authors in [26] duplicate the player sets into two sets. In the first set, each player updates its decision by best response to maximize its individual payoff in (4.36). In another set, each player charge prices according to the first order necessary condition (4.35). The existence and stability of a Nash equilibrium is guaranteed using supermodular game theory. In [39], the divisible resource allocation is addressed by proportional auction scheme. The efficiency of this auction is determined by the cost function, since it is related to the Lagrange multiplier, which is used to relax the global constraint in the corresponding social welfare maximization problem. To maximize the social welfare, the cost function is carefully designed by comparing the first order necessary conditions of local payoff function with those of social utility function. More discussion on pricing to improve equilibrium efficiency can be found in [42].

The intuition behind why pricing allows to improve the efficiency of Nash equilibria is that it introduces an (implicit) message passing between players such that the original interior equilibrium is driven to the Pareto frontier. In a word, the discrepancy of equilibrium states between the social optimum and selfish behavior is compensated by the price scheme.

4.6 Dynamics of Gradient Algorithms

When people propose a gradient algorithm to solve an optimization problem, one of the basic questions to be answered is whether the algorithm will converge to the desired equilibria. We have already provided several convergence results when the gradient is immediately and accurately available to the decision-makers. However, in practice, information is often delayed and sometimes distorted. To study such information limitations, it is often useful to study the properties of the corresponding differential equation under delays and perturbations. For this purpose, we next introduce the Lyapunov stability theory, which is widely used in control theory.

Let $x = 0$ be an equilibrium point for

$$\frac{dx(t)}{dt} = \vartheta\left(x(t)\right) \tag{4.37}$$

and $B \subset \mathbb{R}^n$ be a region containing 0. Let $V : B \to \mathbb{R}$ be a continuously differentiable function such that $V(x) > 0$, $\forall x \neq 0$ and $V(0) = 0$. There are the following conditions for various notions of stability.

(1) If $\frac{dV(x(t))}{dt} \leq 0$, $\forall x \in B$, then the equilibrium is stable and $V(x)$ is called a Lyapunov function.
(2) In addition, if $\frac{dV(x(t))}{dt} < 0$, $\forall x \in B \setminus \{0\}$, then the equilibrium is asymptotically stable.
(3) In addition to (1) and (2) above, if V is radially unbounded *i.e.* $V(x) \to \infty$ as $x \to \infty$ then the equilibrium is globally asymptotically stable.

Note that the above theorem also holds if the equilibrium is a nonzero $\hat{x}$. In this case, consider a system with state $y = x - \hat{x}$ and the results hold immediately.

4.6.1 Connection between Lyapunov Functions and Objective Functions

In this section we first start to consider

$$\min_{x\in\mathbb{R}^n} f(x), \tag{4.38}$$

where the function $f : \mathbb{R}^n \to \mathbb{R}$ is differentiable and strictly convex. One of the simplest methods for solving (4.38) is the gradient descent algorithm (4.39), which attempts to maximize the decrease of the objective function in each iteration by updating the current iterate in the opposite direction of the gradient of f.

$$x^{(k+\Delta)} = x^{(k)} - \alpha\nabla f(x) \tag{4.39}$$

Let us use the gradient flow in (4.40) to approximate the sequence $\left\{x^{(k)}\right\}$ generated by (4.39).

$$\frac{dx}{dt} = \lim_{\Delta\to 0^+} \frac{x^{(k+\Delta)} - x^{(k)}}{\Delta} = -\alpha\nabla f(x). \tag{4.40}$$

When people study the dynamics of gradient method, Lyapunov functions are usually adopted to study the stability of stationary point (local or global optimum) to which the gradient algorithm may converge. However, there is no common method to construct a Lyapunov function, whose existence is only sufficient to guarantee the stability of a stationary point. A Lyapunov function can be regarded as the "energy" of a system. If one can prove that the energy along the considered dynamics is continuously decreasing, the system would finally settle down at the lowest-energy state. When we study the convergence of gradient method in an optimization setting, the convex cost function $f(x)$ will keep decreasing until it reaches a stationary point of the gradient dynamics, which coincides with the minimum of $f(x)$. Thus, one intuition of selecting Lyapunov function for $\dot{x} := \frac{dx}{dt} := \vartheta\,(x(t))$ is to associate it with the objective function. More precisely, choose

$$V(x(t)) := f(x^*) - f(x(t)) \tag{4.41}$$

as the Lyapunov candidate, where x^* is one optimum solution of (4.38). It is straightforward to see that $V(x) \geq 0$, $\forall x \in X$ and $V(x) = 0$ iff $x = x^*$. Furthermore $V(x)$ is nondecreasing along the trajectories of (4.40) by showing

$$\frac{dV(x(t))}{dt} = \sum_{i=1}^{n} \frac{\partial V(x)}{\partial x_i} \cdot \frac{dx_i}{dt} = -\alpha\left\|\nabla f(x)\right\|^2 \leq 0 \tag{4.42}$$

with $\frac{dV(x(t))}{dt} = 0$ iff $x = x^*$. It follows that (4.41) is a Lyapunov function. According to Lyapunov stability theory, the unique minimum x^* is globally asymptotically stable.

For a concrete example, Kelly *et al.* [31] study the stability of dual-based flow control algorithm in communication networks, where the Lyapunov function coincides the concave objective function of dual network problem. For the stability of primal algorithm in network resource allocation [2], [11], the Lyapunov function is the same as the associated strictly concave objective function.

Similar like the direct connection between Lyapunov function and objective function in convex optimization problems, in some non-cooperative game, say potential game, its Lyapunov function is just the potential of the game. Let's consider a strategic non-cooperative game $\Xi = \left\{\mathscr{V}, \{\mathscr{S}_v\}_{v\in\mathscr{V}}, \{w_v\}_{v\in\mathscr{V}}\right\}$, where $\mathscr{V}$ is the set of $|\mathscr{V}|$ players; $\mathscr{S}_v$ is the set of strategies of player v and w_v is its payoff function.

Gradient based algorithm are usually adopted for each player to follow to reach a Nash equilibrium,

$$\frac{dx_v}{dt} = \alpha \frac{\partial w_v\left(x\left(t\right)\right)}{\partial x_v}, \ \forall v \in \mathscr{V}, \tag{4.43}$$

where $\alpha > 0$ is the stepsize.

One interesting property of potential games is that the Lyapunov function for the gradient system (4.43) is just the (scaled version) of potential function $\Phi(x)$ given in (4.32). Let's consider the following candidate Lyapunov function [61]:

$$V(x) = \Phi_{\max} - \Phi(x),$$

where $\Phi(x^*) := \Phi_{\max}$ denotes the maximum value of the potential Φ over $\mathscr{S}$. It is easy to show the positiveness of $V(x)$, $\forall x \in \mathscr{S}$ except $x = x^*$. Following (4.32), $V(x(t))$ is non-decreasing along the trajectories of the system (4.43),

$$\frac{dV\left(x\left(t\right)\right)}{dt} = -\nabla_x^T \Phi\left(x\right) \left(\frac{d}{dt}x\left(t\right)\right) = -\alpha \left\|\nabla_x \Phi\left(x\right)\right\|^2 \leq 0. \tag{4.44}$$

By LaSalle's invariance principle[5] and (4.44) the trajectories of (4.43) converge to the largest invariant set

$$\left\{x \in \mathscr{S} : \frac{dV\left(x\left(t\right)\right)}{dt} = 0\right\}. \tag{4.45}$$

Since the set in (4.45) contains only the unique Nash equilibrium of the game Ξ if $\Phi(x)$ is strictly concave, the dynamics (4.43) converges to such an equilibrium asymptotically. In a potential game, although each player v selfishly maximizes its own payoff $w_v(x)$, it implicitly maximizes an imaginary objective, the potential function. Thus, $V(x) = \Phi_{\max} - \Phi(x)$ can be used to measure the "energy" accumulated along the trajectory (4.43).

By the discussions above, the convexity (concavity) plays an essential role in finding a candidate Lyapunov function. The connection between Lyapunov function and objective function may be invalidated in some cases. For example, the integration of gradient dynamics with respect to the variables does not result in a common objective function or the potential function in a non-cooperative game is hard to find. One may seek Lyapunov function by a more general method, the Krasovskii's method.

[5] To be introduced afterwards.

4.6.2 Krasovskii's Method

Let us use a saddle point problem to illustrate Krasovskii's method: find x^* and λ^* such that

$$L(x^*,\lambda) \leq L(x^*,\lambda^*) \leq L(x,\lambda^*),\ \forall x \in X,\ \forall \lambda \in M,$$

where $L: X \times M \to \mathbb{R}$ is a strictly convex-concave function. X and M are closed convex sets in $\mathbb{R}^n$ and $\mathbb{R}^m$. The saddle point function can for example be a Lagrangian

$$L(x,\lambda) = f_0(x) + \lambda^T f(x)$$

of a convex programming problem

$$x^* \in \arg\min \{f_0(x) \,|\, f_i(x) \leq 0, i = 1,\ldots,m,\ \ x \in X\}.$$

Assuming that the function $L(x,\lambda)$ is differentiable, the sufficient and necessary conditions to be a saddle point are

$$x^* = P_X(x^* - \alpha \nabla_x L(x^*,\lambda^*)), \tag{4.46a}$$

$$\lambda^* = P_M(\lambda^* + \alpha \nabla_\lambda L(x^*,\lambda^*)) \tag{4.46b}$$

where $P_X(\cdot)$ and $P_M(\cdot)$ are the projection on sets X and M, respectively. The difference between the left and right side of of (4.46), which is equal to zero at the point x^*, λ^* and non-zero at an arbitrary point x, λ, specifies a mapping of the set $\mathbb{R}^n \times \mathbb{R}^m$ into itself. The resultant space can be viewed as a vector field with the fixed point x^*, λ^*. If $X = \mathbb{R}^n$, $M = \mathbb{R}^m$, this problem is written as the system of

$$\frac{dx}{dt} = -\alpha \nabla_x L(x,\lambda) \tag{4.47a}$$

$$\frac{d\lambda}{dt} = \alpha \nabla_\lambda L(x,\lambda) \tag{4.47b}$$

where α is the step-size. It can be checked that (4.47) admits a unique solution.

Obviously, $L(x,\lambda)$ is not a Lyapunov function of (4.47) since (4.47) accounts for the decrease in one variable x and increase in another variable λ. One alternative Lyapunov function is constructed by Krasovskii's method,

$$V(z) = \dot{z}^T A \dot{z}, \tag{4.48}$$

where $z = [x,\lambda]^T$, $A = \frac{1}{2} diag\left(\alpha^{-1}\right)$. $V(z)$ is positive except the equilibrium $z^* = [x^*,\lambda^*]^T$. It can be computed that the time-derivative of (4.48) along the trajectories of (4.47) results in

$$\dot{V}(z) = -\dot{z}^T \begin{bmatrix} \frac{\partial^2 L}{\partial x^2} & 0 \\ 0 & -\frac{\partial^2 L}{\partial \lambda^2} \end{bmatrix} \dot{z} \leq 0. \tag{4.49}$$

$\dot{V}(z) = 0$ happens only at the equilibrium point $z^* = [x^*,\lambda^*]^T$. Hence, the trajectory of dynamics (4.47) will tend to the saddle point of (x^*,λ^*) asymptotically.

Krasovskii's method is quite general and it can also be used to construct a Lyapunov function for the gradient based algorithms (4.40) and (4.43). Recently it has been used for studying stability of gradient algorithms for network resource allocation based on convex optimization [19]. Since the time-derivative of Lyapunov function (4.48) contains the second order global information, it can be used to prove the global asymptotic stability of the unique Nash equilibrium in a non-cooperative game [58], [1], [70]. However, imposing conditions on the Jacobian of pseudo-gradient [58]-[70] may be sometimes conservative since it ignores local structure of each player's payoff. By exploiting the local structure of payoff in [17], the same stability of gradient algorithm can be proved by a quadratic Lyapunov function without imposing global conditions proposed by [1].

4.6.3 Non-strictly Convex Problem

So far, we only discussed the stability of gradient algorithm for strictly concave/convex problem. For non-strictly convex/concave objective function, there may exist multiple equilibria. Then the corresponding gradient algorithm is no more guaranteed to converge to one of the equilibria asymptotically, see [12], [38]. We will invoke LaSalle's invariance principle to determine the convergence of gradient algorithms to multiple equilibria.

Theorem 4.8 (LaSalle's invariance principle [32]). *Consider the differential equation in (4.37). Let $V : B \to \mathbb{R}$ be a radially unbounded, continuoustly differentiable, positive definite function such that $\dot{V}(x) \leq 0$ for all $x \in B$. Let $\mathscr{E}$ be the set of points in B such that $\dot{V}(x) = 0$. Let $\mathscr{M}$ be the largest invariant set[6] in $\mathscr{E}$. Then, every solution of (4.37) starting in B tends to $\mathscr{M}$ as $t \to \infty$.*

LaSalle's invariance principle implies that any equilibrium point of a gradient algorithm is in invariant set $\mathscr{M}$. Also the domain of attraction[7] of an equilibrium point is also an invariant set. However, we do not know whether all elements in invariant set are preferred in terms of stability and optimality. The key is to ensure that the invariant set includes the global optimal equilibria exclusively. Basically, the invariant set also gives us some insight to construct a Lyapunov function. Recall that the invariant set should include all global optimum for the gradient algorithm to converge to. One possible way of constructing Lyapunov function is to connect the time-derivative of Lyapunov function with KKT conditions, *i.e.* any equilibrium satisfying KKT conditions should be in the largest invariant set $\mathscr{M}$. After that, one may argue that the trajectories of gradient algorithm converge to the desired equilibrium in $\mathscr{M}$. They may also converge to non-equilibria in $\mathscr{M}$ if there are any.

[6] A set $\mathscr{I}$ is an invariant set for a dynamic system $\frac{dx(t)}{dt} = \vartheta(x(t))$ if every trajectory $x(t)$ which starts from a point in $\mathscr{I}$ remains in $\mathscr{I}$ for all time. For example, any equilibrium point is an invarient set.

[7] Let $\varphi(t;x)$ be the solution of (4.37) that starts at initial state x at time $t = 0$. Then, the domain of attraction is defined as the set of all points x such that $\varphi(t;x)$ is defined for $t \geq 0$ and $\lim_{t\to\infty} \varphi(t;x) = 0$..

To achieve optimality, one usually establish conditions to exclude those undesired points in the set $\mathscr{M}$. This is done by studying the dynamics or properties of a reduced system, which is obtained by substituting the elements in $\mathscr{M}$ into original gradient systems.

Recently, a primal-dual algorithm was proposed in [73] to solve a non-strictly concave problem, which comes from uplink resource allocation problem in OFDM networks. The authors first established some properties that the elements in the largest invariant set should satisfy. Based on those properties they reduced the original primal-dual gradient-based algorithm into a set of linear differential equations. By studying stability of the reduced linear system, they equivalently excluded the undesired elements in $\mathscr{M}$. It should be noted that LaSalle's invariance principle can also be used to argue the uniqueness of Lagrange multiplier to which a primal-dual algorithm converges, although the Lagrange multiplier satisfying KKT conditions can be multiple, see [16] for network resource allocation example.

4.7 Conclusions

We have attempted to provide a tutorial overview of distributed optimization and games for decision-making in networked systems. Starting with a review of first-order methods for convex optimization, we have discussed how distributed optimization mechanisms can be designed using mathematical decomposition, networked optimization, and game theory. While decomposition methods result in very efficient computations, they typically require central coordination of subsystems. Networked optimization techniques, on the other hand, remove the central coordination and subsystems communicate and cooperate only with their nearest neighbors to find the global optimum. Finally, techniques from non-cooperative games allow to eliminate the overhead traffic for coordination of subsystems altogether, but will in general not be able to find the global optimum (unless a pricing mechanism is introduced that encourages entities to strive towards the common good).

Naturally, a book chapter like this has to be selective in scope. Many interesting and useful ideas and results had to be left out. This includes, for example, optimization of systems with stochastic parameters or noise, and analysis of decomposition and networked optimization under information delay. For the material that we have presented, our focus has been on basic concepts and ideas rather than the very latest extensions and generalizations. Although many of the key results are now almost half a century old, they have proven to be very useful in a wide range of applications, from resource allocation in communication systems to wide-area control of infrastructures. It is our firm belief that these techniques will play an increasingly important role in engineering, as the systems that we build become more and more networked and interconnected, and as the requirements on their performance and resource-efficiency continue to increase.

References

1. Alpcan, T., Basar, T., Dey, S.: A power control game based on outage probabilities for multicell wireless data networks. IEEE transactions on wireless communications 5(4) (2006)
2. Alpcan, T., Fan, X., Basar, T., Arcak, M., Wen, J.T.: Power control for multicell CDMA wireless networks: A team optimization approach. Wireless Networks 14(5), 647–657 (2008)
3. Arrow, K.J., Hurwicz, L.: Decentralization and Computation in Resource Allocation. In: Essays in Economics and Econometrics. University of North Carolina Press, Rayleigh (1960)
4. Benders, J.F.: Partitioning procedures for solving mixed-variables programming. Numerische Matematik 4, 238–252 (1962)
5. Bertsekas, D.P.: Nonlinear Programming. Athena Scientific, Belmont (1999)
6. Bertsekas, D.P., Tsitsiklis, J.N.: Parallel and distributed computation: numerical methods. Prentice-Hall, Englewood Cliffs (1989)
7. Border, K.C.: Fixed point theorems with applications to economics and game theory. Cambridge Univ. Pr., Cambridge (1989)
8. Boyd, S., Diaconis, P., Xiao, L.: Fastest mixing markov chain on a graph. SIAM Review 46, 667–689 (2004)
9. Boyd, S.P.: Course material for ee364b, stanford university (2007), `http://www.stanford.edu/class/ee364b/`
10. Boyd, S.P., Vandenberghe, L.: Convex Optimization. Cambridge University Press, Cambridge (2004)
11. Chen, L., Low, S.H., Doyle, J.C.: Joint congestion control and media access control design for wireless ad hoc networks. In: Proceedings of IEEE Infocom, pp. 2212–2222 (2005)
12. Chen, M., Ponec, M., Sengupta, S., Li, J., Chou, P.A.: Utility maximization in peer-to-peer systems. In: Proc. ACM Sigmetrics (2008)
13. Chiang, M.: Balancing transport and physical layers in wireless multihop networks: Jointly optimal congestion control and power control. IEEE Journal on Selected Areas in Communications 23, 104–116 (2005)
14. Cohen, G.: Optimization by decomposition and coordination: a unified approach. IEEE Transactions on Automatic Control 23(2), 222–232 (1978)
15. Dantzig, G.B., Wolfe, P.: Decomposition principle for linear programs. Operations Research 8, 101–111 (1960)
16. Eryilmaz, A., Srikant, R.: Joint congestion control, routing and mac for stability and fairness in wireless networks. IEEE Journal on Selected Areas in Communications 24, 1514–1524 (2006)
17. Fan, X., Alpcan, T., Arcak, M., Wen, T.J., Basar, T.: A passivity approach to game-theoretic CDMA power control. Automatica 42(11), 1837–1847 (2006)
18. Fawal, H.E., Georges, D., Bornard, G.: Optimal control of complex irrigation systems via decomposition-coordination and the use of augmented lagrangian. In: IEEE International Conference on Systems, Man, and Cybernetics, San Diego, CA, pp. 3874–3879 (November 1998)
19. Feijer, D., Paganini, F.: Krasovskiis Method in the Stability of Network Control. In: Proceedings of the 2009 conference on American Control Conference, pp. 3292–3297. Institute of Electrical and Electronics Engineers Inc. (2009)
20. Fudenberg, D., Tirole, J.: Game theory. MIT Press, Cambridge (1991)

21. Geoffrion, A.M.: Elements of large-scale mathematical programming I–II. Management Science 16, 652–691 (1970)
22. Goodman, D., Mandayam, N.: Power control for wireless data. IEEE Personal Communications 7, 48–54 (2000)
23. Hastings, W.K.: Monte carlo sampling methods using markov chains and their applications. Biometrika 57, 97–109 (1970)
24. Hiriart-Urruty, J.-B., Lemaréchal, C.: Fundamentals of Convex Analysis. Springer, Heidelberg (2001)
25. Holmberg, K.: Primal and dual decomposition as organizational design: price and/or resource directive decomposition. In: Design models for hierarchical organizations: computation, information and decentralization, pp. 61–92. Kluwer Academic Publishers, Dordrecht (1995)
26. Huang, J., Berry, R.A., Honig, M.L.: Distributed interference compensation for wireless networks. IEEE Journal on Selected Areas in Communications 24, 1074–1084 (2006)
27. Johansson, B., Keviczky, T., Johansson, M., Johansson, K.H.: Subgradient methods and consensus algorithms for solving convex optimization problems. In: 47th IEEE Conference on Decision and Control, CDC 2008, pp. 4185–4190 (2008)
28. Johansson, B., Rabi, M., Johansson, M.: A randomized incremental subgradient method for distributed optimization in networked systems. SIAM Journal on Optimization 20(3), 1157–1170 (2009)
29. Johari, R., Mannor, S., Tsitsiklis, J.N.: Efficiency loss in a network resource allocation game: The case of elastic supply. Mathematics of Operations Research 29, 407–435 (2004)
30. Kelly, F.: Charging and rate control for elastic traffic. European transactions on Telecommunications 8(1), 33–37 (1997)
31. Kelly, F.P., Maulloo, A.K., Tan, D.K.H.: Rate control for communication networks: shadow prices, proportional fairness and stability. Journal of the Operational Research society 49(3), 237–252 (1998)
32. Khalil, H.K.: Nonlinear systems, 3rd edn (2002)
33. Koshal, J., Nedic, A., Shanbhag, U.V.: Distributed multiuser optimization: Algorithms and error analysis. In: Proceedings of the 48th IEEE Conference on Decision and Control, held jointly with the 2009 28th Chinese Control Conference. CDC/CCC 2009., pp. 4372–4377 (2009)
34. Koutsoupias, E., Papadimitriou, C.: Worst-case equilibria. Computer Science Review 3(2), 65–69 (2009)
35. Larsson, T., Patriksson, M., Strömberg, A.-B.: Ergodic primal convergence in dual subgradient schemes for convex programming. Mathematical Programming 86, 283–312 (1999)
36. Lasaulce, S., Debbah, M., Altman, E., de Cachan, E.N.S., Cachan, F.: Methodologies for analyzing equilibria in wireless games. IEEE Signal Processing Magazine 26(5), 41–52 (2009)
37. Lasdon, L.S.: Optimization Theory for Large Systems. Macmillan Co., New York (1970)
38. Lin, X., Shroff, N.B.: Utility maximization for communication networks with multipath routing. IEEE Transactions on Automatic Control 51(5) (2006)
39. Maheswaran, R., Basar, T.: Efficient signal proportional allocation (ESPA) mechanisms: Decentralized social welfare maximization for divisible resources. IEEE Journal on Selected Areas in Communications 24(5) (2006)
40. Marden, J.R., Arslan, G., Shamma, J.S.: Cooperative control and potential games. IEEE Transactions on Systems, Man, and Cybernetics, Part B: Cybernetics 39(6), 1393–1407 (2009)

41. Marden, J.R., Wierman, A.: Overcoming limitations of game-theoretic distributed control. In: 48th IEEE Conference on Decision and Control (2009)
42. Mazumdar, R.R., Courcoubetis, C.A., Duffield, N., Kesidis, G., Odlyzko, A., Srikant, R., Walrand, J., Cosman, P.: Guest Editorial Price-Based Access Control and Economics of Networking. IEEE Journal on Selected Areas in Communications 24(5) (2006)
43. Monderer, D., Shapley, L.S.: Potential games. Games and economic behavior 14, 124–143 (1996)
44. Nedic, A., Ozdaglar, A.: Distributed subgradient methods for multi-agent optimization. IEEE Transactions on Automatic Control 54(1), 48–61 (2009)
45. Nedic, A., Bertsekas, D.P.: Incremental subgradient methods for nondifferentiable optimization. SIAM J. on Optimization 12(1), 109–138 (2001)
46. Nedić, A., Ozdaglar, A.: Approximate primal solutions and rate analysis for dual subgradient methods. SIAM Journal on Optimization 19(4), 1757–1780 (2008)
47. Nesterov, Y.: A method for unconstrained convex minimization problem with the rate of convergence $o(1/k^2)$. Doklady AN SSSR (translated as Soviet Math. Docl.) 269, 543–547 (1983)
48. Nesterov, Y.: Smooth minimization of non-smooth functions. Mathematical Programming 103(1), 127–152 (1995)
49. Nesterov, Y.: Introductory Lectures on Convex Optimization: A Basic Course. Springer, Netherlands (2003)
50. Olfati-Saber, R., Fax, J.A., Murray, R.M.: Consensus and cooperation in networked multi-agent systems. Proceedings of the IEEE 95(1), 215–233 (2007)
51. Olfati-Saber, R., Fax, J.A., Murray, R.M.: Consensus and cooperation in networked multi-agent systems. Proceedings of the IEEE 95(1), 215–233 (2007)
52. Olshevsky, A., Tsitsiklis, J.N.: Convergence rates in distributed consensus and averaging. In: Proceedings of IEEE CDC (2006)
53. Polyak, B.: Introduction to Optimization. Optimization Software (1987)
54. Rabbat, M., Nowak, R.: Distributed optimization in sensor networks. In: Third International Symposium on Information Processing in Sensor Networks, IPSN 2004, pp. 20–27 (2004)
55. Ram, S.S., Nedic, A., Veeravalli, V.V.: Incremental stochastic subgradient algorithms for convex optimization. SIAM Journal on Optimization 20(2), 691–717 (2009)
56. Rantzer, A.: Using game theory for distributed control engineering. Department of Automatic Control, Lund University, Sweden, Tech. Rep. ISRN LUTFD2/TFRT–7620–SE (July 2008)
57. Rockafellar, R.T.: Convex Analysis. Princeton University Press, Princeton (1996)
58. Rosen, J.B.: Existence and uniqueness of equilibrium points for concave n-person games. Econometrica: Journal of the Econometric Society 33(3), 520–534 (1965)
59. Saraydar, C.U., Mandayam, N.B., Goodman, D.J.: Efficient power control via pricing in wireless data networks. IEEE transactions on Communications 50(2), 291–303 (2002)
60. Schizas, I.D., Ribeiro, A., Giannakis, G.B.: Consensus in ad hoc wsns with noisy links part i: Distributed estimation of deterministic signals. IEEE Transactions on Signal Processing 56(1), 350–364 (2008)
61. Scutari, G., Barbarossa, S., Palomar, D.P.: Potential games: A framework for vector power control problems with coupled constraints. In: Proceedings 2006 IEEE International Conference on Acoustics, Speech and Signal Processing, ICASSP 2006, vol. 4 (2006)
62. Shor, N.Z.: Minimization methods for non-differentiable functions. Springer, Heidelberg (1985)

63. Shum, K.W., Leung, K.K., Sung, C.W.: Convergence of iterative waterfilling algorithm for Gaussian interference channels. IEEE Journal on Selected Areas in Communications 25(6), 1091–1100 (2007)
64. Sung, C.W., Leung, K.K.: A generalized framework for distributed power control in wireless networks. IEEE Transactions on Information Theory 51(7), 2625 (2005)
65. Takahara, Y.: Multilevel approach to dynamic optimization. Technical Report SRC-50-C-64-18, Systems Research Center, Case Western Reserve University (1964)
66. Topkis, D.M.: Supermodularity and complementarity. Princeton Univ. Pr., Princeton (1998)
67. Walter, R.: Principles of mathematical analysis. McGraw-Hill, New York (1976)
68. Xiao, L., Boyd, S.: Fast linear iterations for distributed averaging. Systems & Control Letters 53(1), 65–78 (2004)
69. Yang, B., Feng, G., Guan, X.: Noncooperative random access game via pricing in ad hoc networks. In: 2007 46th IEEE Conference on Decision and Control, pp. 5704–5709 (2007)
70. Yang, B., Feng, G., Shen, Y., Long, C., Guan, X.: Channel-Aware Access for Cognitive Radio Networks. IEEE transactions on vehicular technology 58(7), 3726–3737 (2009)
71. Yang, B., Shen, Y., Johansson, M., Guan, X.: Threshold-based Multichannel Access with Energy Constraint. In: ICC 2010 IEEE International Conference on Communications (2010)
72. Yates, R.D.: A framework for uplink power control in cellular radio systems. IEEE Journal on Selected Areas in Communications 13(7), 1341–1347 (1995)
73. Zhang, X., Chen, L., Huang, J., Chen, M., Zhao, Y.P.: Distributed and optimal reduced primal-dual algorithm for uplink OFDM resource allocation. In: 2009 Joint 48th IEEE Conference on Decision and Control (CDC) and 28th Chinese Control Conference, CCC 2009 (2009)

Chapter 5
Decentralized Model Predictive Control

Alberto Bemporad and Davide Barcelli

Abstract. Decentralized and distributed model predictive control (DMPC) addresses the problem of controlling a multivariable dynamical process, composed by several interacting subsystems and subject to constraints, in a computation and communication efficient way. Compared to a centralized MPC setup, where a global optimal control problem must be solved on-line with respect to all actuator commands given the entire set of states, in DMPC the control problem is divided into a set of local MPCs of smaller size, that cooperate by communicating each other a certain information set, such as local state measurements, local decisions, optimal local predictions. Each controller is based on a partial (local) model of the overall dynamics, possibly neglecting existing dynamical interactions. The global performance objective is suitably mapped into a local objective for each of the local MPC problems.

This chapter surveys some of the main contributions to DMPC, with an emphasis on a method developed by the authors, by illustrating the ideas on motivating examples. Some novel ideas to address the problem of hierarchical MPC design are also included in the chapter.

5.1 Introduction

Most of the procedures for analyzing and controlling dynamical systems developed over the last decades rest on the common presupposition of *centrality*. Centrality means that all the information available about the system is collected at a single location, where all the calculations based on such information are executed. Information includes both *a priori* information about the dynamical model of the system available off-line, and *a posteriori* information about the system response gathered

Alberto Bemporad
Department of Mechanical and Structural Engineering, University of Trento
e-mail: bemporad@ing.unitn.it

Davide Barcelli
Department of Information Engineering, University of Siena, Italy
e-mail: barcelli@dii.unisi.it

A. Bemporad, M. Heemels, and M. Johansson: Networked Control Systems, LNCIS 406, pp. 149–178.
springerlink.com

by different sensors on-line. When considering large-scale systems the presupposition of centrality fails because of the lack of a centralized information-gathering system or of centralized computing capabilities. Typical examples of such systems are power networks, water networks, urban traffic networks, cooperating vehicles, digital cellular networks, flexible manufacturing networks, supply chains, complex structures in civil engineering,and many others. In such systems the centrality assumption often fails because of geographical separation of components (spatial distribution), as the costs and the reliability of communication links cannot be neglected. Moreover, technological advances and reduced cost of microprocessors provide a new force for distributed computation. Hence the current trend for *decentralized* decision making, distributed computations, and hierarchical control.

Several new challenges arise when addressing a decentralized setting, where most of the existing analysis and control design methodologies cannot be directly applied. In a distributed control system which employs decentralized control techniques there are several local control stations, where each controller observes only local outputs and only controls local inputs. Besides advantages in controller implementation (namely reduced and parallel computations, reduced communications), a great advantage of decentralization is maintenance: while certain parts of the overall process are interrupted, the remaining parts keep operating in closed-loop with their local controllers, without the need of stopping the overall process as in case of centralized control. Moreover, a partial re-design of the process does not necessarily imply a complete re-design of the controller, as it would instead in case of centralized control. However, all the controllers are involved in controlling the same large-scale process, and is therefore of paramount importance to determine conditions under which there exists a set of appropriate local feedback control laws stabilizing the entire system.

Ideas for decentralizing and hierarchically organizing the control actions in industrial automation systems date back to the 70's [37, 26, 27, 31, 11], but were mainly limited to the analysis of stability of decentralized linear control of interconnected subsystems, so the interest faded. Since the late 90's, because of the advances in computation techniques like convex optimization, the interest in decentralized control raised again [14, 29], and convex formulations were developed, although limited to special classes of systems such as spatially invariant systems [4]. Decentralized control and estimation schemes based on distributed convex optimization ideas have been proposed recently in [30, 20] based on Lagrangean relaxations. Here global solutions can be achieved after iterating a series of local computations and inter-agent communications.

Large-scale multi-variable control problems, such as those arising in the process industries, are often dealt with model predictive control (MPC) techniques. In MPC the control problem is formulated as an optimization one, where many different (and possibly conflicting) goals are easily formalized and state and control constraints can be included. Many results are nowadays available concerning stability and robustness of MPC, see *e.g.* [24]. However, centralized MPC is often unsuitable for control of large-scale networked systems, mainly due to lack of scalability and to maintenance issues of global models. In view of the above considerations, it is

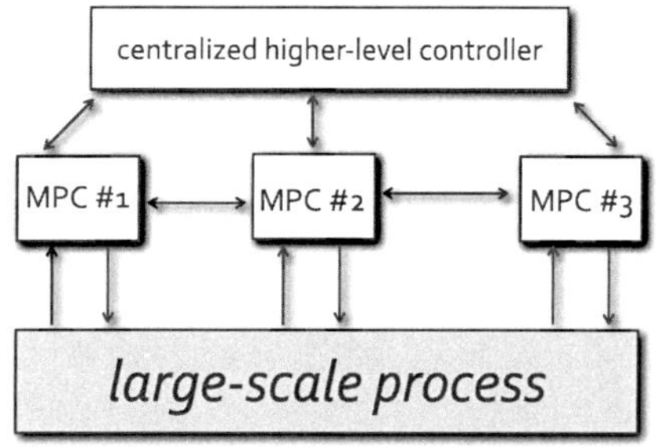

Fig. 5.1 Hierarchical and decentralized/distributed model predictive control of a large-scale process

then natural to look for decentralized or for distributed MPC (DMPC) algorithms, in which the original large-size optimization problem is replaced by a number of smaller and easily tractable ones that work iteratively and cooperatively towards achieving a common, system-wide control objective.

Even though there is not a universal agreement on the distinction between "decentralized" and "distributed", the main difference between the two terms depends on the type of information exchange:

- *decentralized MPC*: Control agents take control decisions independently on each other. Information exchange (such as measurements and previous control decisions) is only allowed before and after the decision making process. There is no negotiation between agents during the decision process. The time needed to decide the control action is not affected by communication issues, such as network delays and loss of packets.
- *distributed MPC*: An exchange of candidate control decisions may also happen during the decision making process, and iterated until an agreement is reached among the different local controllers, in accordance with a given stopping criterion.

In DMPC M subproblems are solved, each one assigned to a different control agent, instead of a single centralized problem. The goal of the decomposition is twofold: first, each subproblem is much smaller than the overall problem (that is, each subproblem has far fewer decision variables and constraints than the centralized one), and second, each subproblem is coupled to only a few other subproblems (that is, it shares variables with only a limited number other subproblems). Although decentralizing the MPC problem may lead to a deterioration of the overall closed-loop performance because of the suboptimality of the resulting control actions, besides computation and communication benefits there are also important operational benefits in using DMPC solutions. For instance local maintenance can be carried out by only stopping the corresponding local MPC controller, while in a centralized MPC approach the whole process should be suspended.

A DMPC control layer is often interacting with a higher-level control layer in a hierarchical arrangement, as depicted in Figure 5.1. The goal of the higher layer is to possibly adjust set-points and constraint specifications to the DMPC layer, based on a *global* (possibly less detailed) model of the entire system. Because of its general overview of the entire process, such a centralized decision layer allows one to reach levels of coordination and performance optimization otherwise very difficult (if not impossible) using a decentralized or distributed action. For a recent survey on decentralized, distributed and hierarchical model predictive control architectures, the reader is referred to the recent survey paper [32].

In a typical DMPC framework the steps performed by the local controllers at each control instant are the following: (*i*) measure local variables and update state estimates, (*ii*) solve the local receding-horizon control problem, (*iii*) apply the control signal for the current instant, (*iv*) exchange information with other controllers. Along with the benefits of a decentralized design, there are some inherent issues that one must face in DMPC: ensuring the asymptotic stability of the overall system, ensure the feasibility of global constraints, quantify the loss of performance with respect to centralized MPC.

5.2 Model Predictive Control

In this section we review the basic setup of linear model predictive control. Consider the problem of regulating the discrete-time linear time-invariant system

$$\begin{cases} x(t+1) = Ax(t) + Bu(t) \\ y(t) = Cx(t) \end{cases} \tag{5.1}$$

to the origin while fulfilling the constraints

$$u_{\min} \leq u(t) \leq u_{\max} \tag{5.2}$$

at all time instants $t \in \mathbb{Z}_{0+}$ where $\mathbb{Z}_{0+}$ is the set of nonnegative integers, $x(t) \in \mathbb{R}^n, u(t) \in \mathbb{R}^m$ and $y(t) \in \mathbb{R}^p$ are the state, input, and output vectors, respectively, and the pair (A,B) is stabilizable. In (5.2) the constraints should be interpreted component-wise and we assume $u_{min} < 0 < u_{max}$.

MPC solves such a constrained regulation problem as described below. At each time t, given the state vector $x(t)$, the following finite-horizon optimal control problem

$$V(x(t)) = \min_U \; x'_{t+N}Px_{t+N} + \sum_{k=0}^{N-1} x'_kQx_k + u'_kRu_k \tag{5.3a}$$

$$\text{s.t.} \; x_{k+1} = Ax_k + Bu_k, \; k = 0,\dots,N-1 \tag{5.3b}$$

$$y_k = Cx_k, \; k = 0,\dots,N \tag{5.3c}$$

$$x_0 = x(t) \tag{5.3d}$$

$$u_{\min} \leq u_k \leq u_{\max}, \; k = 0,\dots,N_u - 1 \tag{5.3e}$$

$$u_k = Kx_k, \; k = N_u,\dots,N-1 \tag{5.3f}$$

Table 5.1 Classification of existing DMPC approaches.

acronym	submodels	constraints	intersampling iterations	broadcast predictions	state constraints	stability constraints	references
ABB	coupled	local inputs	no	no	no	none	[2, 3, 5, 1]
VRW	coupled	local inputs	yes	no	no	none	[35, 34]
MD	coupled	local inputs	yes	yes	no	none	[25]
DM	decoupled	local inputs	no	yes	yes	compatibility	[15]
KBB	decoupled		no	yes	yes	none	[21]
JK	coupled	local inputs	no	yes	yes	compatibility	[17, 12, 18]

is solved, where $U \triangleq \{u_0, \ldots, u_{N_u-1}\}$ is the sequence of future input moves, x_k denotes the predicted state vector at time $t+k$, obtained by applying the input sequence $u_0, \ldots, u_{k-1}$ to model (5.1), starting from $x(t)$. In (5.3) $N > 0$ is the prediction horizon, $N_u \leq N-1$ is the input horizon, $Q = Q' \geq 0$, $R = R' > 0$, $P = P' \geq 0$ are square weight matrices defining the performance index, and K is some terminal feedback gain. As we will discuss below, P, K are chosen in order to ensure closed-loop stability of the overall process.

Problem (5.3) can be recast as a quadratic programming (QP) problem (see *e.g.* [24, 9]), whose solution $U^*(x(t)) \triangleq \{u_0^* \ \ldots \ u_{N_u-1}^*\}$ is a sequence of optimal control inputs. Only the first input

$$u(t) = u_0^* \tag{5.4}$$

is actually applied to system (5.1), as the optimization problem (5.3) is repeated at time $t+1$, based on the new state $x(t+1)$ (for this reason, the MPC strategy is often referred to as *receding horizon* control). The MPC algorithm (5.3)-(5.4) requires that all the n components of the state vector $x(t)$ are collected in a (possibly remote) central unit, where a quadratic program with mN_u decision variables needs to be solved and the solution broadcasted to the m actuators. As mentioned in the introduction, such a centralized MPC approach may be inappropriate for control of large-scale systems, and it is therefore natural to look for decentralized or distributed MPC (DMPC) algorithms.

5.3 Existing Approaches to DMPC

A few contributions have appeared in recent years in the context of DMPC, mainly motivated by applications of decentralized control of cooperating air vehicles [10, 28, 22]. We review in this section some of the main contributions on DMPC, summarized in Table 5.1, that have appeared in the scientific literature. An application of some of the results surveyed in this chapter in a problem of distributed control of power networks with comparisons among DMPC approaches is reported in [13].

In the following sections, we denote by M be the number of local MPC controllers that we want to design, for example $M = m$ in case each individual actuator is governed by its own local MPC controller.

5.3.1 DMPC Approach of Alessio, Barcelli, and Bemporad

In [2, 3, 5, 1] a decentralized MPC design approach for possibly dynamically coupled processes was proposed. A (partial) decoupling assumption only appears in the *prediction* models used by different MPC controllers. The chosen degree of decoupling represents a tuning knob of the approach. Sufficient criteria for analyzing the asymptotic stability of the process model in closed loop with the set of decentralized MPC controllers are provided. If such conditions are not verified, then the structure of decentralization should be modified by augmenting the level of dynamical coupling of the prediction submodels, increasing consequently the number and type of exchanged information about state measurements among the MPC controllers. Following such stability criteria, a hierarchical scheme was proposed to change the decentralization structure on-line by a supervisory scheme without destabilizing the system. Moreover, to cope with the case of a non-ideal communication channel among neighboring MPC controllers, sufficient conditions for ensuring closed-loop stability of the overall closed-loop system when packets containing state measurements may be lost were given. We review here the main ingredients and results of this approach.

5.3.1.1 Decentralized Prediction Models

Consider again process model (5.1). Matrices A, B may have a certain number of zero or negligible components corresponding to a partial dynamical decoupling of the process, especially in the case of large-scale systems, or even be block diagonal in case of total dynamical decoupling. This is the case for instance of independent moving agents each one having its own dynamics and only coupled by a global performance index.

For all $i = 1, \dots, M$, we define $x^i \in \mathbb{R}^{n_i}$ as the vector collecting a subset $\mathscr{I}_{xi} \subseteq \{1, \dots, n\}$ of the state components,

$$x^i = W_i' x = \begin{bmatrix} x_1^i \\ \vdots \\ x_{n_i}^i \end{bmatrix} \in \mathbb{R}^{n_i} \tag{5.5a}$$

where $W_i \in \mathbb{R}^{n \times n_i}$ collects the n_i columns of the identity matrix of order n corresponding to the indices in $\mathscr{I}_{xi}$, and, similarly,

$$u^i = Z_i' u = \begin{bmatrix} u_1^i \\ \vdots \\ u_{m_i}^i \end{bmatrix} \in \mathbb{R}^{m_i} \tag{5.5b}$$

as the vector of input signals tackled by the i-th controller, where $Z_i \in \mathbb{R}^{m \times m_i}$ collects m_i columns of the identity matrix of order m corresponding to the set of indices $\mathscr{I}_{ui} \subseteq \{1, \ldots, m\}$. Note that

$$W_i' W_i = I_{n_i},\ Z_i' Z_i = I_{m_i},\ \forall i = 1, \ldots, M \tag{5.6}$$

where $I_{(\cdot)}$ denotes the identity matrix of order $(\cdot)$. By definition of x^i in (5.5a) we obtain

$$x^i(t+1) = W_i' x(t+1) = W_i' A x(t) + W_i' B u(t) \tag{5.7}$$

An *approximation* of (5.1) is obtained by changing $W_i' A$ in (5.7) into $W_i' A W_i W_i'$ and $W_i' B$ into $W_i' B Z_i Z_i'$, therefore getting the new prediction reduced order model

$$x^i(t+1) = A_i x^i(t) + B_i u^i(t) \tag{5.8}$$

where matrices $A_i = W_i' A W_i \in \mathbb{R}^{n_i \times n_i}$ and $B_i = W_i' B Z_i \in \mathbb{R}^{m_i \times m_i}$ are submatrices of the original A and B matrices, respectively, describing in a possibly approximate way the evolution of the states of subsystem #i.

The size (n_i, m_i) of model (5.8) in general will be much smaller than the size (n, m) of the overall process model (5.1). The choice of the pair (W_i, Z_i) of *decoupling matrices* (and, consequently, of the dimensions n_i, m_i of each submodel) is a tuning knob of the DMPC procedure proposed in the sequel of the paper.

We want to design a controller for each set of moves u^i according to prediction model (5.8) and based on feedback from x^i, for all $i = 1, \ldots, M$. Note that in general different input vectors u^i, u^j may share common components. To avoid ambiguities on the control action to be commanded to the process, we impose that only a subset $\mathscr{I}_{ui}^{\#} \subseteq \mathscr{I}_{ui}$ of input signals computed by controller #i is actually applied to the process, with the following conditions

$$\bigcup_{i=1}^{M} I_{ui}^{\#} = \{1, \ldots, m\} \tag{5.9a}$$

$$I_{ui}^{\#} \cap I_{uj}^{\#} = \emptyset,\ \forall i, j = 1, \ldots, M,\ i \neq j \tag{5.9b}$$

Condition (5.9a) ensures that all actuators are commanded, condition (5.9b) that each actuator is commanded by only one controller. For the sake of simplicity of notation, since now on we assume that $M = m$ and that $I_{ui}^{\#} = i$, $i = 1, \ldots, m$, *i.e.*, that each controller #i only controls the ith input signal. As observed earlier, in general $\mathscr{I}_{x^i} \cap \mathscr{I}_{x^j} \neq \emptyset$, meaning that controller #i may partially share the same feedback information with controller #j, and $\mathscr{I}_{ui} \cap \mathscr{I}_{uj} \neq \emptyset$. This means that controller #i may take into account the effect of control actions that are actually decided by another controller #j, $i \neq j$, $i, j = 1, \ldots, M$, which ensures a certain degree of cooperation.

The designer has the flexibility of choosing the pairs (W_i, Z_i) of decoupling matrices, $i = 1, \ldots, M$. A first guess of the decoupling matrices can be inspired by the intensity of the dynamical interactions existing in the model. The larger (n_i, m_i) the smaller the model mismatch and hence the better the performance of the overall-closed loop system. On the other hand, the larger (n_i, m_i) the larger is the communication and computation efforts of the controllers, and hence the larger the sampling time of the controllers. An example of model decomposition is given later in Section 5.4.1.

5.3.1.2 Decentralized Optimal Control Problems

In order to exploit submodels (5.8) for formulating local finite-horizon optimal control problems that lead to an overall closed-loop stable DMPC system, let the following assumptions be satisfied (these will be relaxed in Theorem 5.5):

Assumption 5.1. *Matrix A in (5.1) is strictly Hurwitz*[1].

Assumption 5.1 restricts the strategy and stability results of DMPC to processes that are open-loop asymptotically stable, leaving to the controller the mere role of optimizing the performance of the closed-loop system.

Assumption 5.2. *Matrix A_i is strictly Hurwitz, $\forall i = 1,\ldots,M$.*

Assumption 5.2 restricts the degrees of freedom in choosing the decentralized models. Note that if $A_i = A$ for all $i = 1,\ldots,M$ is the only choice satisfying Assumption 5.2, then no decentralized MPC can be formulated within this framework. For all $i = 1,\ldots,M$ consider the following infinite-horizon constrained optimal control problems

$$V_i(x(t)) = \min_{\{u_k^i\}_{k=0}^{\infty}} \sum_{k=0}^{\infty} (x_k^i)'W_i'QW_i x_k^i + (u_k^i)'Z_i'RZ_i u_k^i = \tag{5.10a}$$

$$= \min_{u_0^i} (x_1^i)'P_i x_1^i + x^i(t)'W_i'QW_i x^i(t) + (u_0^i)'Z_i'RZ_i u_0^i \tag{5.10b}$$

$$\text{s.t. } x_1^i = A_i x^i(t) + B_i u_0^i \tag{5.10c}$$

$$x_0^i = W_i'x(t) = x^i(t) \tag{5.10d}$$

$$u_{\min}^i \leq u_0^i \leq u_{\max}^i \tag{5.10e}$$

$$u_k^i = 0,\ \forall k \geq 1 \tag{5.10f}$$

where $P_i = P_i' \geq 0$ is the solution of the Lyapunov equation

$$A_i'P_iA_i - P_i = -W_i'QW_i \tag{5.11}$$

that exists by virtue of Assumption 5.2. Problem (5.10) corresponds to a finite-horizon constrained problem with control horizon $N_u = 1$ and linear stable prediction model. Note that only the local state vector $x^i(t)$ is needed to solve Problem (5.10).

[1] While usually a matrix A is called *Hurwitz* if all its eigenvalues have strictly negative real part (continuous-time case), in this paper we say that a matrix A is *Hurwitz* if all the eigenvalues λ_i of A are such that $|\lambda_i| < 1$ (discrete-time case).

At time t, each controller MPC $\#i$ measures (or estimates) the state $x^i(t)$ (usually corresponding to local and neighboring states), solves problem (5.10) and obtains the optimizer

$$u_0^{*i} = [u_0^{*i1}, \ldots, u_0^{*ii}, \ldots, u_0^{*im_i}]' \in \mathbb{R}^{m_i} \tag{5.12}$$

In the simplified case $M = m$ and $I_{ui}^{\#} = i$, only the i-th sample of u_0^{*i}

$$u_i(t) = u_0^{*ii} \tag{5.13}$$

will determine the i-th component $u_i(t)$ of the input vector actually commanded to the process at time t. The inputs u_0^{*ij}, $j \neq i$, $j \in \mathscr{I}_{ui}$ to the neighbors are discarded, their only role is to provide a better prediction of the state trajectories x_k^i, and therefore a possibly better performance of the overall closed-loop system.

The collection of the optimal inputs of all the M MPC controllers,

$$u(t) = [u_0^{*11} \ \ldots \ u_0^{*ii} \ \ldots \ u_0^{*mm}]' \tag{5.14}$$

is the actual input commanded to process (5.1). The optimizations (5.10) are repeated at time $t+1$ based on the new states $x^i(t+1) = W_i'x(t+1)$, $\forall i = 1, \ldots, M$, according to the usual receding horizon control paradigm. The degree of coupling between the DMPC controllers is dictated by the choice of the decoupling matrices (W_i, Z_i). Clearly, the larger the number of interactions captured by the submodels, the more complex the formulation (and, in general, the solution) of the optimization problems (5.10) and hence the computations performed by each control agent. Note also that a higher level of information exchange between control agents requires a higher communication overhead. We are assuming here that the submodels are constant at all time steps.

5.3.1.3 Convergence Properties

As mentioned in the introduction, one of the major issues in decentralized RHC is to ensure the stability of the overall closed-loop system. The non-triviality of this issue is due to the fact that the prediction of the state trajectory made by MPC $\#i$ about state $x^i(t)$ is in general not correct, because of partial state and input information and of the mismatch between u^{*ij} (desired by controller MPC $\#i$) and u^{*jj} (computed and applied to the process by controller MPC $\#j$).

The following theorem, proved in [1, 2], summarizes the closed loop convergence results of the proposed DMPC scheme using the cost function $V(x(t)) \triangleq \sum_{i=1}^{M} V_i(W_i'x(t))$ as a Lyapunov function for the overall system.

Theorem 5.3. *Let Assumptions 5.1, 5.2 hold and define P_i as in* (5.11) $\forall i = 1, \ldots, M$. *Define*

$$\begin{aligned} \Delta u^i(t) &\triangleq u(t) - Z_i u_0^{*i}(t), & \Delta x^i(t) &\triangleq (I - W_iW_i')x(t) \\ \Delta A^i &\triangleq (I - W_iW_i')A, & \Delta B^i &\triangleq B - W_iW_i'BZ_iZ_i' \end{aligned} \tag{5.15}$$

Also, let

$$\Delta Y^i(x(t)) \triangleq W_i W_i'(A\Delta x^i(t) + BZ_i Z_i' \Delta u^i(t)) + \Delta A^i x(t) + \Delta B^i u(t) \tag{5.16a}$$

and

$$\Delta S^i(x(t)) \triangleq \left(2(A_i W_i' x(t) + B_i u_0^{*i}(t))' + \Delta Y^i(x(t))' W_i\right) P_i W_i' \Delta Y^i(x(t)) \tag{5.16b}$$

If the condition

$$(i)\ x' \left(\sum_{i=1}^{M} W_i W_i' Q W_i W_i' \right) x - \sum_{i=1}^{M} \Delta S^i(x) \geq 0,\ \forall x \in \mathbb{R}^n \tag{5.17a}$$

is satisfied, or the condition

$$(ii)\ x' \left(\sum_{i=1}^{M} W_i W_i' Q W_i W_i' \right) x - \alpha x' x - \sum_{i=1}^{M} \Delta S^i(x) + \sum_{i=1}^{M} u_0^{*i}(x)' Z_i' R Z_i u_0^{*i}(x) \geq 0,$$
$$\forall x \in \mathbb{R}^n \tag{5.17b}$$

is satisfied for some scalar $\alpha > 0$, then the decentralized MPC scheme defined in (5.10)–(5.14) in closed loop with (5.1) is globally asymptotically stable.

By using the explicit MPC results of [9], each optimizer function $u_0^{*i} : \mathbb{R}^n \mapsto \mathbb{R}^{m_i}$, $i = 1, \ldots, M$, can be expressed as a piecewise affine function of x

$$u_0^{*i}(x) = F_{ij} x + G_{ij} \quad \text{if} \quad H_{ij} x \leq K_{ij}, \quad j = 1, \ldots, n_{ri} \tag{5.18}$$

Hence, both condition (5.17a) and condition (5.17b) are a composition of quadratic and piecewise affine functions, so that *global stability* can be tested through linear matrix inequality relaxations [19] (cf. the approach of [16]).

As $u_{\min} < 0 < u_{\max}$, there exists a ball around the origin $x = 0$ contained in one of the regions, say $\{x \in \mathbb{R}^n : H_{i1} x \leq K_{i1}\}$, such that $G_{i1} = 0$. Therefore, around the origin both (5.17a) and (5.17b) become a quadratic form $x'(\sum_{i=1}^{M} E_i)x$ of x, where matrices E_i can be easily derived from (5.15), (5.16) and (5.17). Hence, *local stability* of (5.10)–(5.14) in closed loop with (5.1) can be simply tested by checking the positive semidefiniteness of the square $n \times n$ matrix $\sum_{i=1}^{M} E_i$. Note that, depending on the degree of decentralization, in order to satisfy the sufficient stability criterion one may need to set $Q > 0$ in order to dominate the unmodeled dynamics arising from the terms ΔS^i.

In the absence of input constraints, Assumptions 5.1, 5.2 can be removed to extend the previous DMPC scheme to the case where (A, B) and (A_i, B_i) may not be Hurwitz, although stabilizable.

Theorem 5.4 ([1, 3]). *Let the pairs (A_i, B_i) be stabilizable, $\forall i = 1, \ldots, M$. Let Problem (5.10) be replaced by*

$$V_i(x(t)) = \min_{\{u_k^i\}_{k=0}^{\infty}} \sum_{k=0}^{\infty} (x_k^i)' W_i' Q W_i x_k^i + (u_k^i)' Z_i' R Z_i u_k^i = \tag{5.19a}$$

$$= \min_{u_0^i} (x_1^i)' P_i x_1^i + x^i(t)' W_i' Q W_i x^i(t) + (u_0^i)' Z_i' R Z_i u_0^i \tag{5.19b}$$

$$s.t.\ x_1^i = A_i x^i(t) + B_i u_0^i \tag{5.19c}$$

$$x_0^i = W_i' x(t) = x^i(t) \tag{5.19d}$$

$$u_k^i = K_{LQ_i} x_k^i,\ \forall k \geq 1 \tag{5.19e}$$

where $P_i = P_i' \geq 0$ *is the solution of the Riccati equation*

$$W_i' Q W_i + K_{LQ_i}' Z_i' R Z_i K_{LQ_i} + (A_i + B_i K_{LQ_i})' P_i (A_i + B_i K_{LQ_i}) = P_i \tag{5.20}$$

and $K_{LQ_i} = -(Z_i' R Z_i + B_i' P_i B_i)^{-1} B_i' P_i A_i$ *is the corresponding local LQR feedback. Let* $\Delta Y^i(x(t))$ *and let* $\Delta S^i(x(t))$ *be defined as in* (5.16), *in which* P_i *is defined as in* (5.20).

If condition (5.17a) *is satisfied, or condition* (5.17b) *is satisfied for some scalar* $\alpha > 0$, *then the decentralized MPC scheme defined in (5.19), (5.14) in closed-loop with (5.1) is globally asymptotically stable.*

So far we assumed that the communication model among neighboring MPC controllers is faultless, so that each MPC agent successfully receives the information about the states of its corresponding submodel. However, one of the main issues in networked control systems is the unreliability of communication channels, which may result in data packet dropout.

A sufficient condition for ensuring convergence of the DMPC closed-loop in the case packets containing measurements are lost for an arbitrary but upper-bounded number N of consecutive time steps was proved in [1, 5]. The underlying operating assumption is that if the actual number of lost packets exceeds the given N, the decentralized controllers are turned off and $u = 0$ is applied persistently, so that a number of packet drops larger than N is not considered. The results shown here are based on formulation (5.10) and rely on the open-loop asymptotic stability Assumptions 5.1 and 5.2. The issue is still non-trivial, as if a set of measures for subsystem i is lost, this would affect not only the trajectory of subsystem i because of the improper control action u^i, but, due to the dynamical coupling, also the trajectories of subsystems $j \in J$, where $J = \{j \mid i \in \mathscr{I}_{xj} \cup \mathscr{I}_{uj}\}$, and thus the closed-loop stability of the overall system may be endangered.

By relying on open-loop stability, setting $u(t) = 0$ is a natural choice for backup input moves when no state measurements are available because of a communication blackout. Different backup options may be considered, such as solving (5.10) by replacing $x^i(t)$ with an estimate obtained through model (5.8) and the available measurements. Of course whether one or the other approach is better strongly depends on the amount of model mismatch introduced by the decentralized modeling.

The next theorem, proved in [1], provides conditions for asymptotic closed-loop stability of decentralized MPC under packet loss, generalizing and unifying the results of [2, 3].

Theorem 5.5. *Let N be a positive integer such that no more than N consecutive steps of channel transmission blackout can occur. Assume $u(t) = 0$ is applied when no packet is received. Let Assumptions 5.1, 5.2 hold and $\forall i = 1,\dots,M$ define P_i as in* (5.11), *$\Delta u^i(t)$, $\Delta x^i(t)$, ΔA^i, ΔB^i as in* (5.15), *$\Delta Y^i(x(t))$ as in* (5.16a),

$$\Delta S^i_j(x) \triangleq [2(A_i W_i' x + B_i u_0^{*i}(x))' W_i' + \Delta Y^i(x)'](A^{j-1})' W_i P_i W_i' A^{j-1} \Delta Y^i(x) \quad (5.21)$$

and let

$$\xi_i(x) \triangleq A_i W_i' x + B_i u_0^{*i}(x)$$

If the condition

$$(i) \sum_{i=1}^{M} \left(x' W_i W_i' Q W_i W_i' x + \xi_i(x)'(P_i - W_i'(A^{j-1})' W_i P_i W_i' A^{j-1} W_i)\xi_i(x) - \Delta S^i_j(x) \right)$$
$$\geq 0, \quad \forall x \in \mathbb{R}^n,\ \forall j = 1,\dots,N \quad (5.22a)$$

is satisfied, or the condition

$$(ii) \sum_{i=1}^{M} \left(x' W_i W_i' Q W_i W_i' x + \xi_i(x)'(P_i - W_i'(A^{j-1})' W_i P_i W_i' A^{j-1} W_i)\xi_i(x) + \right.$$
$$\left. -\Delta S^i_j(x) + u_0^{*i}(x)' Z_i' R Z_i u_0^{*i}(x) \right) \alpha x' x \geq 0, \qquad \forall x \in \mathbb{R}^n,\ \forall j = 1,\dots,N \quad (5.22b)$$

is satisfied for some scalar $\alpha > 0$, then the decentralized MPC scheme defined in (5.10)–(5.14) in closed loop with (5.1) is globally asymptotically stable under packet loss.

Note again that around the origin the conditions in (5.22) become a quadratic form to be checked positive semidefinite, so local stability of (5.10)–(5.14) in closed loop with (5.1) under packet loss can be tested for any arbitrary fixed N. Note also that conditions (5.22) are a generalization of (5.17), as for $j = 1$ (no packet drop) matrix $P_i - W_i'(A^{j-1})' W_i P_i W_i' A^{j-1} W_i = P_i - P_i = 0$.

5.3.1.4 Extension to Set-Point Tracking

Consider the following discrete-time linear global process model

$$\begin{cases} z(t+1) = Az(t) + Bv(t) + F_d(t) \\ h(t) = Cz(t) \end{cases} \quad (5.23)$$

where $z \in \mathbb{R}^n$ is the state vector, $v \in \mathbb{R}^m$ is the input vector, $y \in \mathbb{R}^p$ is the output vector, $F_d \in \mathbb{R}^d$ is a vector of measured disturbances. Let A satisfy Assumption 5.1 and assume F_d is constant. The considered set-point tracking problem is that of h tracking a given reference value $r \in \mathbb{R}^p$, despite the presence of F_d. In order to recast the problem as a regulation problem, assume steady-state vectors $z_r \in \mathbb{R}^n$ and $v_r \in \mathbb{R}^m$ exist solving the static problem

$$\begin{cases} (I-A)z_r = Bv_r + F_d \\ r = Cz_r \end{cases} \tag{5.24}$$

and let $x = z - z_r$ and $u = v - v_r$. Input constraints $v_{\min} \leq v \leq v_{\max}$ are mapped into constraints $v_{\min} - v_r \leq u \leq v_{\max} - v_r$[2].

Proposition 5.1. *Under the global coordinate transformation* (5.24), *the process* (5.23) *under the decentralized MPC law* (5.10)–(5.14) *is such that* $h(t)$ *converges asymptotically to the set-point r, either under the assumption of Theorem 5.3 or, in the presence of packet drops, of Theorem 5.5.*

Note that problem (5.24) is solved in a *centralized* way. Defining local coordinate transformations v_{ir}, z_{ir} based on submodels (5.8) would not lead, in general, to offset-free tracking, due to the mismatch between global and local models. This is a general observation one needs to take into account in decentralized tracking. Note also that both v_r and z_r depend on F_d as well as r, so problem (5.24) should be solved each time the value of F_d or r change and retransmitted to each controller.

5.3.2 DMPC Approach of Jia and Krogh

In [17, 12] the system under control is composed by a number of unconstrained linear discrete-time subsystems with decoupled input signals, described by the equations

$$\begin{bmatrix} x_1(k+1) \\ \vdots \\ x_M(k+1) \end{bmatrix} = \begin{bmatrix} A_{11} & \dots & A_{1M} \\ \vdots & \ddots & \vdots \\ A_{M1} & \dots & A_{MM} \end{bmatrix} \begin{bmatrix} x_1(k) \\ \vdots \\ x_M(k) \end{bmatrix} + \begin{bmatrix} B_1 & & 0 \\ & \ddots & \\ 0 & & B_M \end{bmatrix} \begin{bmatrix} u_1(k) \\ \vdots \\ u_M(k) \end{bmatrix} \tag{5.25}$$

The effect of dynamical coupling between neighboring states is modeled in prediction through a disturbance signal v, for instance the prediction model used by controller $\#j$ is

$$x_j(k+i+i|k) = A_{jj}x_j(k+i|k) + B_j + u_j(k+i|k) + K_jv_j(k+i|k) \tag{5.26}$$

where $K_j = [A_{j1} \; \dots \; A_{j,j-1} \; A_{j,j+1} \; \dots \; A_{jM}]$. The information exchanged between control agents at the end of each sample step is the entire prediction of the local state vector. In particular, controller $\#j$ receives the signal

[2] In case $v_r \notin [v_{\min}, v_{\max}]$, perfect tracking under constraints is not possible, and an alternative is to set

$$\begin{aligned} \left[\begin{smallmatrix} z_r \\ v_r \end{smallmatrix}\right] = & \arg\min \left\| \left[\begin{smallmatrix} I-A & -B \\ C & 0 \end{smallmatrix}\right] \left[\begin{smallmatrix} z_r \\ v_r \end{smallmatrix}\right] - \left[\begin{smallmatrix} F_d \\ r \end{smallmatrix}\right] \right\| \\ & \text{s.t.} \quad v_{\min} \leq v_r \leq v_{\max} \end{aligned}$$

$$v_j(k+i|k) = \begin{bmatrix} x_1(k+i|k-1) \\ \vdots \\ x_{j-1}(k+i|k-1) \\ x_{j+1}(k+i|k-1) \\ \vdots \\ x_M(k+i|k-1) \end{bmatrix}$$

where i is the prediction time index, from the other MPC controllers at the end of the previous time step $k-1$. The signal $v_j(k+i|k)$ is used by controller #j at time k to estimate the effect of the neighboring subsystem dynamics in (5.26).

Under certain assumptions of the model matrix A, closed-loop stability is proved by introducing a contractive constraint on the norm of $x_j(k+1|k)$ in each local MPC problem, which the authors prove to be a recursively feasible constraint.

The authors deal with state constraints in [18] by proposing a min-max approach, at the price of a possible conservativeness of the approach.

5.3.3 DMPC Approach of Venkat, Rawlings, and Wright

In [35, 36, 34] the authors propose distributed MPC algorithm based on a process of negotiations among DMPC agents. The adopted prediction model is

$$\begin{cases} x_{ii}(k+1) = A_{ii}x_{ii}(k) + B_{ii}u_i(k) & \text{(local prediction model)} \\ x_{ij}(k+1) = A_{ij}x_{ij}(k) + B_{ij}u_j(k) & \text{(interaction model)} \\ y_i(k) = \sum_{j=1}^{M} C_{ij}x_{ij}(k) \end{cases}$$

The effect of the inputs of subsystem #j on subsystem #i is modeled by using an "interaction model". All interaction models are assumed stable, and constraints on inputs are assumed decoupled (*e.g.*, input saturation).

Starting from a multiobjective formulation, the authors distinguish between a "communication-based" control scheme, in which each controller #i is optimizing his own local performance index Φ_i, and a "cooperation-based" control scheme, in which each controller #i is optimizing a weighted sum $\sum_{j=1}^{M} \alpha_j \Phi_j$ of all performance indices, $0 \le \alpha_j \le 1$. As performance indices depend on the decisions taken by the other controllers, at each time step k a sequence of iterations is taken before computing and implementing the input vector $u(k)$. In particular, within each sampling time k, at every iteration p the previous decisions $u_{j\neq i}^{p-1}$ are broadcast to controller #i, in order to compute the new iterate u_i^p. With the communication-based approach, the authors show that if the sequence of iterations converges, it converges to a Nash equilibrium. With the cooperation-based approach, convergence to the optimal (centralized) control performance is established. In practical situations the process sampling interval may be insufficient for the computation time required for convergence of the iterative algorithm, with a consequent loss of performance. Nonetheless, closed-loop stability is not compromised: as it is achieved even though the convergence of the iterations is not reached. Moreover, all iterations are plantwide feasible, which naturally increases the applicability of the approach including a certain robustness to transmission faults.

5.3.4 DMPC Approach of Dunbar and Murray

In [15] the authors consider the control of a special class of dynamically decoupled continuous-time nonlinear subsystems

$$\dot{x}_i(t) = f_i(x_i(t), u_i(t))$$

where the local states of each model represent a position and a velocity signal

$$x_i(t) = \begin{bmatrix} q_i(t) \\ \dot{q}_i(t) \end{bmatrix}$$

State vectors are only coupled by a global performance objective

$$L(x,u) = \sum_{(i,j)\in\mathscr{E}_0} \omega \|q_i - q_j + d_{ij}\|^2 + \omega \|q_\Sigma - q_d\|^2 + \nu \|\dot{q}\|^2 + \mu \|u\|^2 \tag{5.27}$$

under local input constraints $u_i(t) \in U$, $\forall i = 1,\dots,M$, $\forall t \geq 0$. In (5.27) $\mathscr{E}_0$ is the set of pair-wise neighbors, d_{ij} is the desired distance between subsystems i and j, $q_\Sigma = (q_1 + q_2 + q_3)/3$ is the average position of the leading subsystems 1,2,3, and $q_d = (q_1^c + q_2^c + q_3^c)/3$ the corresponding target.

The overall integrated cost (5.27) is decomposed in distributed integrated cost functions

$$L_i(x_i, x_{-i}, u_i) = L_i^x(x_i, x_{-i}) + \gamma\mu \|u_i\|^2 + L^d(i)$$

where $x_{-i} = (x_{j1}, \dots, x_{jk})$ collects the states of the neighbors of agent subsystem #i, $L_i^x(x_i, x_{-i}) = \sum_{j\in N_i} \frac{\gamma\omega}{2} \|q_i - q_j + d_{ij}\|^2 + \gamma\nu \|\dot{q}_i\|^2$, and

$$L^d(i) = \begin{cases} \gamma\omega \|q_\Sigma - q_d\|^{2/3} & i \in \{1,2,3\} \\ 0 & \text{otherwise} \end{cases}$$

It holds that

$$L(x,u) = \frac{1}{\gamma} \sum_{i=1}^{N} L_i(x_i, x_{-i}, u_i)$$

Before computing DMPC actions, neighboring subsystems broadcast in a synchronous way their states, and each agent transmits and receives an "assumed" control trajectory $\hat{u}_i(\tau;t_k)$. Denoting by $u_i^p(\tau;t_k)$ the control trajectory predicted by controller #i, by $u_i^*(\tau;t_k)$ the optimal predicted control trajectory, by T the prediction horizon, and by $\delta \in (0,T]$ the update interval, the following DMPC performance index is minimized

$$
\begin{aligned}
&\min_{u_i^p} \quad J_i(x_i(t_k), x_{-i}(t_k), u_i^p(\cdot;t_k)) \\
&= \min_{u_i^p} \quad \int_{t_k}^{t_k+T} L_i(x_i^p(s;t_k), \hat{x}_{-i}(s;t_k), u_i^p(s;t_k))ds + \gamma \|x_i^p(t_k+T;t_k) - x_i^C\|_{P_i}^2 \\
&\qquad \text{s.t.} \quad \dot{x}_i^p(\tau;t_k) = f_i(x_i^p(\tau;t_k), u_i^p(\tau;t_k)) \\
&\qquad\qquad \dot{\hat{x}}_i^p(\tau;t_k) = f_i(\hat{x}_i^p(\tau;t_k), \hat{u}_i^p(\tau;t_k)) \\
&\qquad\qquad \dot{\hat{x}}_{-i}^p(\tau;t_k) = f_{-i}(\hat{x}_{-i}^p(\tau;t_k), \hat{u}_{-i}^p(\tau;t_k)) \\
&\qquad\qquad u_i^p(\tau;t_k) \in U \\
&\qquad\qquad \|x_i^p(\tau;t_k) - \hat{x}_i(\tau;t_k)\| \leq \delta^2 \kappa \\
&\qquad\qquad x_i^p(t_k+T;t_k) \in \Omega_i(\varepsilon_i)
\end{aligned}
$$

The second last constraint is a "compatibility" constraint, enforcing consistency between what agent #i plans to do and what its neighbors believe it plans to do. The last constraint is a terminal constraint.

Under certain technical assumptions, the authors prove that the DMPC problems are feasible at each update step k, and under certain bounds on the update interval δ convergence to a given set is also proved. Note that closed-loop stability is ensured by constraining the state trajectory predicted by each agent to stay close enough to the trajectory predicted at the previous time step that has been broadcasted. The main drawback of the approach is the conservativeness of the compatibility constraint.

5.3.5 DMPC Approach of Keviczy, Borrelli, and Balas

Dynamically decoupled submodels are also considered in [21], where the special nonlinear discrete-time system structure

$$x_{k+1}^i = f^i(x_k^i, u_k^i)$$

is assumed, subject to local input and state constraints $x_k^i \in \mathscr{X}^i$, $u_k^i \in \mathscr{U}^i$, $i = 1, \ldots, M$. Subsystems are coupled by the cost function

$$l(\tilde{x}, \tilde{u}) = \sum_{i=1}^{N_v} l^i(x^i, u^i, \tilde{x}^i, \tilde{u}^i)$$

and by the global constraints

$$g^{i,j}(x^i, u^i, x^j, u^j) \leq 0, \ (i, j) \in \mathscr{A}$$

where $\mathscr{A}$ is a given set. Each local MPC controller is based on the optimization of the following problem

$$\min_{\tilde{U}_t} \sum_{k=0}^{N-1} l(\tilde{x}_{k,t}, \tilde{u}_{k,t}) + l_N(\tilde{x}_{N,t}) \tag{5.28a}$$

$$\text{s.t.} \quad x^i_{k+1,t} = f^i(x^i_{k,t}, u^i_{k,t}) \tag{5.28b}$$

$$x^i_{k,t} \in \mathscr{X}^i, \quad u^i_{k,t} \in \mathscr{U}^i, \; k = 1, \ldots, N-1 \tag{5.28c}$$

$$x^i_{N,t} \in \mathscr{X}^i_f \tag{5.28d}$$

$$x^j_{k+1,t} = f^j(x^j_{k,t}, u^j_{k,t}), (i,j) \in \mathscr{A} \tag{5.28e}$$

$$x^j_{k,t} \in \mathscr{X}^j, \quad u^j_{k,t} \in \mathscr{U}^j, (i,j) \in \mathscr{A} \; k = 1, \ldots, N-1 \tag{5.28f}$$

$$x^j_{N,t} \in \mathscr{X}^j_f, (i,j) \in \mathscr{A} \tag{5.28g}$$

$$g^{i,j}(x^i_{k,t}, u^i_{k,t}, x^j_{k,t}, u^j_{k,t}) \le 0, (i,j) \in \mathscr{A} \; k = 1, \ldots, N-1 \tag{5.28h}$$

$$x^i_{0,t} = x^i_t, \tilde{x}^i_{0,t} = \tilde{x}^i_t \tag{5.28i}$$

where (5.28b)–(5.28d) are the local model and constraints of the agent, (5.28e)–(5.28g) are the model and constraints of the neighbors, and (5.28h) represent interaction constraints of agent $\#i$ with its own neighbors.

The information exchanged among the local MPC agents are the neighbors' current states, terminal regions, and local models and constraints. As in (5.13), only the optimal input $u^i_{0,t}$ computed by controller $\#i$ is applied; the remaining inputs $u^j_{k,t}$ are completely discarded, as they are only used to enhance the prediction.

Stability is analyzed for the problem without coupling constraints (5.28h), under the assumption that the following inequality holds

$$\sum_{k=1}^{N-1} 2\|Q(x^{j,j}_{k,t} - x^{j,i}_{k,t})\|_p + \|R(u^{j,j}_{k,t} - u^{j,i}_{k,t})\|_p \le \|Qx^i_t\|_p + \|Qx^j_t\|_p +$$
$$\|Q(x^i_t - x^j_t)\|_p + \|Ru^{i,i}_{0,t}\|_p + \|Ru^{j,i}_{0,t}\|_p$$

where $\|Qx\|_2 \triangleq x'Qx$, and $\|Qx\|_1$, $\|Qx\|_\infty$ are the standard q and ∞ norm, respectively.

5.3.6 DMPC Approach of Mercangöz and Doyle

The distributed MPC *and* estimation problems are considered in [25] for square plants (the number of inputs equals the number of outputs) perturbed by noise, whose local prediction models are

$$\begin{cases} x_i(k+1) = A_i x_i(k) + B_i u_i(k) + \sum_{j=1}^{M} B_j u_j(k) + w_i(k) \\ y_i(k) = C_i x_i(k) + v_i(k) \end{cases} \tag{5.29}$$

A distributed Kalman filter based on the local submodels (5.29) is used for state estimation. The DMPC approach is similar to Venkat et al.'s "communication-based" approach, although only first moves $u_j(k)$ are transmitted and assumed frozen in prediction, instead of the entire optimal sequences. Only constraints on local inputs are handled by the approach. Although general stability and convergence results are not proved in [25], experimental results on a four-tank system are reported to show the effectiveness of the approach.

5.3.7 DMPC Approach of Magni and Scattolini

Another interesting approach to decentralized MPC for nonlinear systems has been formulated in [23]. The problem of regulating a nonlinear system affected by disturbances to the origin is considered under some technical assumptions of regularity of the dynamics and of boundedness of the disturbances. Closed-loop stability is ensured by the inclusion in the optimization problem of a contractive constraint. The considered class of functions and the absence of information exchange between controllers leads to some conservativeness of the approach.

5.4 Example of Decentralized Temperature Control in a Railcar

5.4.1 Example Description

In this section we test the DMPC approach of Alessio, Barcelli, and Bemporad for decentralized control of the temperature in different passenger areas in a railcar [5]. The system is schematized in Figure 5.2. Each passenger area has its own heater and air conditioner but its thermal dynamics interacts with surrounding areas (neighboring passenger areas, external environment, antechambers) directly or through windows, walls and doors. Passenger areas are composed by a table and the associated four seats. Temperature sensors are located in each four-seat area, in each antechamber, and along the corridor. The goal of the controller is to adjust each passenger area to its own temperature set-point to maximize passenger comfort. Temperature sensors may be wired or wireless, in the latter case we assume that information packets may be dropped, because of very low power transmission, simplified transmission protocols, and no acknowledgement and retransmission and because of time-varying communication disturbances due for example to passengers' electronic equipment.

Let $2N$ be the number of four-seat areas ($N = 8$ in Figure 5.2), N the number of corridor partitions, and 2 the number of antechambers. Under the assumption of perfectly mixed fluids in each jth volume, $j = 1, \dots, n$ where $n = 3N + 2$, the heat transmission equations by conduction lead to the linear model

$$\frac{dT_j(\tau)}{d\tau} = \sum_{i=0}^{n} Q_{ij}(\tau) + Q_{uj},\ Q_{ij}(\tau) = \frac{S_{ij}K_{ij}(T_i(\tau) - T_j(\tau))}{C_j L_{ij}},\ j = 1, \dots, n \quad (5.30)$$

where $T_j(\tau)$ is the temperature of volume $\#j$ at time $\tau \in \mathbb{R}$, $T_0(\tau)$ is the ambient temperature outside the railcar, $Q_{ij}(\tau)$ is heat flow due to the temperature difference $T_i(\tau) - T_j(\tau)$ with the neighboring volume $\#i$, S_{ij} is the contact surface area, Q_{uj} is the heat flow of heater $\#j$, K_{ij} is the thermal coefficient that depends on the materials, $C_j = K_c^j V_j$ is the (material dependent) heat capacity coefficient K_c^j times the fluid

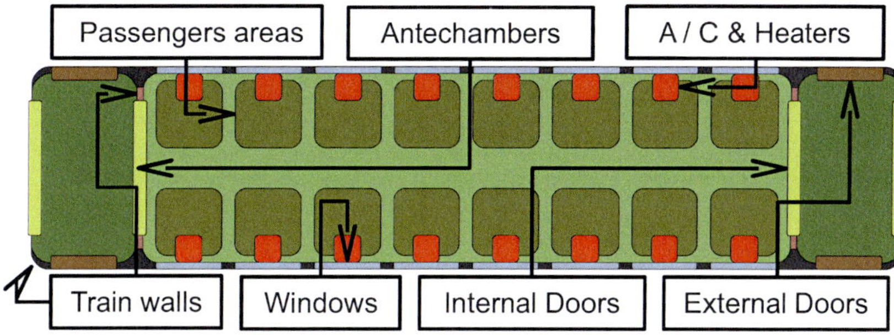

Fig. 5.2 Physical structure of the railcar

volume V_j, and L_{ij} is the thickness of the separator between the two volumes $\#i$ and $\#j$. We assume that $Q_{ij}(\tau) = 0$ for all volumes i, j that are not adjacent, $\forall \tau \in \mathbb{R}$. Hence, the process can be modeled as a linear time-invariant continuous-time system with state vector $z \in \mathbb{R}^{3N+2}$ and input vector $v \in \mathbb{R}^{2N}$

$$\begin{cases} \dot{z}(\tau) = A_c z(\tau) + B_c v(\tau) + F T_0(\tau) \\ h(\tau) = C z(\tau) \end{cases} \tag{5.31}$$

where $F \in \mathbb{R}^n$ is a constant matrix, $T_0(\tau)$ is treated as a piecewise constant measured disturbance, and $C \in \mathbb{R}^{p \times n}$ is such that $h \in \mathbb{R}^p$ contains the components of z corresponding to the temperatures of the passenger seat areas, $p = 2N$. Since we assume that the thermal dynamics are relatively slow compared to the sampling time T_s of the decentralized controller we are going to synthesize, we use first-order Euler approximation to discretize dynamics (5.31) without introducing excessive errors:

$$\begin{cases} z(t+1) = Az(t) + Bv(t) + F_d T_0(t) \\ h(t) = Cz(t) \end{cases} \tag{5.32}$$

where $A = I + A_c T_s$, $B = B_c T_s$, and $F_d = F T_s$. We assume that A is asymptotically stable, as an inheritance of the asymptotic stability of matrix A_c.

In order to track generic temperature references $r(t)$, we adopt the coordinate shift defined by (5.24). The next step is to decentralize the resulting global model. The particular topology of the railcar suggests a decomposition of model (5.1) as the cascade of four-seat areas. There are two kinds of four-seat areas, namely (*i*) the ones next to the antechambers, and (*ii*) the remaining ones. Besides interacting with the external environment, the areas of type (*i*) interact with another four-seat-area, an antechamber, and a section of the corridor, while the areas of type (*ii*) only with the four-seat areas at both sides and the corresponding section of the corridor. Note that the decentralized models overlap, as they share common states and inputs. The decoupling matrices Z_i are chosen so that in each subsystem only the first component of the computed optimal input vector is actually applied to the process.

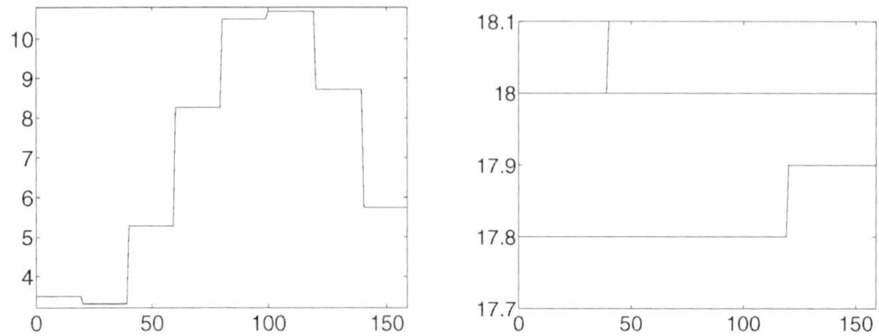

Fig. 5.3 Exogenous signals used in the reported simulations

As a result, each submodel has 5 states and 2 or 3 inputs, depending whether it describes a seat area of type (*i*) or (*ii*), which is considerably simpler than the centralized model (5.1) with 26 states and 16 inputs.

We apply the DMPC approach (5.10) with

$$Q = 2\begin{bmatrix} 10^2 I_{16} & 0 \\ 0 & I_{10} \end{bmatrix},\ R = 10^5 I_{16}, v_{\min} = -0.03\text{ W},\ v_{\max} = 0.03\text{ W},\ T_s = 9\text{ min} \tag{5.33}$$

where $v_{\min}$ is the lower bound on the heat flow subtracted by the air-conditioners, and $v_{\max}$ is the maximum heating power of the heaters (with a slight abuse of notation we denoted by $v_{\min}$, $v_{\max}$ the entries of the corresponding lower and upper bound vectors of $\mathbb{R}^{16}$). Note that the first sixteen diagonal elements of matrix Q correspond to the temperatures of the four-seat areas. It is easy to check that with the parameters in (5.33) condition (5.17a) for local stability is satisfied. For comparison, a centralized MPC approach (5.3) with the same weights, horizon, and sampling time as in (5.33) is also designed. The associated QP problem has 16 optimization variables and 32 constraints, while the complexity of each DMPC controller is either 2 (or 3) variables and 4 (or 6) constraints. The DMPC approach is in fact largely scalable: for longer railcars the complexity of the DMPC controllers remains the same, while the complexity of the centralized MPC problem grows with the increased model size. Note also that, even if a centralized computation is taken, the DMPC approach can be immediately parallelized.

5.4.2 Simulation Results

We investigate different simulation outcomes depending on four ingredients: *i*) type of controller (centralized/decentralized), *ii*) packet-loss probability, *iii*) change in reference values, *iv*) changes of external temperature (acting as a measured disturbance). Figure 5.3 shows the external temperature and reference scenarios used in all simulations.

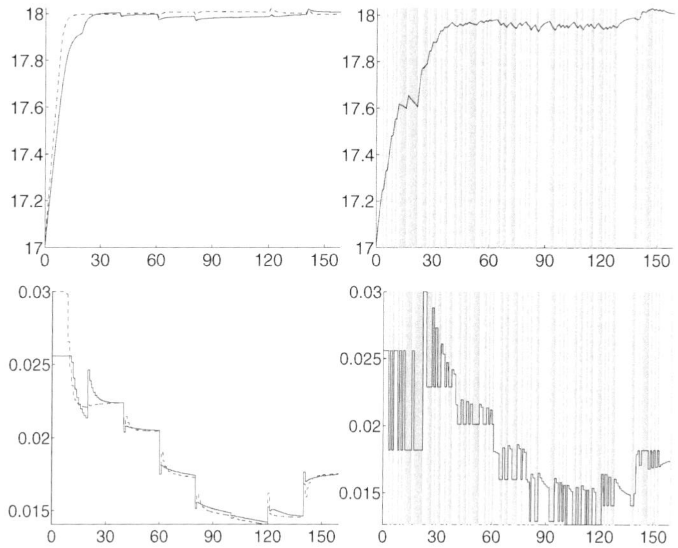

Fig. 5.4 Comparison between centralized MPC (dashed lines) and decentralized MPC (continuous lines): output h_1 (upper plots) and input v_1 (lower plots). Gray areas denote packet drop intervals

In order to compare closed-loop performances in different simulation scenarios, the following performance index

$$J = \Sigma_{t=1}^{N_{\text{sim}}} (z(t) - r(t))'Q(z(t) - r(t)) + (v(t) - v_r)'R(v(t) - v_r) \qquad (5.34)$$

is defined, where $N_{sim} = 160$ (one day) is the total number of simulation steps.

The initial condition is 17°C for all seat-area temperatures, except for the antechamber, which is 15°C. Note that the steady-state value of antechamber temperatures is not relevant for the posed control goals. The closed-loop trajectories of centralized MPC feedback vs. decentralized MPC with no packet-loss are shown in Figure 5.4.2 (we only show the first state and input for clarity). In both cases the temperatures of the four-seat areas converge to the set-point asymptotically. Figure 5.4.2 shows the temperature vector $h(t)$ tracking the time-varying reference $r(t)$ in the absence of packet-loss, where the coordinate transformation (5.24) is repeated after each set-point and external temperature change.

To simulate packet loss, we assume that the probability of losing a packet depends on the state of a Markov chain with N states (see Figure 5.6). We parameterize with the probability parameter p, $0 \le p \le 1$ the probabilities associated with the Markov chain: the Markov chain is in the jth state if $j-1$ consecutive packets have been lost. The probability of losing a further packet is $1-p$ in every state of the chain, except for the $(N+1)$th state where no packet can be lost any more.

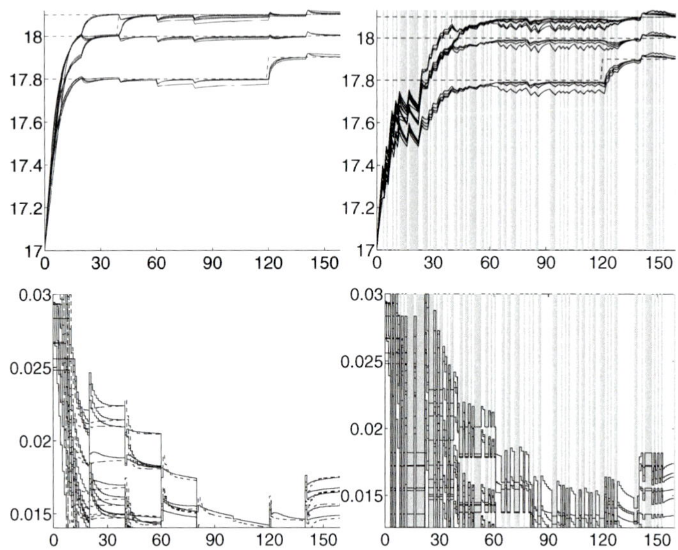

Fig. 5.5 Decentralized MPC results. Upper plots: output variables h (continuous lines) and references r (dashed lines). Lower plots: command inputs v. Gray areas denote packet drop intervals

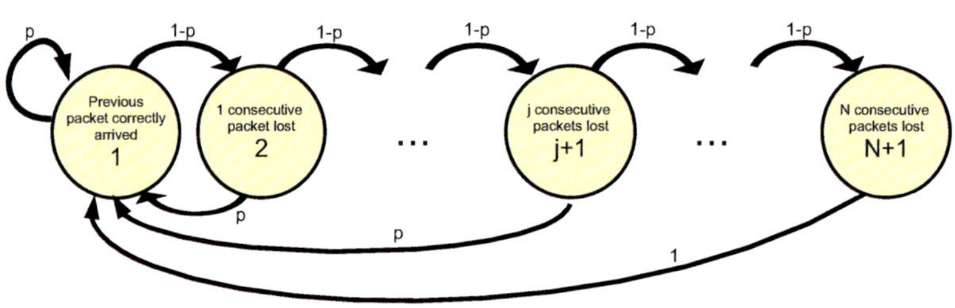

Fig. 5.6 Markov chain model of packet-loss probability

Let π be the stationary probability vector of the Markov chain of Figure 5.6, obtained through the one-step probability matrix

$$P = \begin{bmatrix} p & 1-p & 0 & \cdots & 0 \\ p & 0 & 1-p & \cdots & 0 \\ \vdots & \vdots & \vdots & \ddots & \vdots \\ p & 0 & 0 & \cdots & 1-p \\ 1 & 0 & 0 & \cdots & 0 \end{bmatrix}$$

by solving

$$\begin{cases} \pi' = \pi' P \\ \sum_{i=1}^{N} \pi_i = 1 \end{cases}$$

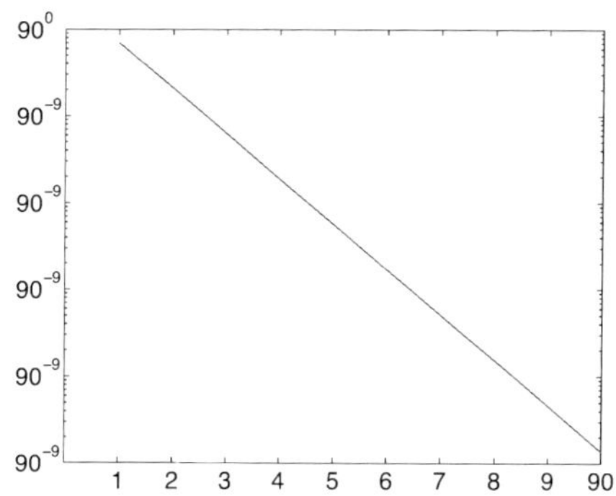

Fig. 5.7 Markov chain packet-loss probability with $N = 10$ and $p = 0.7$

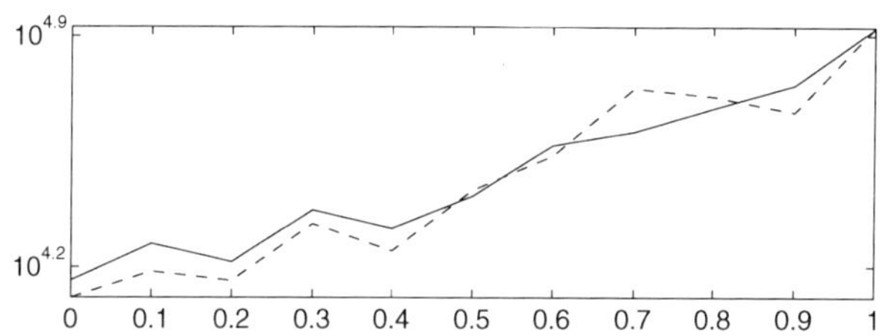

Fig. 5.8 Performance indices of Centralized MPC (dashed line) and Decentralized MPC (solid line)

Recalling the meaning of the Markov chain nodes, the steady-state probability π_i of the ith state is the probability of loosing consecutively exactly $i-1$ packets. The packet-loss discrete probability is shown in Figure 5.7 when $N = 10$ maximum consecutive packet-losses are possible and $p = 0.7$.

Figure 5.7 highlights the exponential decrease of the stationary probabilities as a function of consecutive packets lost. Such a probability model is confirmed by the experimental results on relative frequencies of packet failure burst length observed in [38]. Note that our model assumes that the probability of losing a packet is null after N packets, hence satisfying the assumption of an upper-bound on the number of consecutive drops (as mentioned earlier, we can assume for instance that if $k > N$ consecutive packets are lost, the control loops are shut down). The simulation results obtained with $p = 0.5$ are shown in Figure 5.4.2 and Figure 5.4.2.

In case of packet loss, we also compare the performance of centralized vs. decentralized MPC. Note that in case packet loss occurs also on the communication channel between the point computing the coordinate shift and the decentralized controllers, the last received coordinate shift is kept. The stability condition (5.22a) of Theorem 5.5 was tested and proved satisfied for values of j up to 160.

Figure 5.8 shows that the performance index J defined in Eq. (5.34) increases as the packet-loss probability grows, implying performance to deteriorate due to the conservativeness of the backup control action $u = 0$ (that is, $v = v_r$). The results of Figure 5.8 are averaged over 10 simulations per probability sample. As a general consideration, centralized MPC dominates over the decentralized, although for certain values of p the average performance of decentralized MPC is slightly better, probably due to the particular packet loss sequences that have realized. However, the loss of performance due to decentralization, with regard to the present example, is largely negligible.

The simulations were run on a MacBook Air 1.86 GHz running Matlab R2008a under OS X 10.5.6 and the Hybrid Toolbox for Matlab [7]. The average CPU time for solving the centralized QP problem associated with (5.3) is 6.0ms (11.9ms in the worst case). For the decentralized case, the average CPU time for solving the QP problem associated with (5.10) is 3.3ms (7.4ms in the worst case). Although the decrease of CPU time is only a few milliseconds, we remark that for increasing N the complexity of DMPC remains constant, while the complexity of centralized MPC would grow with N. To quantify this aspect consider that, if one thinks to the explicit form of the MPC controllers [9], the number of regions of the centralized MPC is upper bounded by 3^{16}, while in decentralized case by 3^2 for submodels with two inputs and by 3^3 for submodels with three inputs.

Note that the reference vectors v_r, z_r are computed globally in all simulations. In this example the complexity of such a static calculation is negligible with respect to solving the QP problems. Moreover, the communication burden is also negligible, as new reference vectors are transmitted individually to each MPC agent only when set-point and disturbances change.

5.5 Hierarchical MPC

5.5.1 Problem Description

In this section we provide some novel ideas on how to possibly couple a DMPC layer with a higher centralized (hybrid) MPC layer in the hierarchical setting of Figure 5.1. The idea is to design a centralized hybrid MPC controller to achieve global coordination, namely to enforce global constraints (linear, logical, mixed linear & logical) and to optimize a global objective (such as an economically-driven objective). To achieve the goal, we need an abstract (hybrid) model of the underlying closed-loop dynamics, without resorting to a global dynamical model of the process and to the need of full state feedback.

Under the assumption that the higher MPC layer runs in real-time at a lower sampling frequency than the underlying DMPC layer, a sensible choice is to use the static global model

$$y(k+1) = G(1)u(k)$$

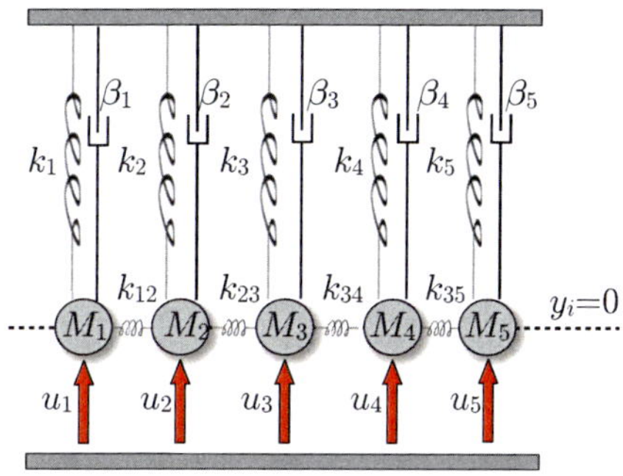

Fig. 5.9 Hierarchical and decentralized MPC of system composed by five dynamically-coupled masses moving vertically

as a centralized abstract model, where $G(1)$ is the DC-gain of (5.1), and k represents the sampling step of the higher-level MPC controller.

Then, the centralized higher-level MPC controller solves the following problem

$$\begin{aligned} \min_{u(k)} \ & f(y(k+1) - r_d(k), u(k)) \\ \text{s.t. } & g(u(k), y(k), r_d(k)) \leq 0 \end{aligned}$$

and defines

$$r(k) = G(1)u(k)$$

as the current setpoints for the DMPC layer. Extensions of these ideas, including quantitative ways of choosing a suitable sampling time for the higher control layer, have been formulated very recently in [6].

5.5.2 *Illustrative Example*

Consider the system depicted in Figure 5.9, composed by five dynamically-coupled masses moving vertically. The dynamics of each mass #i is described by the dynamics

$$M_i\ddot{y}_i = u_i - \beta_i\dot{y}_i - k_iy_i - \underbrace{k_{ij}(y_i - y_j)}_{j=i-1,i+1} \tag{5.35}$$

where $M_i = 5$ [kg], $\beta_i = 0.1$ [kg/s], $k_i = 1$ [kg/s^2], $k_{ij} = 0.5$ [kg/s^2]. For local prediction purposes, each DMPC controller #i neglects the velocities of the neighboring masses, $\dot{y}_{i-1} = \dot{y}_{i+1} = 0$, and therefore only considers $y_i, \dot{y}_i, y_{i-1}, y_{i+1}$ as local states.

After discretizing the dynamics (5.35) with sampling time $T_L = 0.25$ [s], each DMPC agent solves the following on-line optimal control problem

$$\min_{u_i(k),\ldots,u_i(k+N_u-1)} \sum_{j=0}^{N_y-1} (y_i(k+j)-r_i(k))^2+0.1\sum_{j=0}^{N_u-1} (u_i(k+j)-u_i(k+j-1))^2+10^4\varepsilon^2$$

$$\begin{array}{lll}\text{s.t.} & u^i_{\min} \le u_i(k+j) \le u^i_{\max}(k) & j=0,\ldots,N_u-1 \\ & y^i_{\min}-\varepsilon \le y_i(k+j) \le y^i_{\max}+\varepsilon & j=0,\ldots,N \\ & u_i(k+j)=u_i(k+N_u-1) & j=N_u,\ldots,N_y-1\end{array} \tag{5.36}$$

where $u^i_{\min}=y^i_{\min}=0$ [m], $y^i_{\max}=2$ [m], $N_y=20$ is the prediction horizon, $N_u=4$ is the control horizon, ε is a slack variable used to soften output constraints to prevent the possible infeasibility of the quadratic program associated with (5.36). The input bounds $u^i_{\max}(k)$ are decided at multiple of the higher-level sampling time $T_H=3$ [s] by a centralized hybrid MPC, together with the local setpoint vector $r(k)$. The hybrid MPC controller [8, 33] is designed to enforce the following constraints:

(i) at most K_u inputs can be over a certain threshold $u_{\lim}=0.7$ [m], $u_i(k)\ge u_{\lim}$;
(ii) set-point changes are bounded by a quantity Δ_r=0.5 [m]

$$|G(1)u(k)-y_i(k)| \le \Delta_r \tag{5.37}$$

The logical "at most" constraint (i) is enforced by defining auxiliary binary inputs $u_{\ell i}(k)\in\{0,1\}$, $i=1,\ldots,5$, and by setting

$$u^i_{\max}(k)=\begin{cases} y^i_{\max} & \text{if } u_{\ell i}(k)=1 \\ u_{\lim} & \text{if } u_{\ell i}(k)=0\end{cases} \tag{5.38}$$

By letting $r_d(k)=[0.2\ 0.5\ 0.75\ 1\ 0.75]'$ [m] be the vector of desired vertical positions of the masses, every $T_H/T_L=12$ steps the hybrid MPC controller solves the following problem

$$\min_{u(k)} \|G(1)u(k)-r_d(k)\|^2+\sum_{i=1}^{5}(u_{\ell i}-1)^2$$

subject to the linear constraints defined by (5.37) and the mixed-integer reformulation of constraint (5.38) [8] to determine the reference vector $r(k)=G(1)u(k)$ for the DMPC layer, and the input upper limit $u^i_{\max}(k)$, which is set either to $y^i_{\max}$ or to $u_{\lim}$, depending on the logical constraint.

We simulate the hierarchical control system from initial positions $y_1(0)=1$, $y_2(0)=0.4$, $y_3(0)=0.7$, $y_4(0)=0.8$, $y_5(0)=0.2$ and null velocities. The closed-loop results are reported in Figure 5.10. Note that in the absence of the higher-level hybrid MPC controller the red, purple, cyan, and blue input force signals ($u_i(k)$) overpass the limit threshold $u_{\lim}(k)$ (Figure 5.10(a)). When the hybrid MPC controller is used with $K_u=3$, we obtain the plots depicted in Figure 5.10(b), where only the red, purple, and cyan input signals are set greater than $u_{\lim}(k)$. For $K_u=2$, we obtain the plots depicted in Figure 5.10(c), where only the red and purple input signals get above $u_{\lim}(k)$.

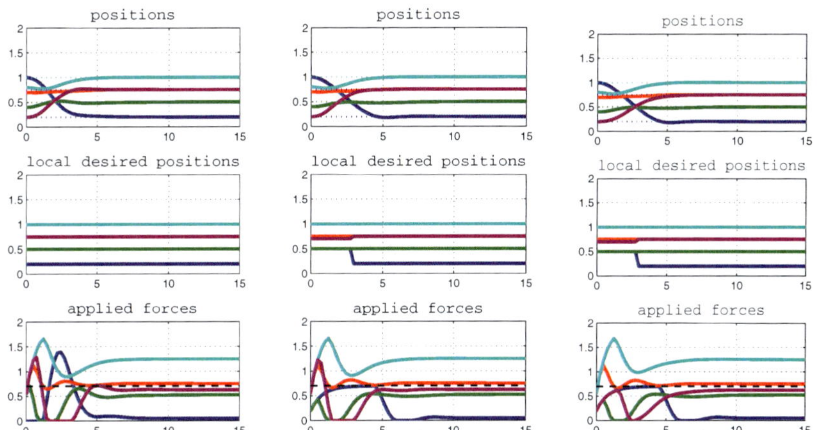

(a) DMPC with $r(k) = r_d(k)$ (no hybrid MPC) (b) hierarchical MPC, $K_u = 3$ (c) hierarchical MPC, $K_u = 2$

Fig. 5.10 Hierarchical and decentralized MPC results for the five-mass system. The limit u_{lim} is shown as a dashed black line

Both the linear DMPC and the hybrid MPC controllers were implemented in Simulink using the Hybrid Toolbox for MATLAB [7].

5.6 Conclusions

In this chapter we have surveyed different approaches to the problem of controlling a distributed process through the cooperation of multiple decentralized model predictive controllers. Each controller is based on a submodel of the overall process, and different submodels may share common states and inputs, to possibly decrease modeling errors in case of dynamical coupling, and to increase the level of cooperativeness of the controllers. The DMPC approach is suitable for control of large-scale systems subject to constraints: the possible loss of global optimal performance is compensated by the gain in controller scalability, reconfigurability, and maintenance. Although a few contributions have been given in the last few years, the DMPC theory is not yet mature and homogenous. In this chapter we have tried to highlight similarities and differences among the various approaches that have been proposed, a little step towards the consolidation of a general theoretical framework for DMPC design.

Open research topics in DMPC include: systematic ways to decompose the model into local submodels, when this is not obvious from the physics of the process, determining the optimal model decomposition (*i.e.*, the best achievable closed-loop performance) for a given channel capacity and computer power available to the control agents; better awareness of DMPC algorithms of the communication efforts, especially when operating over wireless sensor networks (for instance, to save battery energy and hence increase the device life span); stochastic DMPC formulations to take into account imperfect communication in a less conservative way than robust approaches; output-feedback DMPC by using suitable complementary decentralized estimation schemes; hierarchical MPC schemes, for instance combining centralized hybrid MPC and decentralized linear MPC; design better distributed MPC algorithms by taking into account the progress in distributed optimization approaches (*e.g.*, to handle coupled input and state constraints), as described in Chapter 3 of this book.

Acknowledgment

The authors acknowledge the help of Bruce Krogh, Dong Jia, Francesco Borrelli, Richard Murray, William Dunbar, Aswin Venkat, James Rawlings, Mehmet Mercangoz, and Francis Doyle III for sharing copies of their presentations related to their papers on DMPC. Mikael Johansson is also acknowledged for pointing out the packet loss results of [38].

References

1. Alessio, A., Barcelli, D., Bemporad, A.: Decentralized model predictive control of dynamically-coupled linear systems. Technical report. Available upon request from the authors
2. Alessio, A., Bemporad, A.: Decentralized model predictive control of constrained linear systems. In: Proc. European Control Conf., Kos, Greece, pp. 2813–2818 (2007)
3. Alessio, A., Bemporad, A.: Stability conditions for decentralized model predictive control under packet dropout. In: Proc. American Contr. Conf., Seattle, WA, pp. 3577–3582 (2008)
4. Bamieh, B., Paganini, F., Dahleh, M.A.: Distributed control of spatially invariant systems. IEEE Trans. Automatic Control 47(7), 1091–1107 (2002)
5. Barcelli, D., Bemporad, A.: Decentralized model predictive control of dynamically-coupled linear systems: Tracking under packet loss. In: 1st IFAC Workshop on Estimation and Control of Networked Systems, Venice, Italy, pp. 204–209 (2009)
6. Barcelli, D., Bemporad, A., Ripaccioli, G.: Hierarchical multi-rate control design for constrained linear systems. In: Proc. 49th IEEE Conf. on Decision and Control, Atlanta, Georgia USA (to appear, 2010)
7. Bemporad, A.: Hybrid Toolbox – User's Guide (December 2003), `http://www.dii.unisi.it/hybrid/toolbox`
8. Bemporad, A., Morari, M.: Control of systems integrating logic, dynamics, and constraints. Automatica 35(3), 407–427 (1999)
9. Bemporad, A., Morari, M., Dua, V., Pistikopoulos, E.N.: The explicit linear quadratic regulator for constrained systems. Automatica 38(1), 3–20 (2002)
10. Borrelli, F., Keviczky, T., Fregene, K., Balas, G.J.: Decentralized receding horizon control of cooperative vechicle formations. In: Proc. 44th IEEE Conf. on Decision and Control and European Control Conf., Sevilla, Spain, pp. 3955–3960 (2005)
11. Callier, F.M., Chan, W.S., Desoer, C.A.: Input-output stability theory of interconnected systems using decomposition techniques. IEEE Trans. Circuits and Systems 23(12), 714–729 (1976)

12. Camponogara, E., Jia, D., Krogh, B.H., Talukdar, S.: Distributed model predictive control. IEEE Control Systems Magazine, 44–52 (February 2002)
13. Damoiseaux, A., Jokic, A., Lazar, M., van den Bosch, P.P.J., Hiskens, I.A., Alessio, A., Bemporad, A.: Assessment of decentralized model predictive control techniques for power networks. In: 16th Power Systems Computation Conference, Glasgow, Scotland (2008)
14. D'Andrea, R.: A linear matrix inequality approach to decentralized control of distributed parameter systems. In: Proc. American Contr. Conf., pp. 1350–1354 (1998)
15. Dunbar, W.B., Murray, R.M.: Distributed receding horizon control with application to multi-vehicle formation stabilization. Automatica 42(4), 549–558 (2006)
16. Grieder, P., Morari, M.: Complexity reduction of receding horizon control. In: Proc. 42nd IEEE Conf. on Decision and Control, Maui, Hawaii, USA, pp. 3179–3184 (2003)
17. Jia, D., Krogh, B.: Distributed model predictive control. In: Proc. American Contr. Conf., Arlington, VA, pp. 2767–2772 (2001)
18. Jia, D., Krogh, B.: Min-max feedback model predictive control for distributed control with communication. In: Proc. American Contr. Conf., pp. 4507–4512. Anchorage, Alaska (2002)
19. Johannson, M., Rantzer, A.: Computation of piece-wise quadratic Lyapunov functions for hybrid systems. IEEE Trans. Automatic Control 43(4), 555–559 (1998)
20. Johansson, B., Keviczky, T., Johansson, M., Johansson, K.H.: Subgradient methods and consensus algorithms for solving convex optimization problems. In: Proc. 47th IEEE Conf. on Decision and Control, Cancun, Mexico, pp. 4185–4190 (2008)
21. Keviczky, T., Borrelli, F., Balas, G.J.: Decentralized receding horizon control for large scale dynamically decoupled systems. Automatica 42(12), 2105–2115 (2006)
22. Li, W., Cassandras, C.G.: Stability properties of a receding horizon contoller for cooperating UAVs. In: Proc. 43th IEEE Conf. on Decision and Control, Paradise Island, Bahamas, pp. 2905–2910 (2004)
23. Magni, L., Scattolini, R.: Stabilizing decentralized model predictive control of nonlinear systems. Automatica 42(7), 1231–1236 (2006)
24. Mayne, D.Q., Rawlings, J.B., Rao, C.V., Scokaert, P.O.M.: Constrained model predictive control: Stability and optimality. Automatica 36(6), 789–814 (2000)
25. Mercangöz, M., Doyle III., F.J.: Distributed model predictive control of an experimental four-tank system. Journal of Process Control 17(3), 297–308 (2007)
26. Michel, A.N.: Stability analysis of interconnected systems. SIAM J. Contr. 12, 554–579 (1974)
27. Michel, A.N., Rasmussen, R.D.: Stability of stochastic composite systems. IEEE Trans. Automatic Control 21, 89–94 (1976)
28. Richards, A., How, J.P.: Decentralized model predictive control of cooperating UAVs. In: Proc. 43rd IEEE Conf. on Decision and Control, Paradise Island, Bahamas (2004)
29. Rotkowitz, M., Lall, S.: A characterization of convex problems in decentralized control. IEEE Trans. Automatic Control 51(2), 1984–1996 (2006)
30. Samar, S., Boyd, S., Gorinevsky, D.: Distributed estimation via dual decomposition. In: Proc. European Control Conf., Kos, Greece, pp. 1511–1516 (2007)
31. Sandell, N.R., Varaiya, P., Athans, M., Safonov, M.G.: Survey of decentralized control methods for large scale systems. IEEE Trans. Automatic Control 23(2), 108–128 (1978)
32. Scattolini, R.: Architectures for distributed and hierarchical model predictive control – a review. Journal of Process Control 19, 723–731 (2009)
33. Torrisi, F.D., Bemporad, A.: HYSDEL — A tool for generating computational hybrid models. IEEE Trans. Contr. Systems Technology 12(2), 235–249 (2004)

34. Venkat, A.N., Hiskens, I.A., Rawlings, J.B., Wright, S.J.: Distributed MPC strategies with application to power system automatic generation control. IEEE Transactions on Control Systems Technology 16(6), 1192–1206 (2008)
35. Venkat, A.N., Rawlings, J.B., Wright, J.S.: Stability and optimality of distributed model predictive control. In: Proc. 44th IEEE Conf. on Decision and Control and European Control Conf., Seville, Spain (2005)
36. Venkat, A.N., Rawlings, J.B., Wright, J.S.: Implementable distributed model predictive control with guaranteed performance properties. In: Proc. American Contr. Conf., Minneapolis, MN, pp. 613–618 (2006)
37. Wang, S., Davison, E.J.: On the stabilization of decentralized control systems. IEEE Trans. Automatic Control 18(5), 473–478 (1973)
38. Willig, A., Mitschke, R.: Results of bit error measurements with sensor nodes and casuistic consequences for design of energy-efficient error control schemes. In: Römer, K., Karl, H., Mattern, F. (eds.) EWSN 2006. LNCS, vol. 3868, pp. 310–325. Springer, Heidelberg (2006)

Chapter 6
Decentralized Control

John Swigart and Sanjay Lall

Abstract. Decentralized control has been a large area of open research for over forty years. To cover every aspect of it would require a vast knowledge of applied mathematics and considerable time. Consequently, in this tutorial we will attempt to restrict our attention to optimal control of systems which are linear with Gaussian random noise and disturbances, where the objective is a quadratic cost function. This encompasses a very general, and commonly encountered, class of systems. While the results herein will be aimed at this class, much of our discussion may be applied more generally to other problems. Most of the results in this chapter are, of course, not new and can be found in the references.

To begin our discussion we will highlight some of the key features of decentralized control with a few motivating examples. From there, we will address what will be called static systems, and show that decentralized problems of this nature admit tractable solutions. Our discussion will then turn to the class of dynamic problems, involving feedback. Decentralized feedback problems in general are known to be difficult. Nevertheless, there exist some problems for which optimal solutions may be obtained. We will end our discussion with some methods for solving these types of problems.

6.1 Motivating Examples

Control problems come in all shapes and sizes. At the lowest level we classify them as either centralized or decentralized. In a *centralized* problem, we tend to think of such systems as involving a single decision maker. This may be because there is only

John Swigart
Department of Aeronautics and Astronautics, Stanford University, Stanford, CA 94305, USA
e-mail: jswigart@stanford.edu

Sanjay Lall
Department of Electrical Engineering and Department of Aeronautics and Astronautics, Stanford University, Stanford, CA 94305, USA
e-mail: lall@stanford.edu

A. Bemporad, M. Heemels, and M. Johansson: Networked Control Systems, LNCIS 406, pp. 179–201.
springerlink.com

one system involved, or because multiple systems might communicate their sensor measurements to a central processor which interprets the data of every system and relays decisions back.

By contrast, *decentralized* problems are basically defined as any system which is not centralized. Intuitively, one might think of this as multiple systems in which each system has its own processing unit and makes its own decisions based on its own measurements.

To illustrate the types of decentralized problems that one might encounter and some of the questions that we might ask about these systems, consider the following examples.

6.1.1 Vehicle Spacing

Consider the following vehicle control problem. There are N vehicles in a line, with vehicle i located at position q_i, shown in Figure 6.1. Suppose each vehicle is displaced a distance x_i from its original unit spacing. Each vehicle has sensors which measure the relative displacements of its neighbors plus noise; for example,

$$y_2 = \begin{bmatrix} x_2 - x_1 \\ x_3 - x_2 \end{bmatrix} + \begin{bmatrix} w_3 \\ w_4 \end{bmatrix}$$

Fig. 6.1 Vehicle Spacing Example

Within this framework, it may be possible for vehicles to communicate their sensor data to other vehicles. Several such *information constraints* are possible, including:

- Every vehicle receives the output of every sensor
- Every vehicle sees only its own sensor data
- Each vehicle i receives the sensor data of vehicles $i-1$, i, and $i+1$

While the first information structure would be considered a centralized structure, the last two information patterns are called *decentralized*.

Based on whatever information is available, each vehicle chooses u_i to move from x_i to $x_i + u_i$. The questions that we might consider here include:

- Is there a strategy that will restore unit spacing between the vehicles?
- If not, is there a strategy which minimizes the mean square relative position error?

$$\mathrm{E} \sum_{i=1}^{N-1} (x_{i+1} - x_i)^2$$

- Can we trade-off the position error with the mean square distance traveled?

$$\mathrm{E}\sum_{i=1}^{N-1}(x_{i+1}-x_i)^2+\lambda\sum_{i=1}^{N}u_i^2 \tag{6.1}$$

- Are such optimal policies linear, and can we easily compute them?

With a centralized information structure, answers to these types of questions are straightforward, and optimal policies can be readily obtained. However, under decentralized information constraints, these questions become far more difficult to answer. We will return to answer these questions later in this chapter.

6.1.2 Witsenhausen's Counterexample

To illustrate some of the difficulties surrounding decentralized problems, which were alluded to in the previous example, consider the following example problem posed by Witsenhausen in 1968 [21]. The block diagram in Figure 6.2 admits controllers of the form

$$u_1=\gamma_1(x_0) \qquad u_2=\gamma_2(x_1+v)$$

where x_0 and v are independent random inputs with a zero mean Gaussian distribution. Thus, this problem has a decentralized information structure, because player 1 and player 2 have access to different measurements. The objective is to minimize the cost function

$$\mathrm{E}(k^2u_1^2+x_2^2)$$

In other words, player 1 attempts to keep u_1 small while player 2 attempts to estimate x_1 given a corrupted signal of it. In the case where $v=0$, we see that player 2 can perfectly estimate x_1, regardless of player 1's decision; this implies that the cost function can be made equal to zero. However, when noise is present at v, the problem is far more difficult.

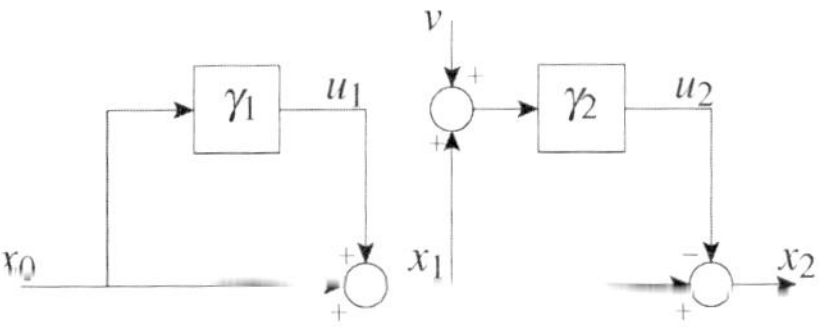

Fig. 6.2 Witsenhausen's Counterexample

Without going into the details of this problem, it can be shown that this problem admits nonlinear controllers γ_1, γ_2 whose cost is less than that of any possible linear controllers. In other words, every linear pair γ_1, γ_2 is a strictly suboptimal policy. Moreover, the optimal nonlinear policy is still unknown.

This example is a classical result which demonstrates the difficulties arising from decentralization. Thus, the question becomes: Is every decentralized control problem difficult to solve? If not, under what conditions can decentralized control problems be solved? These are the questions that we focus on in this chapter.

6.2 Static Problems

Let us return to the vehicle spacing problem considered in Section 6.1.1. Suppose there are 4 vehicles, and 3 sensor outputs

$$\begin{aligned} y_1 &= x_2 - x_1 + w_1 \\ y_2 &= x_3 - x_2 + w_2 \\ y_3 &= x_4 - x_3 + w_3 \end{aligned}$$

Let us restrict our attention to finding the best linear policy for the cost function in (6.1), where each vehicle only receives information about its neighbors. We will later show that such a linear policy is in fact optimal. Linear controllers correspond to matrices K, such that

$$u = Ky$$

In this setting, decentralized linear controllers correspond to sparse matrices of the form

$$\begin{bmatrix} u_1 \\ u_2 \\ u_3 \\ u_4 \end{bmatrix} = \begin{bmatrix} K_{11} & & \\ K_{21} & K_{22} & \\ & K_{32} & K_{33} \\ & & K_{43} \end{bmatrix} \begin{bmatrix} y_1 \\ y_2 \\ y_3 \end{bmatrix}$$

The set of matrices with this sparsity structure is a subspace S, called the *information constraint* . Letting H be the matrix

$$H = \begin{bmatrix} -1 & 1 & & & \\ & -1 & 1 & & \\ & & & \ddots & \\ & & & -1 & 1 \end{bmatrix}$$

this problem can be represented by the block diagram in Figure 6.3. Here, $e = H(x + u)$ is the final position error, $y = Hx + w$ are the sensor outputs, and $u = Ky$ are the position updates.

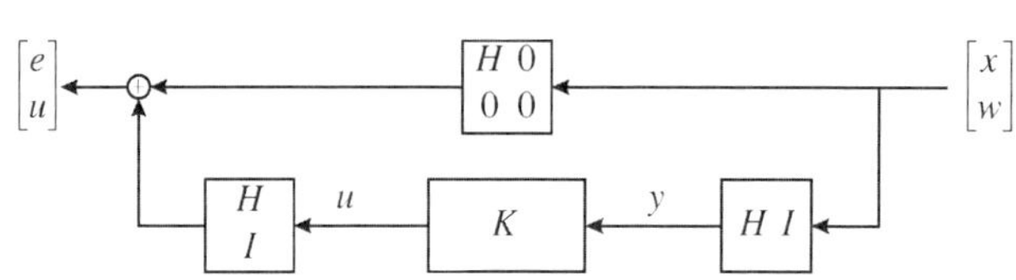

Fig. 6.3 Vehicle Block Diagram

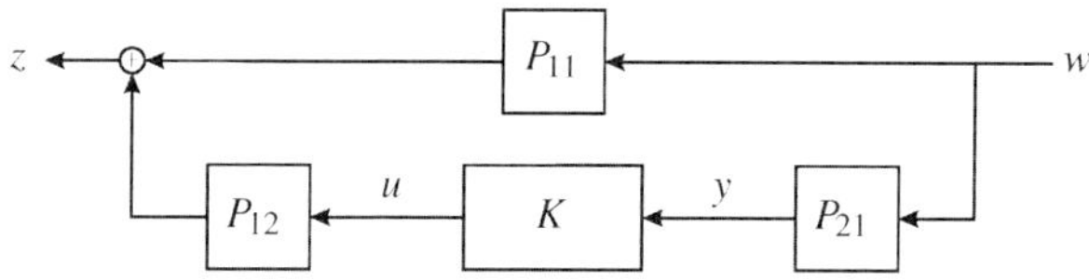

Fig. 6.4 General Static Problem Structure

The problem considered here is a special case of a more general class of problems, depicted in Figure 6.4. In this general framework, we assume that w is Gaussian, and we attempt to minimize the mean square norm of z (that is, $\mathrm{E}\|z\|^2$), subject to the constraint $K \in S$. Here, P_{11}, P_{12}, P_{21} are matrices, though, in general, they could be more general operators.

From the block diagram, we have

$$z = (P_{11} + P_{12}KP_{21})w$$

If $w \sim \mathcal{N}(0, I)$, then the mean square norm of z is

$$\mathrm{E}\|z\|^2 = \|P_{11} + P_{12}KP_{21}\|_F^2$$

where the Frobenius norm satisfies

$$\|A\|_F^2 = \sum_{i,j} A_{ij}^2 = \operatorname{trace} AA^T$$

Consequently, the optimization of the general one-step decentralized control problem can be written as

$$\begin{array}{ll} \text{minimize} & \|P_{11} + P_{12}KP_{21}\|_F^2 \\ \text{subject to} & K \in S \end{array} \tag{6.2}$$

Note that this is a linearly constrained convex quadratic program. As a result, there exist optimization techniques which can efficiently solve this problem, even when K has millions of variables [3].

Lemma 1. *Suppose P_{11}, P_{12}, P_{21} are matrices of appropriate dimensions, and S is a subspace. Define A and b so that*

$$A = P_{21}^T \otimes P_{12} \qquad b = -\operatorname{vec} P_{11}$$

and choose C such that

$$C \operatorname{vec} K = 0$$

if and only if $K \in S$. Then, K is optimal in (6.2) if and only if

$$x = \operatorname{vec} K$$

and x, λ satisfy

$$\begin{bmatrix} A^T A & C^T \\ C & 0 \end{bmatrix} \begin{bmatrix} x \\ \lambda \end{bmatrix} = \begin{bmatrix} A^T b \\ 0 \end{bmatrix}$$

Proof. Using the definition of the Frobenius norm, (6.2) is equivalent to

$$\begin{array}{ll} \text{minimize} & \|A \operatorname{vec} K - b\|_2^2 \\ \text{subject to} & C \operatorname{vec} K = 0 \end{array}$$

The result follows from the KKT conditions for this problem. ∎

6.2.1 Solution of the Multi-vehicle Problem

Returning to our sample problem of Section 6.1.1, suppose that x and w are independent Gaussian vectors, with distributions

$$x \sim \mathcal{N}(0, \Sigma_x) \qquad w \sim \mathcal{N}(0, \Sigma_w)$$

Using the general framework above, the cost in (6.1) can be written as

$$\mathrm{E}\|e\|_2^2 + \lambda\, \mathrm{E}\|u\|_2^2 = \|P_{11} + P_{12} K P_{21}\|_F^2$$

where

$$P_{11} = \begin{bmatrix} H\Sigma_x^{\frac{1}{2}} & 0 \\ 0 & 0 \end{bmatrix} \quad P_{12} = \begin{bmatrix} H \\ \lambda^{\frac{1}{2}} I \end{bmatrix} \quad P_{21} = \begin{bmatrix} H\Sigma_x^{\frac{1}{2}} & \Sigma_w^{\frac{1}{2}} \end{bmatrix}$$

Let $\Sigma_x = 0.1I$ and assume $\Sigma_w \approx 0$, indicating a high signal-to-noise ratio. By varying the parameter λ, the optimal trade-off curves for both centralized and decentralized solutions are plotted in Figure 5(a), when there are three vehicles, and in Figure 5(b), when there are ten vehicles. In these figures, the bottom-right ends of the curves represent the *do-nothing* policies, which is why the centralized and decentralized costs are the same. Conversely, the upper left corners represent the minimum mean square error (MMSE) policies. In the three vehicle case, it is possible to achieve zero mean square error, though the decentralized policy requires more effort than

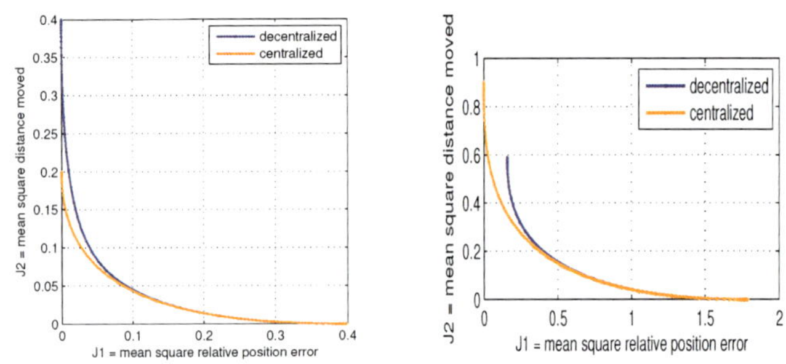

(a) Three Vehicles (b) Ten Vehicles

Fig. 6.5 Optimal Trade-off Curves

the centralized one. However, in the ten vehicle problem, no policy will achieve zero error.

This result can be formalized by noting that the mean square error can be written as

$$\mathrm{E}\|e\|_2 = \|H(KH+I)\Sigma_x^{\frac{1}{2}}\|_F^2 + \|HK\Sigma_w^{\frac{1}{2}}\|_F^2$$

Thus, in this noise-free case, the mean square error is zero if and only if

$$H(KH+I) = 0$$

The final question to address for this problem is the form of the optimal MMSE policies. Using Lemma 1, the optimal decentralized K for the three vehicle problem has the form

$$u = \begin{bmatrix} 1 & 0 \\ 0 & 0 \\ 0 & -1 \end{bmatrix} y$$

which implies that the new position of the vehicles are given by

$$q^{\text{new}} = \begin{bmatrix} 0 & 1 & 0 \\ 0 & 1 & 0 \\ 0 & 1 & 0 \end{bmatrix} q + \begin{bmatrix} -1 \\ 0 \\ 1 \end{bmatrix} + Kw$$

Hence, in the optimal MMSE policy, vehicle 2 does not move and vehicles 1 and 3 position themselves one unit away from 2. Contrast this policy with the centralized solution wherein vehicle 2 moves to the centroid of the three vehicles and the other vehicles position themselves relative to 2.

As the number of vehicles increases, it can be shown that the optimal decentralized policy tends toward positioning the vehicles at

$$q_i^{\text{new}} = \frac{1}{10}(3q_{i-1} + 4q_i + 3q_{i+1}) + \text{noise term}.$$

6.2.2 Nonlinear Policies

Let us generalize these results for the one-step decentralized problem by considering nonlinear policies. Suppose we have a two-player problem where each player receives information y_i and uses the policy $\gamma_i(y_i)$ to make decision u_i. The objective function that they must minimize is of the form

$$J = \int c(\gamma_1(y_1), \gamma_2(y_2), x)\, d\mu$$

where $c(u_1, u_2, x)$ is a convex quadratic function and μ is a probability density on x, y_1, y_2. Thus, we let

$$c(u_1,u_2,x) = \begin{bmatrix} u_1 \\ u_2 \\ x \end{bmatrix} \begin{bmatrix} M_{11} & M_{12} & M_{13} \\ M_{12}^T & M_{22} & M_{23} \\ M_{13}^T & M_{23}^T & M_{33} \end{bmatrix} \begin{bmatrix} u_1 \\ u_2 \\ x \end{bmatrix}$$

where $M \geq 0$.

Theorem 2. *Suppose $M \geq 0$, and F_i is the set of Borel measurable functions mapping $\mathbb{R}^{p_i} \to \mathbb{R}^{m_i}$ for $i = 1,2$. Let $J : F_1 \times F_2 \to \mathbb{R}$ satisfy*

$$J(\gamma_1,\gamma_2) = \int \begin{bmatrix} \gamma_1(y_1) \\ \gamma_2(y_2) \\ x \end{bmatrix}^T \begin{bmatrix} M_{11} & M_{12} & M_{13} \\ M_{12}^T & M_{22} & M_{23} \\ M_{13}^T & M_{23}^T & M_{33} \end{bmatrix} \begin{bmatrix} \gamma_1(y_1) \\ \gamma_2(y_2) \\ x \end{bmatrix} d\mu$$

where μ is a probability distribution of x, y_1, y_2. Then, (γ_1^*, γ_2^*) satisfies

$$J(\gamma_1^*,\gamma_2^*) \leq J(\gamma_1,\gamma_2) \qquad \text{for all } \gamma_1 \in F_1, \gamma_2 \in F_2$$

if and only if

$$M_{11}\gamma_1^*(y_1) = -M_{13}\,\mathrm{E}(x|y_1) - M_{12}\,\mathrm{E}(\gamma_2^*(y_2)|y_1) \tag{6.3}$$

$$M_{22}\gamma_2^*(y_2) = -M_{23}\,\mathrm{E}(x|y_2) - M_{12}^T\,\mathrm{E}(\gamma_1^*(y_1)|y_2) \tag{6.4}$$

Proof. An outline of the proof is as follows. Since J is convex, and sufficiently smooth, optimal points are those at which the directional derivatives are zero. Proceeding formally, to find player 1's optimal policy, we take a functional derivative.

$$\begin{aligned} J(\gamma_1 + h\Gamma, \gamma_2) &= J(\gamma_1,\gamma_2) \\ &\quad + 2h \int \Gamma(y_1)^T \big(M_{11}\gamma_1(y_1) + M_{12}\gamma_2(y_2) + M_{13}x\big) \\ &\qquad\qquad \times \mathrm{Prob}(x,y_1,y_2)\,dx\,dy_1 dy_2 \\ &\quad + h^2 \ldots \\ &= J(\gamma_1,\gamma_2) \\ &\quad + 2h \int \Gamma(y_1)^T \big(M_{11}\gamma_1(y_1) + M_{12}\,\mathrm{E}(\gamma_2(y_2)|y_1) + M_{13}\,\mathrm{E}(x|y_1)\big) \\ &\qquad\qquad \times \mathrm{Prob}(y_1)\,dy_1 \\ &\quad + h^2 \ldots \end{aligned}$$

Consequently, the functional derivative is the term linear in h above. Thus, player 1's policy is optimal if and only if

$$\int \Gamma(y_1)^T \big(M_{11}\gamma_1(y_1) + M_{12}\,\mathrm{E}(\gamma_2(y_2)|y_1) + M_{13}\,\mathrm{E}(x|y_1)\big)\,\mathrm{Prob}(y_1)\,dy_1 = 0$$

for all $\Gamma \in F_1$, which holds if and only if

$$M_{11}\gamma_1(y_1) = -M_{12}\,\mathrm{E}(\gamma_2(y_2)|y_1) - M_{13}\,\mathrm{E}(x|y_1)$$

The same argument, applied to player 2, produces (6.4). ■

Notice in Theorem 2 that the optimal strategies involve estimating the decision that the other player will make. This is commonly referred to as *second-guessing* .

Consider the following problem: two players are provided with some measurements y_1 and y_2 of a random variable x. Their goal is then to choose numbers u_1 and u_2 which minimize

$$\mathrm{E}(u_1 + u_2 - x)^2$$

That is, they try to pick values whose sum is close to x. Clearly, in the centralized case, this problem is trivial. However, in the decentralized case in which player 1 only has access to y_1 and player 2 only has access to y_2, what is the optimal strategy? Is there an optimal linear strategy? We can answer that with the following lemma.

Lemma 3. *Suppose J satisfies the conditions of Theorem 2, where*

$$M = \begin{bmatrix} I & I & -I \\ I & I & -I \\ -I & -I & I \end{bmatrix}$$

Also, suppose that y_1 and y_2 satisfy

$$\begin{aligned} y_1 &= A_1 x + w_1 \\ y_2 &= A_2 x + w_2 \end{aligned}$$

where x, w_1, w_2 are independent random vectors with normal distributions

$$\begin{aligned} x &\sim \mathcal{N}(0, \Sigma_x) & \Sigma_x &\geq 0 \\ w_i &\sim \mathcal{N}(0, \Sigma_{w_i}) & \Sigma_{w_i} &> 0 \qquad i = 1, 2 \end{aligned}$$

Then, the optimal γ_1, γ_2 which minimizes J is linear.

Proof. From the definitions of y_1 and y_2, it is straightforward to show that

$$\begin{aligned} \mathrm{E}(x|y_i) &= H_i y_i \qquad i = 1, 2 \\ \mathrm{E}(y_2|y_1) &= A_2 H_1 y_1 \\ \mathrm{E}(y_1|y_2) &= A_1 H_2 y_2 \end{aligned}$$

where

$$H_i = \Sigma_x A_i^T (A_i \Sigma_x A_i^T + \Sigma_{w_i})^{-1} \qquad i = 1, 2$$

From Theorem 2, the optimality conditions become

$$\begin{aligned} \gamma_1(y_1) &= \mathrm{E}(x|y_1) - \mathrm{E}(\gamma_2(y_2)|y_1) \\ \gamma_2(y_2) &= \mathrm{E}(x|y_2) - \mathrm{E}(\gamma_1(y_1)|y_2) \end{aligned}$$

We will now show that linear controllers satisfy these equations. Suppose that γ_1, γ_2 satisfy

$$\gamma_1(y_1) = K_1 y_1 \qquad \gamma_2(y_2) = K_2 y_2$$

Substituting into the optimality conditions yields

$$K_1 y_1 = H_1 y_1 - K_2 A_2 H_1 y_1$$
$$K_2 y_2 = H_2 y_2 - K_1 A_1 H_2 y_2$$

Since these expressions are true for every y_1, y_2, this implies that

$$\begin{bmatrix} K_1 & K_2 \end{bmatrix} \begin{bmatrix} I & A_1 H_2 \\ A_2 H_1 & I \end{bmatrix} = \begin{bmatrix} H_1 & H_2 \end{bmatrix}$$

or, equivalently,

$$\begin{bmatrix} K_1 & K_2 \end{bmatrix} \left(\begin{bmatrix} \Sigma_{w_1} & 0 \\ 0 & \Sigma_{w_2} \end{bmatrix} + \begin{bmatrix} A_1 \\ A_2 \end{bmatrix} \Sigma_x \begin{bmatrix} A_1^T & A_2^T \end{bmatrix} \right) = \Sigma_x \begin{bmatrix} A_1^T A_2^T \end{bmatrix} \tag{6.5}$$

Since the matrix on the left side is invertible, then there exist matrices K_1, K_2 which satisfy the optimality conditions. Hence, there exist optimal linear policies γ_1, γ_2. ■

Let us briefly examine the optimal controller obtained in (6.5). It can be shown that the optimal K in this problem is in fact the optimal centralized MMSE policy, which has the form

$$u = \mathrm{E}(x \mid y_1, y_2) = K_1 y_1 + K_2 y_2$$

Thus, since the objective of our decentralized problem is to estimate x with the sum of u_1 and u_2, the optimal decentralized policy is to let $u_1 = K_1 y_1$ and $u_2 = K_2 y_2$.

Notice of course that Lemma 3 applies more generally [8]. In fact, this same method of proof can be used to show that our multiple vehicle problem has an optimal linear solution, as we had assumed. For static problems in general, optimal decentralized controllers satisfy some type of second-guessing criterion. In most cases, linear controllers may be found, via convex optimization, which satisfy these optimality conditions.

6.3 Dynamic Problems

In Section 6.2, we only considered static/one-step problems; that is, problems in which players simultaneously make a single decision, and the cost is some function of the resulting state and the decisions. These problems had the nice feature that they could be written as convex optimization problems which involved sparsity constraints on the controllers.

Unfortunately, the dynamic case, which we will now discuss, is considerably more difficult. What makes it harder is the addition of feedback in the dynamics; that is, the decisions made by some players may affect the measurements that other players receive. The general framework looks like Figure 6.6. Contrast this block diagram with Figure 6.4 for the static problem. Thus, we see that the static problem is a special case of the dynamic problem where $P_{22} = 0$.

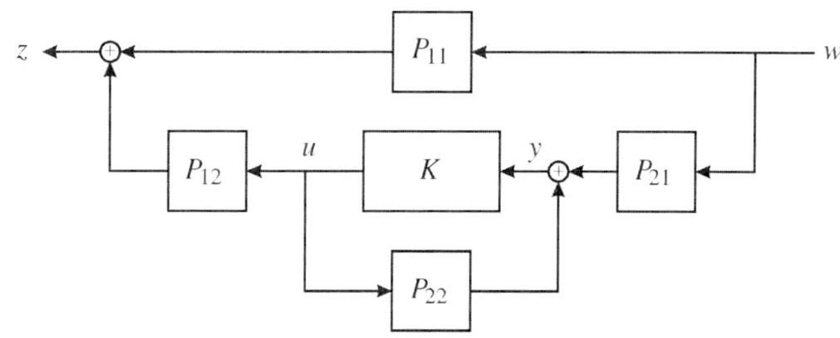

Fig. 6.6 General Feedback Structure

In this feedback framework, the mapping $w \mapsto z$ may be written as

$$z = \left(P_{11} + P_{12}K(I - P_{22}K)^{-1}P_{21}\right)w$$

whenever $(I - P_{22}K)^{-1}$ exists. This is the well-posedness condition, and it is sufficient to assume that P_{22} is strictly proper. As a result, the control problem that we wish to solve is

$$\begin{array}{ll} \text{minimize} & \|P_{11} + P_{12}K(I - P_{22}K)^{-1}P_{21}\| \\ \text{subject to} & K \text{ is stabilizing} \\ & K \in S \end{array} \tag{6.6}$$

where S is a subspace representing the decentralized information structure.

Some example subspaces would be block diagonal constraints, so that $K \in S$ if and only if

$$K = \begin{bmatrix} K_1 & & & \\ & K_2 & & \\ & & K_3 & \\ & & & K_4 \end{bmatrix}$$

Here, S would be considered fully decentralized, where each player makes decisions based solely on their own measurements. Another type of subspace S might involve both sparsity and delay constraints; for instance, requiring K to be of the form

$$K = \begin{bmatrix} K_{11} & DK_{12} & & \\ DK_{21} & K_{22} & DK_{23} & \\ & DK_{32} & K_{33} & DK_{34} \\ & & DK_{43} & K_{44} \end{bmatrix}$$

where D is a delay operator. In this case, S allows communication between neighboring players with a communication delay of D.

Note that in (6.6), if the constraint $K \in S$ is removed, the problem becomes centralized, and solutions can be readily obtained.

However, with the subspace constraint in place, no known solutions exist for the general problem. The Witsenhausen counterexample of Section 6.1.2 is sufficient evidence of this. As a result, much research in decentralized control is aimed toward characterizing cases where solutions to (6.6) can be found [2, 9, 5, 6, 7, 1, 20].

For convenience, let $G = P_{22}$. Suppose that P is stable, and define

$$h(K) = -K(I - GK)^{-1}$$

This mapping is bijective. Hence, using the change of variables, $Q = h(K)$, results in the equivalent problem

$$\begin{aligned} \text{minimize} \quad & \|P_{11} - P_{12}QP_{21}\| \\ \text{subject to} \quad & Q \text{ is stable} \\ & h(Q) \in S \end{aligned} \tag{6.7}$$

Although the objective function is now affine in Q and every stable Q produces a stabilizing K, the constraint $h(Q) \in S$ is no longer convex.

6.3.1 Quadratic Invariance

Since this is a large area of open research, we will restrict our attention to a class of systems which are shown to admit convex problems in (6.7). This set of systems, which has been shown to be quite general, are classified by a property known as *quadratic invariance* [13] .

A subspace S is called quadratically invariant under G if

$$KGK \in S \qquad \text{for all } K \in S$$

The result can be stated in the following theorem.

Theorem 4. *Suppose*

- *$\mathscr{U}$ and $\mathscr{Y}$ are Banach spaces*
- *$G : \mathscr{U} \to \mathscr{Y}$ is compact*
- *$S \subset L(\mathscr{Y}, \mathscr{U})$ is a closed subspace*
- *Let $M = \{K \in L(\mathscr{Y}, \mathscr{U})\ ;\ (I - GK) \text{ is invertible}\}$*

Then the subspace S is quadratically invariant under G if and only if

$$h(S \cap M) = S \cap M$$

Proof. We provide an outline of the proof [14]. If S is quadratically invariant, then we'll show that for all $K \in S$ and all $n \geq 0$,

$$K(GK)^n \in S$$

By definition, this holds for $n = 0, 1$. Hence, suppose it holds for some $n \geq 1$, and consider the fact that

$$2K(GK)^{n+1} = \big(K + K(GK)^n\big)G\big(K + K(GK)^n\big) - KGK - K(GK)^n GK(GK)^n$$

Since S is a subspace, each term on the right is in S, and consequently, $K(GK)^{n+1} \in S$. By induction this holds for all n.

Now, consider the resolvent set $\rho(GK) = \{\lambda \in \mathbb{C} \mid (\lambda I - GK) \text{ is invertible}\}$. For sufficiently large $\lambda \in \mathbb{C}$, $\lambda \in \rho(GK)$ and

$$K\left(I - \frac{GK}{\lambda}\right)^{-1} = K + \frac{KGK}{\lambda} + \frac{K(GK)^2}{\lambda^2} + \dots$$

Since each term $K(GK)^n \in S$ and S is a closed subspace, we have

$$K(\lambda I - GK)^{-1} \in S$$

Without going into the details, there exists an analytic function $q : \rho(GK) \to \mathbb{C}$ which satisfies $q(\lambda) = 0$ whenever $K(\lambda I - GK)^{-1} \in S$. Since $q(\lambda) = 0$ for sufficiently large λ, it follows that $q(\lambda) = 0$ for all λ in the unbounded connected component of $\rho(GK)$. Since G is compact, GK has a countable spectrum, and the result follows. ∎

Using Theorem 4, if S is quadratically invariant under G, then (6.7) is equivalent to

$$\begin{aligned} \text{minimize} \quad & \|P_{11} - P_{12}QP_{21}\| \\ \text{subject to} \quad & Q \text{ is stable} \\ & Q \in S \end{aligned} \tag{6.8}$$

This is now a convex program and can be efficiently solved. Note that quadratic invariance is an algebraic condition; hence, it applies to continuous and discrete time systems as well as any norm. It also applies to systems represented by transfer functions though some additional technical conditions and a slightly different proof are needed. In addition, it can be shown that systems which are quadratically invariant have optimal linear solutions [10]; hence, the convex problem above can be solved to find a linear policy which is optimal with respect to every nonlinear policy. More recently, quadratic invariance has also been extended to unstable systems and nonlinear systems [12].

Quadratic invariance is a powerful result, since it unifies and generalizes many known classes of tractable problems which had previously been discovered. Such constraint classes include systems on strings and arrays, group symmetries, Hermitian symmetries, one-step-delay problems, funnel-causal systems, and partially-nested systems. We will highlight a few of these results in the following sections.

6.3.2 Skyline Information Structures

Quadratic invariance is one of the most general classes of problems which is currently known. To illustrate the scope of this class, we will provide some examples of quadratically invariant systems.

Suppose that G and S have the following sparsity patterns

$$G \sim \begin{bmatrix} \bullet & \circ & \circ & \circ & \circ \\ \bullet & \bullet & \circ & \circ & \circ \\ \bullet & \bullet & \bullet & \circ & \circ \\ \bullet & \bullet & \bullet & \bullet & \circ \\ \bullet & \bullet & \bullet & \bullet & \bullet \end{bmatrix} \qquad S = \left\{ K \mid K \sim \begin{bmatrix} \circ & \circ & \circ & \circ & \circ \\ \circ & \bullet & \circ & \circ & \circ \\ \circ & \bullet & \circ & \circ & \circ \\ \bullet & \bullet & \bullet & \circ & \circ \\ \bullet & \bullet & \bullet & \circ & \bullet \end{bmatrix} \right\}$$

where the empty circles represent those elements of the matrix constrained to be zero. This is known as a vertical skyline information structure. For a finite horizon problem, this corresponds to any causal plant and a controller that receives measurements which are not in the same order that they were made. This would correspond to problems involving multiple sensors with different delays, or systems with packet delays/drops. In problems of this nature, S is quadratically under G since

$$KGK \sim \begin{bmatrix} \circ & \circ & \circ & \circ & \circ \\ \circ & \bullet & \circ & \circ & \circ \\ \circ & \bullet & \circ & \circ & \circ \\ \bullet & \bullet & \bullet & \circ & \circ \\ \bullet & \bullet & \bullet & \circ & \bullet \end{bmatrix}$$

Hence, we can find solutions to problems of this form via convex optimization (6.8). In an analogous fashion, we can show that a horizontal skyline information structure is also quadratically invariant under this G. A sample horizontal skyline structure for S might be

$$S = \left\{ K \mid K \sim \begin{bmatrix} \bullet & \circ & \circ & \circ & \circ \\ \bullet & \bullet & \circ & \circ & \circ \\ \bullet & \circ & \circ & \circ & \circ \\ \bullet & \bullet & \bullet & \circ & \circ \\ \bullet & \bullet & \bullet & \bullet & \bullet \end{bmatrix} \right\}$$

Such a system is an interesting case in which measurements can be forgotten in later time steps.

To illustrate quadratic invariance in the vertical skyline case, consider the following numerical example [12]. Suppose G is given by

$$G(s) = \begin{bmatrix} \frac{1}{s+1} & 0 & 0 & 0 & 0 \\ \frac{1}{s+1} & \frac{1}{s-1} & 0 & 0 & 0 \\ \frac{1}{s+1} & \frac{1}{s-1} & \frac{1}{s+1} & 0 & 0 \\ \frac{1}{s+1} & \frac{1}{s-1} & \frac{1}{s+1} & \frac{1}{s+1} & 0 \\ \frac{1}{s+1} & \frac{1}{s-1} & \frac{1}{s+1} & \frac{1}{s+1} & \frac{1}{s-1} \end{bmatrix}$$

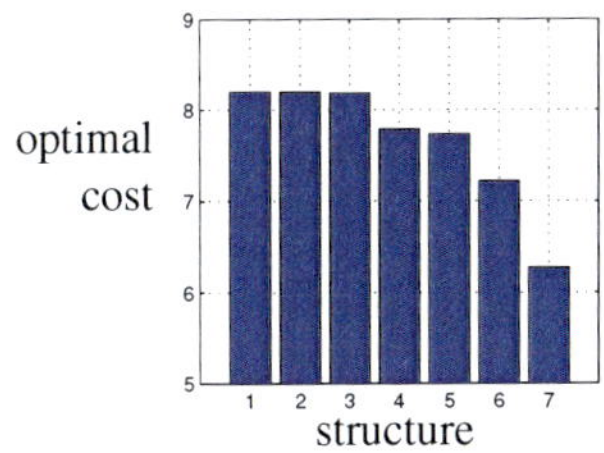

Fig. 6.7 Optimal Vertical Skyline Costs

Let the P matrices be defined as follows.

$$P_{11} = \begin{bmatrix} G & 0 \\ 0 & 0 \end{bmatrix} \qquad P_{12} = \begin{bmatrix} G \\ I \end{bmatrix} \qquad P_{21} = \begin{bmatrix} G & I \end{bmatrix} \qquad P_{22} = G$$

Suppose we have the following sequence of quadratically invariant sparsity structures

$$\begin{bmatrix} \circ & \circ & \circ & \circ & \circ \\ \circ & \bullet & \circ & \circ & \circ \\ \circ & \bullet & \circ & \circ & \circ \\ \circ & \bullet & \circ & \circ & \circ \\ \circ & \bullet & \circ & \circ & \bullet \end{bmatrix} \begin{bmatrix} \circ & \circ & \circ & \circ & \circ \\ \circ & \bullet & \circ & \circ & \circ \\ \circ & \bullet & \circ & \circ & \circ \\ \circ & \bullet & \circ & \circ & \circ \\ \bullet & \bullet & \circ & \circ & \bullet \end{bmatrix} \begin{bmatrix} \circ & \circ & \circ & \circ & \circ \\ \circ & \bullet & \circ & \circ & \circ \\ \circ & \bullet & \circ & \circ & \circ \\ \bullet & \bullet & \circ & \circ & \circ \\ \bullet & \bullet & \circ & \circ & \bullet \end{bmatrix} \begin{bmatrix} \circ & \circ & \circ & \circ & \circ \\ \circ & \bullet & \circ & \circ & \circ \\ \circ & \bullet & \circ & \circ & \circ \\ \bullet & \bullet & \circ & \circ & \circ \\ \bullet & \bullet & \bullet & \circ & \bullet \end{bmatrix} \begin{bmatrix} \circ & \circ & \circ & \circ & \circ \\ \circ & \bullet & \circ & \circ & \circ \\ \circ & \bullet & \circ & \circ & \circ \\ \bullet & \bullet & \bullet & \circ & \circ \\ \bullet & \bullet & \bullet & \circ & \bullet \end{bmatrix} \begin{bmatrix} \bullet & \circ & \circ & \circ & \circ \\ \bullet & \bullet & \circ & \circ & \circ \\ \bullet & \bullet & \bullet & \circ & \circ \\ \bullet & \bullet & \bullet & \bullet & \circ \\ \bullet & \bullet & \bullet & \bullet & \bullet \end{bmatrix}$$

Notice that these information structures become increasingly more informative; that is, there is more information for making decisions with each successive structure. By solving the resulting convex optimization problem (6.8) for each structure, we obtain the following histogram of optimal $\mathscr{H}_2$ costs. As expected, as the amount of information increases, the optimal cost drops.

6.3.3 Control of Networks

As another example, consider the networked control problem shown in Figure 6.8. In this system, there are three subsystems connected in a line. The plants are connected with a propagation delay of p, and the controllers are restricted to communicate with a delay of c. In other words, G and K have the following structures.

$$G = \begin{bmatrix} * & D_p* & D_{2p}* \\ D_p* & * & D_p* \\ D_{2p}* & D_p* & * \end{bmatrix} \qquad K = \begin{bmatrix} * & D_c* & D_{2c}* \\ D_c* & * & D_c* \\ D_{2c}* & D_c* & * \end{bmatrix}$$

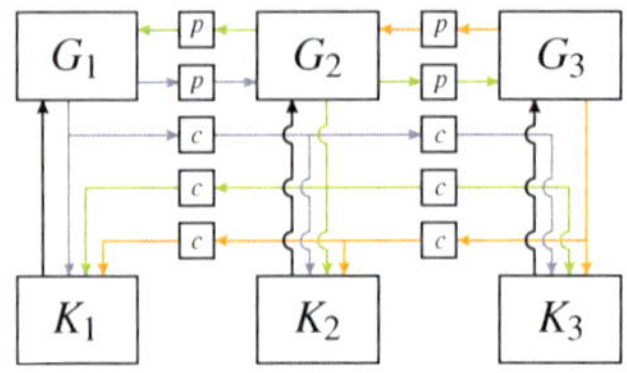

Fig. 6.8 Sample Network Problem

Consequently, if $c \leq p$, then this system is quadratically invariant since

$$KGK = \begin{bmatrix} * & D_r* & D_{2r}* \\ D_r* & * & D_r* \\ D_{2r}* & D_r* & * \end{bmatrix}$$

where $r = \min\{c, p\}$. As an example, consider the finite horizon problem with $p = 2$ and $c = 1$. This problem has G and K with the following sparsity structures.

$$G \sim \left[\begin{array}{cccc|cccc|cccc}
\bullet & \circ & \circ & \circ & \circ & \circ & \circ & \circ & \circ & \circ & \circ & \circ \\
\bullet & \bullet & \circ & \circ & \circ & \circ & \circ & \circ & \circ & \circ & \circ & \circ \\
\bullet & \bullet & \bullet & \circ & \bullet & \circ & \circ & \circ & \circ & \circ & \circ & \circ \\
\bullet & \bullet & \bullet & \bullet & \bullet & \bullet & \circ & \circ & \circ & \circ & \circ & \circ \\
\hline
\circ & \circ & \circ & \circ & \bullet & \circ & \circ & \circ & \circ & \circ & \circ & \circ \\
\circ & \circ & \circ & \circ & \bullet & \bullet & \circ & \circ & \circ & \circ & \circ & \circ \\
\bullet & \circ & \circ & \circ & \bullet & \bullet & \bullet & \circ & \bullet & \circ & \circ & \circ \\
\bullet & \bullet & \circ & \circ & \bullet & \bullet & \bullet & \bullet & \bullet & \bullet & \circ & \circ \\
\hline
\circ & \circ & \circ & \circ & \circ & \circ & \circ & \circ & \bullet & \circ & \circ & \circ \\
\circ & \circ & \circ & \circ & \circ & \circ & \circ & \circ & \bullet & \bullet & \circ & \circ \\
\circ & \circ & \circ & \circ & \bullet & \circ & \circ & \circ & \bullet & \bullet & \bullet & \circ \\
\circ & \circ & \circ & \circ & \bullet & \bullet & \circ & \circ & \bullet & \bullet & \bullet & \bullet
\end{array}\right]
\quad
K \sim \left[\begin{array}{cccc|cccc|cccc}
\bullet & \circ & \circ & \circ & \circ & \circ & \circ & \circ & \circ & \circ & \circ & \circ \\
\bullet & \bullet & \circ & \circ & \bullet & \circ & \circ & \circ & \circ & \circ & \circ & \circ \\
\bullet & \bullet & \bullet & \circ & \bullet & \bullet & \circ & \circ & \bullet & \circ & \circ & \circ \\
\bullet & \bullet & \bullet & \bullet & \bullet & \bullet & \bullet & \circ & \bullet & \bullet & \circ & \circ \\
\hline
\circ & \circ & \circ & \circ & \bullet & \circ & \circ & \circ & \circ & \circ & \circ & \circ \\
\bullet & \circ & \circ & \circ & \bullet & \bullet & \circ & \circ & \bullet & \circ & \circ & \circ \\
\bullet & \bullet & \circ & \circ & \bullet & \bullet & \bullet & \circ & \bullet & \bullet & \circ & \circ \\
\bullet & \bullet & \bullet & \circ & \bullet & \bullet & \bullet & \bullet & \bullet & \bullet & \bullet & \circ \\
\hline
\circ & \circ & \circ & \circ & \circ & \circ & \circ & \circ & \bullet & \circ & \circ & \circ \\
\circ & \circ & \circ & \circ & \bullet & \circ & \circ & \circ & \bullet & \bullet & \circ & \circ \\
\bullet & \circ & \circ & \circ & \bullet & \bullet & \circ & \circ & \bullet & \bullet & \bullet & \circ \\
\bullet & \bullet & \circ & \circ & \bullet & \bullet & \bullet & \circ & \bullet & \bullet & \bullet & \bullet
\end{array}\right]$$

As a result, we see that KGK has the same structure as K; hence, S is quadratically invariant under G.

These results apply to more general networks [11], like the lattice network in Figure 6.9. In this lattice network, the system is quadratically invariant if $c \leq p$.

In general networked systems, there are two underlying graph structures, one representing the interconnection of the plants, and the other representing the allowable communication channels between controllers. Ignoring delays, it can be shown that the controller subspace is quadratically invariant if and only if the transitive closure of the plant graph is contained in the transitive closure of the controller graph [17]. Here, the term *transitive closure* refers to the set of paths of any length in the graph. Interestingly, this does not require the two graphs to be equal. For instance, consider Figure 6.10. These graph structures satisfy the conditions for quadratic invariance, implying that optimal controllers communicating with the graph structure in Figure 6.10 can be found.

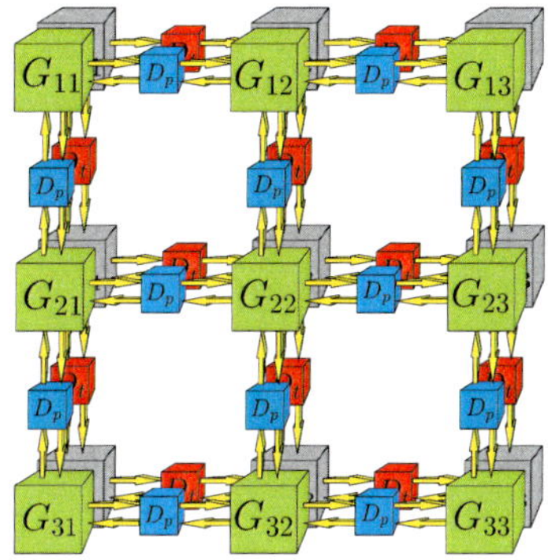

Fig. 6.9 Two-dimensional Lattice

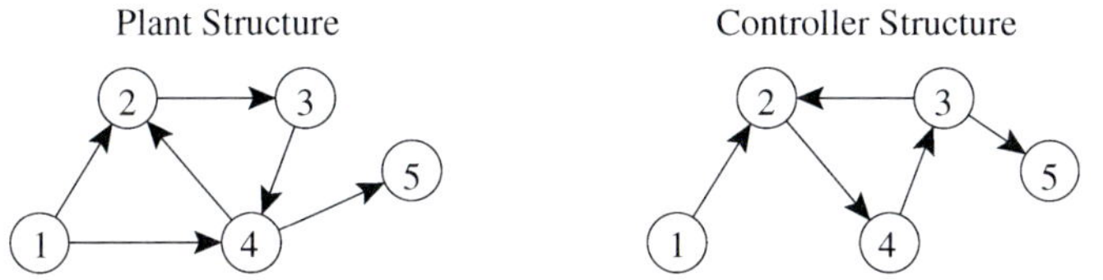

Fig. 6.10 Multiple Graphs Example

6.3.4 Non-convex Systems

In contrast to the above quadratically invariant systems, consider the following systems which remain intractable. Suppose G and S have the following sparsity structure.

$$G \sim \begin{bmatrix} \circ & \circ & \circ \\ \bullet & \circ & \circ \\ \circ & \bullet & \bullet \end{bmatrix} \qquad S = \left\{ K \mid K \sim \begin{bmatrix} \circ & \circ & \circ \\ \bullet & \circ & \circ \\ \circ & \bullet & \bullet \end{bmatrix} \right\}$$

Direct computation shows that

$$KGK \sim \begin{bmatrix} \circ & \circ & \circ \\ \circ & \circ & \circ \\ \bullet & \bullet & \bullet \end{bmatrix}$$

which implies that the system is not quadratically invariant, despite the plant and controller having similar sparsity structures. Consequently, the optimization problem remains non-convex.

In should be clear by now that Witsenhausen's counterexample is not a quadratically invariant system; otherwise, it would have a known solution. In addition, considering the networked control problems of the previous section, suppose that the controllers are fully decentralized; that is, there is no communication between controllers. From the results of Section 6.3.3, this implies that this system is quadratically invariant if and only if the plants are also decoupled, so that the system reduces to independent subsystems. Thus, this supports the well-known fact that decentralized problems, in general, are intractable.

6.3.5 Unstable Plants

Before concluding our discussion on quadratic invariance, we note that the results obtained above can also be applied when the plant is unstable [12].

In this case, one requires any stable and stabilizing controller $K_{\text{nom}} \in S$. Then, using the following change of variables

$$K = K_{\text{nom}} + Q(I - GK_{\text{nom}} + GQ)^{-1}(I - GK_{\text{nom}})$$

and assuming that S is quadratically invariant under G, the optimization problem (6.6) may be equivalently posed as

$$\begin{array}{ll} \text{minimize} & \|T_1 - T_2QT_3\| \\ \text{subject to} & Q \text{ is stable} \\ & Q \in S \end{array}$$

where T_1, T_2, T_3 are constructed from P and K_{nom}.

6.4 Solving the Optimization Problem

While quadratic invariance achieves a convex formulation in (6.8) for previously intractable problems, these convex problems are infinite-dimensional. Thus, actually computing an optimal solution for (6.8) is a non-trivial step.

Several methods exist to find optimal solutions to this problem. The parameter Q itself is a linear stable system, and we can parameterize it via a basis for the impulse response function. In other words, this is similar to solving for the first n coefficients to the impulse response function. Computing a solution in this manner is efficient since it is now a finite-dimensional optimization problem. However, there exist some drawbacks to this approach. Since we must work in a finite basis, the solution is inherently approximate. Moreover, to obtain high precision requires very high-dimensional optimization problems. Lastly, any structure or interpretation of the optimal controller is lost; in particular, determining the dimension of the optimal controller is impossible. Nevertheless, this method is most commonly used due to its efficiency and generality.

Dynamic programming approaches have also been considered [16]. While these have the advantage of being efficient and provide intuition to the optimal solutions, few decentralized control problems seem to lend themselves nicely to solution with this method. Nevertheless, some approximation schemes have been suggested via dynamic programming [4].

Semidefinite programming is another alternative approach. While this approach is efficient and finite-dimensional, it still lacks analytic results which provide important insights to the structure of the optimal controllers. Moreover, only a few special cases have yet been solved with this approach [22, 15, 9].

Another approach has already been alluded to in Lemma 1. That is, in the $\mathscr{H}_2$ optimization problem, the objective function may be *vectorized* to achieve a linearly constrained least-squares type of problem [10]. In fact, in many cases the constraint set may be subsumed into the objective function resulting in an unconstrained $\mathscr{H}_2$ optimization problem, for which well-known solutions exist. The advantage here is that the resulting solution is optimal. However, like the basis parametrization method, the resulting problem can be very high-dimensional, and loses the structure associated with the problem.

This is in contrast to the centralized case, for which there are explicit state-space formulae. Such formulae offer the practical advantages of computational reliability and simplicity, as well as providing understanding and interpretation of the controller structure, such as the separation into controller and estimator as well as determining the state dimension of the optimal controller. It is an open area of research to find general state-space formulae for quadratically invariant systems. One promising approach has been the method of spectral factorization, which we now describe.

6.4.1 Spectral Factorization

The common disadvantage to most techniques described above is the loss of intrinsic structure in the solution. By this, we mean that these approaches seek to solve the optimization problem (6.8) element by element in the parameter Q, rather than optimize over elements in S. This is a subtle point which might be made more clear by an example. Consider the problem (6.8) where S is the set of lower triangular matrices. By vectorizing this problem, one can arrive at the equivalent optimization problem

$$\text{minimize} \qquad \|b - Ax\|_2$$

Here, x is the vectorization of $Q \in S$, where we've dropped all elements of Q constrained to be zero by eliminating columns of $P_{21}^T \otimes P_{12}$ to construct A. While the resulting least-squares problem may now be efficiently solved, reconstructing Q from the optimal x is a very difficult problem, even in this centralized example.

In spectral factorization, we recognize that there exists an optimality condition which must be satisfied for the $\mathscr{H}_2$ optimization problem (6.8), which is given by

$$P_{12}^T P_{11} P_{21}^T + P_{12}^T P_{12} Q P_{21} P_{21}^T \in S^\perp \tag{6.9}$$

where $S^\perp$ is the set of strictly upper triangular matrices in this example. Visually, this equation has the following form.

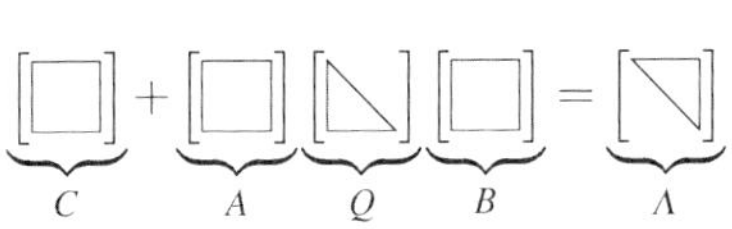

where Λ is some matrix in $S^{\perp}$. Of course, this optimality condition can be solved via vectorization in the same manner as above. However, we can solve for $Q \in S$ as a whole by recognizing that A may be factorized into an upper triangular matrix times a lower triangular matrix, $A = U_A L_A$, and similarly for $B = L_B U_B$. Consequently, we can preserve the structure of (6.9) by multiplying it on the left by U_A^{-1} and on the right by U_B^{-1}. The resulting equation now looks like

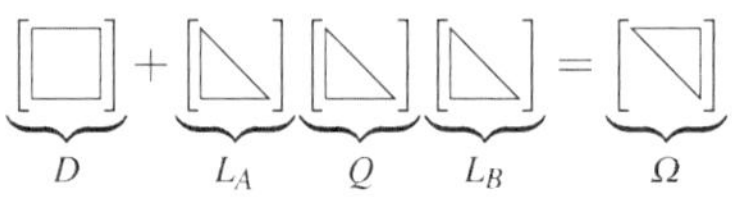

where $\Omega \in S^{\perp}$. Now, by partitioning D into its lower and strictly upper triangular components, $D = L_D + U_D$, we can directly solve for $Q \in S$ as

$$Q = -L_A^{-1} L_D L_B^{-1}$$

Notice that the resulting Q is lower triangular, as desired, and can be computed analytically. At no point during this solution was the structure of Q altered.

Of course, this example is simply a centralized problem. Nevertheless, it highlights some of the key features of spectral factorization. Moreover, recent study has shown that this method can be applied to many decentralized problems.

Consider, for instance, the skylinestructures of Section 6.3.2. In this case, though the constraint set S has changed, the optimality condition may still be expressed as (6.9). While triangular factorizations may no longer work in this setting, there exist other factorizations which allow us to solve for Q while preserving its structure [18].

Similarly, the networked control problems of Section 6.3.3 lend themselves nicely to spectral factorization schemes. Preliminary results have further demonstrated the importance of this method, since the optimal controllers have been shown to separate nicely into control and estimation problems, and the state dimension of the optimal controllers may be obtained [19].

6.4.2 Solution of the Two-Player Problem

To illustrate the effectiveness of the spectral factorization approach described above on a networked control problem, consider the following two-player system in Figure 6.11. In this problem, player 1 may communicate to player 2, but not vice versa. The resulting sparsity constraint is that $K \in S$ if and only if $K_{12} = 0$. The overall dynamics are given by

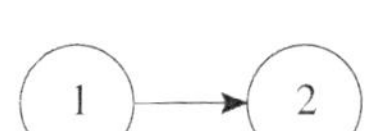

Fig. 6.11 Two-player System

$$\begin{bmatrix} x_1(t+1) \\ x_2(t+1) \end{bmatrix} = \begin{bmatrix} A_{11} & 0 \\ A_{21} & A_{22} \end{bmatrix} \begin{bmatrix} x_1(t) \\ x_2(t) \end{bmatrix} + \begin{bmatrix} B_{11} & 0 \\ B_{21} & B_{22} \end{bmatrix} \begin{bmatrix} u_1(t) \\ u_2(t) \end{bmatrix} + v(t)$$

where x_i, u_i are the state and inputs of player i, and v is exogenous noise. Note that the dynamics of player 1 may propagate to player 2's dynamics, but not vice versa. This represents the simplest possible decentralized and quadratically invariant network structure. By using a spectral factorization approach and without going into too many details, the optimal solutions may be expressed as follows [19].

- Controller 1 has realization

$$\begin{aligned} q_1(t+1) &= A^K q_1(t) + B^K x_1(t) \\ u_1(t) &= -K_{12} q_1(t) - K_{11} x_1(t) \end{aligned}$$

- Controller 2 has realization

$$\begin{aligned} q_2(t+1) &= A^K q_2(t) + B^K x_1(t) \\ u_2(t) &= (J - K_{22}) q_2(t) - K_{21} x_1(t) - J x_2(t) \end{aligned}$$

where K and J are constructed from the solutions of two *different* Riccati equations, and

$$\begin{aligned} A^K &= A_{22} - B_{21} K_{12} - B_{22} K_{22} \\ B^K &= A_{21} - B_{21} K_{11} - B_{22} K_{21} \end{aligned}$$

With the inclusion of q_1 and q_2, the optimal controller is not a static gain, despite the fact that we have state feedback in each subsystem and player two has complete state information. Contrast this result with the classical LQR controller in which the optimal centralized controller would be the static gain K. In fact, both controllers have dynamics, and each has the same number of states as system 2.

It can be shown that q_1 and q_2 in the optimal controllers are in fact the minimum-mean square error estimate of x_2 given the history of x_1's. In other words, letting $\eta(t) = \mathrm{E}\big(x_2(t) \mid x_1(t), \ldots, x_1(0)\big)$ the optimal control policy can be written as

$$\begin{aligned} u_1(t) &= -K_{11} x_1(t) - K_{12} \eta(t) \\ u_2(t) &= -K_{21} x_1(t) - K_{22} \eta(t) + J\big(\eta(t) - x_2(t)\big) \end{aligned}$$

Thus, the optimal policy is, in fact, attempting to perform the optimal centralized policy, though using η instead of x_2. However, there is an additional term in u_2 which represents the error between x_2 and its estimate η. We also see that in the case where x_2 is deterministic, so that $\eta = x_2$, then the optimal distributed controller reduces to the optimal centralized solution, as it should.

6.5 Summary

In this tutorial on decentralized control, we began by illustrating the types of problems that one might consider. Unfortunately, it became readily apparent that optimal decentralized control problems can be very difficult and, in some cases, intractable. As a result, our goal became one of characterizing the types of problems that could be solved efficiently.

To this end, we first discussed static control problems. We were able to show that static decentralized control could be solved via convex optimization, and that linear optimal controllers could be found in most cases. Thus, we generally view static problems as solvable.

In progressing to dynamic problems involving feedback, we discovered that the resulting optimization problems were not convex in general, and the goal became classifying the tractable problems. Perhaps the most general class of problems currently found to admit convex synthesis are those systems which are quadratically invariant. When systems are quadratically invariant, we can show that the optimal solution is linear and may be found via a convex program. Several examples of important problems which are quadratically invariant were provided to illustrate the generality of this result.

Lastly, it was argued that controller synthesis does not end with the formulation of the convex program, since this problem is often infinite-dimensional. Though many techniques are proposed to solve for the optimal controllers, the method of spectral factorization has been shown to provide additional benefits, including computational reliability, insight into the control and estimation structure, and explicit state-space formulae. The solution to an example problem was provided to highlight these advantages.

References

1. Bamieh, B., Voulgaris, P.G.: Optimal distributed control with distributed delayed measurements. In: Proceedings of the IFAC World Congress (2002)
2. Blondel, V.D., Tsitsiklis, J.N.: A survey of computational complexity results in systems and control. Automatica 36(9), 1249–1274 (2000)
3. Boyd, S., Vandenberghe, L.: Convex Optimization. Cambridge University Press, Cambridge (2004)
4. Gupta, V.: Distributed Estimation and Control in Networked Systems. PhD thesis, California Institute of Technology (2006)
5. Ho, Y.-C., Chu, K.C.: Team decision theory and information structures in optimal control problems – Part I. IEEE Transactions on Automatic Control 17(1), 15–22 (1972)
6. Sandell Jr., N., Athans, M.: Solution of some nonclassical LQG stochastic decision problems. IEEE Transactions on Automatic Control 19(2), 108–116 (1974)
7. Qi, X., Salapaka, M., Voulgaris, P., Khammash, M.: Structured optimal and robust control with multiple criteria: A convex solution. IEEE Transactions on Automatic Control 49(10), 1623–1640 (2004)
8. Radner, R.: Team decision problems. Annals of mathematical statistics 33, 857–881 (1962)
9. Rantzer, A.: Linear quadratic team theory revisited. In: Proceedings of American Control Conference, pp. 1637–1641 (June 2006)
10. Rotkowitz, M.: Information structures preserved under nonlinear time-varying feedback. In: Proceedings of the American Control Conference, pp. 4207–4212 (2006)

11. Rotkowitz, M., Cogill, R., Lall, S.: A simple condition for the convexity of optimal control over networks with delays. In: Proceedings of the IEEE Conference on Decision and Control, pp. 6686–6691 (2005)
12. Rotkowitz, M., Lall, S.: A characterization of convex problems in decentralized control. IEEE Transactions on Automatic Control 51(2), 274–286 (2002)
13. Rotkowitz, M., Lall, S.: Decentralized control information structures preserved under feedback. In: Proceedings of the IEEE Conference on Decision and Control, pp. 569–575 (2002)
14. Rotkowitz, M., Lall, S.: Affine controller parameterization for decentralized control over banach spaces. IEEE Transactions on Automatic Control 51(9), 1497–1500 (2006)
15. Scherer, C.W.: Structured finite-dimensional controller design by convex optimization. Linear Algebra and its Applications 351(352), 639–669 (2002)
16. Swigart, J., Lall, S.: Dynamic programming with non-classical information structures. In: Proceedings of Mathematical Theory of Networks and Systems, pp. 457–462 (2008)
17. Swigart, J., Lall, S.: A graph-theoretic approach to distributed control over networks. In: Proceedings of the IEEE Conference on Decision and Control (2009)
18. Swigart, J., Lall, S.: Spectral factorization of non-classical information structures under feedback. In: Proceedings of the American Control Conference, pp. 457–462 (2009)
19. Swigart, J., Lall, S.: An explicit state-space solution for a decentralized two-player optimal linear-quadratic regulator. Submitted to American Control Conference, pp. 457–462 (2010)
20. Vouglaris, P.: Control of nested systems. In: Proceedings of the American Control Conference, vol. 6, pp. 4442–4445 (2000)
21. Witsenhausen, H.S.: A counterexample in stochastic optimum control. SIAM Journal of Control 6(1), 131–147 (1968)
22. Zelazo, D., Mesbahi, M.: $\mathscr{H}_2$ analysis and synthesis of networked dynamic systems. In: Proceedings of the American Control Conference, pp. 2966–2971 (2009)

Chapter 7
Stability and Stabilization of Networked Control Systems

W.P.M.H. Heemels and N. van de Wouw

Abstract. The presence of a communication network in a control loop induces many imperfections such as varying transmission delays, varying sampling/transmission intervals, packet loss, communication constraints and quantization effects, which can degrade the control performance significantly and can even lead to instability. Various techniques have been proposed in the literature for stability analysis and controller design for these so-called networked control systems. The aim of this chapter is to survey the main research lines in a comprehensive manner.

7.1 Introduction

Networked control systems (NCSs) have received considerable attention in recent years. The interest for NCSs is motivated by many benefits they offer such as the ease of maintenance and installation, the large flexibility and the low cost. However, still many issues need to be resolved before all the advantages of wired and wireless networked control systems can be harvested. Part of the solution will be formed by improvements of the employed communication networks and protocols, resulting in increased reliability and reduction of the end-to-end latencies and packet dropouts. However, the solution cannot be obtained in a cost-effective manner by only improving the communication infrastructure. It is important to take a system's perspective to overcome these problems and also develop control algorithms that can deal with communication imperfections and constraints. This latter aspect is

W.P.M.H. Heemels
Eindhoven University of Technology, Department of Mechanical Engineering, P.O. Box 513, 5600 MB Eindhoven, The Netherlands
e-mail: M.Heemels@tue.nl

N. van de Wouw
Eindhoven University of Technology, Department of Mechanical Engineering, P.O. Box 513, 5600 MB Eindhoven, The Netherlands
e-mail: N.v.d.Wouw@tue.nl

A. Bemporad, M. Heemels, and M. Johansson: Networked Control Systems, LNCIS 406, pp. 203–253.
springerlink.com

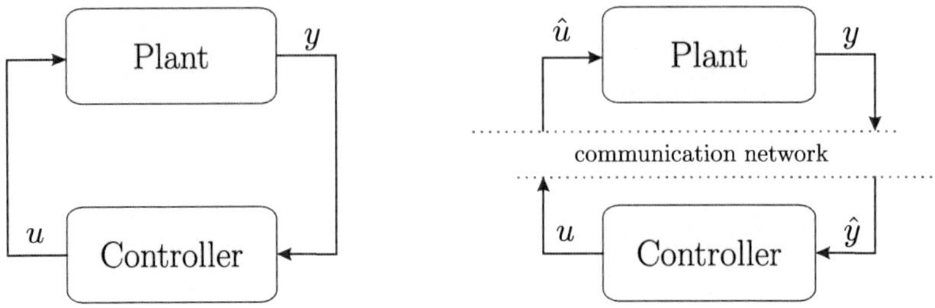

Fig. 7.1 Introduction of a network in a control loop

recognized widely in the control community, as evidenced by the many publications appearing recently, see, *e.g.*, the survey papers [40, 102, 83, 98].

Roughly speaking, the network-induced imperfections and constraints can be categorized in five types:

(i) Variable sampling/transmission intervals;
(ii) Variable communication delays;
(iii) Packet dropouts caused by the unreliability of the network;
(iv) Communication constraints caused by the sharing of the network by multiple nodes and the fact that only one node is allowed to transmit its packet per transmission;
(v) Quantization errors in the signals transmitted over the network due to the finite word length of the packets.

Basically, the introduction of a communication network in a control loop (see Figure 7.1) modifies the external signals (u,y) of the plant and the controller due to these five imperfections. Indeed, the control input $\hat{u}$ going into the plant is no longer equal to the output u of the controller, and the measured output of the plant y is not exactly known by the controller that only has access to a 'networked' version $\hat{y}$ of this output. Each of the imperfections has its own particular effect on the network-induced differences $e_y := \hat{y} - y$ and $e_u := \hat{u} - u$. Obviously, the presence of these network phenomena can degrade the performance of the control loop significantly and can even lead to instability, see, *e.g.*, [10, 14] for an illustrative example. Therefore, it is of importance to understand how these phenomena influence the closed-loop stability and performance properties, preferably in a quantitative manner. Since in any practical communication network all aforementioned network-induced imperfections are present, there is a need for analysis and synthesis methods including all these imperfections. This is especially of importance, considering that the design of a NCS often requires tradeoffs between the different types. For instance, reducing quantization errors (and thus transmitting larger or more packets) typically results in larger transmission delays. To support the designers in making these tradeoffs in the design of the complete NCS (plant, controller and network) in an integral fashion, tools are needed that provide quantitative information on the consequences of each of the possible choices in plant, controller and network design.

Although the NCS field is relatively young, various major research lines are already appearing these days. Unfortunately, much of the available literature on NCS considers only some of above mentioned types of network phenomena, while ignoring the other types. The available results need to be extended and integrated to obtain a framework in which all the network-induced imperfections can be studied simultaneously and tradeoffs can be made. This chapter has the aim to provide an overview of the rapidly growing literature on NCS with a focus on methods for stability analysis that incorporate several of the above mentioned communication imperfections. To a lesser extent we will also discuss the stabilization problem. As such, this chapter strives to form the basis for further research that eventually leads to a practically useful analysis and design framework for control over communication networks.

7.2 Overview of Existing Approaches

A categorization of the available literature on stability analysis of NCSs can be done, firstly, on the basis of the types of network-induced imperfections considered (time-varying sampling intervals, time-varying delays, packet dropouts, communication constraints and quantization as mentioned in the introduction), see Section 7.2.1, and, secondly, on the modeling and analysis approach adopted to study the stability of the NCS under these network-induced imperfections, see Section 7.2.2.

Before categorizing the existing approaches, let us start by noting that two essentially different ways exist to model network-induced uncertainties such as time-varying sampling intervals, time-varying delays and packet dropouts. The first class of models assumes (deterministic) bounds on the delays, sampling intervals and the number of subsequent packet dropouts, without adopting any further assumptions on the possibly random processes behind the generation of, *e.g.*, sequences of delays or packet drops. With some abuse of terminology, we will call this the *deterministic* approach. A second class of models exist in which information about the stochastic nature of these variables is taken into account, provided this additional information is available, which we call the *stochastic* approach. In this overview, we focus mainly on *deterministic* approaches and refer the interested reader to [56, 94, 74, 50, 75, 97] and the references therein for stochastic approaches. One observation is that the cited references for the stochastic approach at present only can handle a *finite* or *countable* number of delays or sampling intervals, while in reality this is often not the case. Fortunately, recently results are appearing that more realistically consider delays and sampling intervals as continuous random variables taking possibly an uncountable number of values, see, for instance, [55, 67, 80, 2, 1, 20].

7.2.1 *The Types of Network-Induced Phenomena*

Many systematic approaches that analyse stability of NCSs consider only one of these network-induced imperfections. Indeed, the effects of quantization are studied in [51, 82, 5, 62, 19, 34, 36], of packet dropouts in [78, 76], of time-varying transmission intervals and delays in [26, 56, 60, 3], and [10, 14, 42, 48, 59, 101, 23, 33], respectively, and of communication constraints in [17, 4, 45, 71, 47].

References that simultaneously consider two types of network-induced imperfections are given in Table 7.1. Moreover, [63] consider imperfections of type (i), (iv), (v), [58, 57, 9, 59, 11] study simultaneously type (i), (ii), (iii), [65] focuses on type (i), (iii), (iv), while [27] studies (ii), (iii) and (v). Also [37, 38, 7, 21] studies three types, namely type (i), (ii), (iv). In addition some of the approaches mentioned in Table 7.1 that study varying sampling intervals and/or varying communication delays can be extended to include type (iii) phenomena as well by modeling dropouts as prolongations of the maximal sampling interval or delay (cf. also Remark 7.18 below). Another subtle though important distinction between existing works incorporating varying delays is whether only small delays or also large delays (delays smaller or larger, respectively, than the sampling interval) are considered. In this chapter we will present methods dealing with both cases.

By recent unifications of the work in [63] and [37, 38] a framework is obtained in [35] that can model and analyze the five imperfections simultaneously. Although certain restrictive assumptions are adopted in [35] (regarding, *e.g.*, the small delay case and the usage of particular quantizers), it is the first framework that includes all five of the mentioned network-induced imperfections.

Table 7.1 References that study NCS with two network-induced imperfections simultaneously

&	(ii)	(iii)	(iv)
(i)	[86, 85, 43]		[22, 6, 66, 91, 92, 81]
(iii)	[28, 32, 12, 53, 100]	-	
(iv)		[46]	-
(v)	[52]	[84]	

7.2.2 *Different Approaches in Modeling/Analysis of NCS*

We distinguish three different approaches towards the modeling, stability analysis and controller synthesis for NCS:

1. Work on the *discrete-time* approach, see e.g [26, 28, 43, 14, 23, 11, 86, 72, 102, 95, 70, 101, 96, 93], has mainly focussed on linear NCS. The first step is to construct discrete-time representations of the sampled-data NCS system (which for linear systems can be done exactly), leading to an uncertain discrete-time system in which the uncertainties appear in an exponential form (due to discretization). The discrete-time modeling approaches can be further categorized by time-driven or event-driven models. In time-driven models the continuous-time model is integrated from sample/transmission time to the next sample/-transmission time, while in event-driven models integration is done from each event time (being control updates times, sample times, etc), see, *e.g.*, [43] for the latter. Here we will mainly focus on time-driven linear NCS models.

Next, to construct models suitable for stability analysis, polytopic overapproximation or embedding techniques are used to capture the exponential uncertainties. Various methods have been proposed to do this (some with fixed approximation error, others with tuning parameters to make the approximation more tight). The resulting polytopic models, possibly also having norm-bounded uncertainty, can then be used in a robust stability analysis, often based on linear matrix inequalities (LMIs), to guarantee the stability of the discrete-time NCS model.

The final step is to guarantee that also the intersample behavior is stable, such that stability of the true sampled-data NCS model can be concluded. This approach allows to consider discrete-time controllers, although by discretizing continuous-time controllers they can be incorporated as well. Typically, this approach is applied to NCS with linear plants and controllers since in that case *exact* discrete-time models can be derived, although recently new results have been obtained that apply to NCS with nonlinear plants and controllers based on approximate discretizations, see [87]. We will discuss the approach for "linear NCS" in more detail in Sections 7.3.2 and 7.4.2;

2. The *sampled-data* approach uses continuous-time models that describe the sampled-data NCS dynamics in the continuous-time domain (so without exploiting any form of discretization) and perform stability analysis and controller synthesis based on these sampled-data NCS models directly. Fridman et al. [25] applied a descriptor system approach to model the sampled-data dynamics of systems with varying sampling intervals in terms of (infinite-dimensional) delay-differential equations (DDEs) and study their stability based on the Lyapunov-Krasovskii functional method. In [99, 100, 27], this approach is used for the stability analysis of NCSs with time-varying delays and constant sampling intervals, using (linear) matrix inequality-based techniques. The recent results in [27] show how varying delays, quantization and dropouts can be formulated in one framework based on DDEs, and stability analysis and H_∞ control design methods, based on LMIs, are presented. However, Mirkin [54] showed that the use of such an approach for digital control systems neglects the piecewise constant nature of the control signal due to the zero-order-hold mechanism thereby introducing conservatism when exploiting such modeling for stability analysis. More specifically, the conservatism is introduced by the fact that the zero-order hold and delay jointly introduce a particular linearly increasing time-varying delay within each control update interval (sometimes indicated by the sawtooth behavior of the delay), whereas in the modeling approach mentioned above it is replaced by an arbitrary bounded time-varying delay.

   An alternative approach, proposed in [58, 60, 59, 86, 85], is based on impulsive DDEs and does take into account the piecewise constant nature of the control signal due to the zero-order-hold mechanism. It has been shown in [54] that this approach is less conservative than the descriptor approach. More specifically, the impulsive DDEs are based on introducing impulses (discontinuous updates) at the moment new information arrives at the controller or the plant. In this manner the true behavior of the underlying NCS is captured. As also noted in [60], the usage of infinite-dimensional DDE models and Lyapunov functionals to analyze the stability of essentially *finite-dimensional* sampled-data NCS does not seem to offer any advantage. The approach in [58, 86] is able to deal simultaneously with time-varying delays and time-varying sampling intervals but does not explicitly include packet dropouts in the model (although they might be considered as variations in the sampling intervals or delays). Here, we will focus mainly on the approach towards the modeling and stability analysis using impulsive DDEs. We call this the *sampled-data* approach, which allows the consideration of discrete-time controllers and nonlinear plants. However, constructive stability conditions have only been obtained for linear NCS. We will discuss this approach in Section 7.3.3;

3. In the so-called *continuous-time* or *emulation* approach, see [92, 91, 65, 38, 37, 17, 66], a continuous-time controller is designed to stabilize the continuous-time plant in the absence of network-induced imperfections. Next, the stability analysis is based on a sampled-data model of the NCS (in the form of a hybrid system) and allows to quantify the level of network-induced uncertainty (in terms of, *e.g.*, the maximal allowable sampling/transmission interval and/or maximal allowable delay) for which the NCS inherits the stability properties from the closed-loop system without the network. This approach is applicable to a wide class of nonlinear NCS, since well-developed tools for the design of (nonlinear) controllers for nonlinear plants can be employed. A drawback is the fact that the controller is formulated in continuous time, whereas for NCS one typically designs the controller in discrete time. We will discuss this approach in detail in Section 7.4.1.

Summarizing, the discrete-time approach considers discrete-time controllers (or discretized continuous-time controllers) and a discrete-time NCS model, while the sampled-data approach also considers discrete-time controllers, but has a continuous-time (sampled-data) NCS model. Finally, the continuous-time (emulation) approach focuses on continuous-time controllers using a continuous-time (sampled-data) NCS model. Within all these different approaches different techniques towards stability analysis are used. While the discrete-time approach uses common quadratic or parameter-dependent Lyapunov functions for the discrete-time model, the continuous-time (emulation) approach uses continuous-time Lyapunov functions constructed by combining separate Lyapunov functions for the network-free closed-loop system on the hand and the network protocol on the other hand (or, alternatively, adopting directly small gain arguments). The sampled-data approaches exploit extensions of Lyapunov-Krasovskii function(al)s. We will discuss these methods in details in the following sections, where we start with NCS without communication constraints and scheduling protocols (Section 7.3) and treat the case with communication constraints in Section 7.4. In this chapter, we will not pay any attention to quantization effects as these are extensively covered in the chapter in this book written by Hideaki Ishii.

7.3 NCS with Delays, Varying Sampling Intervals and Packet Loss

In Section 7.3.1, we discuss a general description of a single-loop NCS with time-varying sampling intervals, delays and packet dropouts. In Section 7.3.2, we discuss a discrete-time approach towards the modeling, stability analysis and controller design for these NCS. Finally, in Section 7.3.3, we present a continuous-time approach towards the modeling and stability analysis for these systems exploiting models in terms of impulsive DDEs.

7.3.1 Description of the NCS

In this section, we present a fairly general description of a NCS including delays larger than the uncertain and time-varying sampling intervals and packet dropouts. It is based on the developments in [11] (see also [12, 14]). We choose this level of generality for the reason that the application of the stability techniques presented later can encompass all these types of network-induced phenomena.

The NCS is depicted schematically in Figure 7.2. It consists of a linear continuous-time plant

$$\dot{x}(t) = Ax(t) + Bu^*(t) \tag{7.1}$$

with $A \in \mathbb{R}^{n \times n}$ and $B \in \mathbb{R}^{n \times m}$, and a discrete-time static time-invariant controller, which are connected over a communication network that induces network delays (τ^{sc} and τ^{ca}). The state measurements ($y(t) = x(t)$) are sampled resulting in the sampling time instants s_k given by:

$$s_0 = 0 \text{ and } s_k = \sum_{i=0}^{k-1} h_i \text{ for } k \geq 1, \tag{7.2}$$

which are non-equidistantly spaced in time due to the time-varying sampling intervals $h_k > 0$. The sequence of sampling instants $s_0, s_1, s_2, \ldots$ is strictly increasing in the sense that $s_{k+1} > s_k$, for all $k \in \mathbb{N}$. The notation $y_k := y(s_k)$ denotes the k-th sampled value of y, $x_k := x(s_k)$ the k-th sampled value of the state and u_k the control value corresponding to $y_k = x_k$. Packet drops may occur (see Figure 7.2) and are modeled by the parameter m_k. This parameter denotes whether or not a packet is dropped:

$$m_k = \begin{cases} 0, \text{ if } y_k \text{ and } u_k \text{ are received} \\ 1, \text{ if } y_k \text{ and/or } u_k \text{ is lost.} \end{cases} \tag{7.3}$$

In (7.3), we make no distinction between packet dropouts that occur in the sensor-to-controller connection and the controller-to-actuator connection in the network. This can be justified by realizing that, for static controllers, the effect of the packet dropouts on the control updates implemented on the plant is the same in both cases. Indeed, for packet dropouts between the sensor and the controller no new control

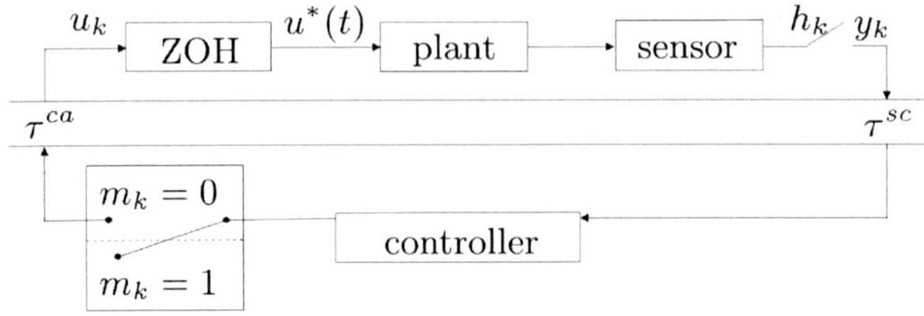

Fig. 7.2 Schematic overview of the NCS with variable sampling intervals, network delays and packet dropouts

update is computed and thus no new control input is sent to the actuator. In the case of packet dropouts between the controller and the actuator no new control update is received by the actuator either. Finally, the zero-order-hold (ZOH) function (in Figure 7.2) is applied to transform the discrete-time control value u_k to a continuous-time control input $u^*(t)$ being the actual actuation signal of the plant.

In the model, both the varying computation time (τ_k^c), needed to evaluate the controller, and the network-induced delays, *i.e.* the sensor-to-controller delay (τ_k^{sc}) and the controller-to-actuator delay (τ_k^{ca}), are taken into account. The sensor is assumed to act in a time-driven fashion (*i.e.*, sampling occurs at the times s_k defined in (7.2)) and both the controller and the actuator act in an event-driven fashion (*i.e.*, responding instantaneously to newly arrived data). Furthermore, we consider that not all the data is used due to packet dropouts and message rejection, *i.e.* the effect that more recent control data is available before older arrives and therefore the older data is neglected. Under these assumptions, all three delays can be captured by a single delay $\tau_k := \tau_k^{sc} + \tau_k^c + \tau_k^{ca}$, see also [68, 102]. To include these effects in the continuous-time model, define the parameter $k^*(t)$ that denotes the index of the most recent control input that is available at time t as $k^*(t) := \max\{k \in \mathbb{N} | s_k + \tau_k \leq t \wedge m_k = 0\}$. The continuous-time model of the plant of the NCS is then given by

$$\dot{x}(t) = Ax(t) + Bu^*(t) \tag{7.4a}$$

$$u^*(t) = u_{k^*(t)}. \tag{7.4b}$$

Here, we assume that the most recent control input remains active at the plant if a packet is dropped. Note that in some NCS setups one also use the different policy to set the control input to 0 in case dropout occurs instead of holding the previous control value, see, *e.g.*, [73].

The delays are assumed to be bounded and contained in the interval $[\tau_{\min}, \tau_{\max}]$, the sampling interval are bounded and contained in the interval $[h_{\min}, h_{\max}]$ and the number of subsequent packet dropouts is upper bounded by $\overline{\delta}$. The latter means that

$$\sum_{v=k-\overline{\delta}}^{k} m_v \leq \overline{\delta}, \tag{7.5}$$

for all $k \in \mathbb{N}$ as this guarantees that from the control inputs $u_{k-\overline{\delta}}, u_{k-\overline{\delta}+1}, \dots, u_k$ at least one is implemented. In summary, the class $\mathscr{S}$ of admissible sequences $\{(s_k, \tau_k, m_k)\}_{k \in \mathbb{N}}$ can be described as follows:

$$\begin{aligned} \mathscr{S} := \Bigg\{ & \{(s_k, \tau_k, m_k)\}_{k\in\mathbb{N}} \,|\, h_{\min} \leq s_{k+1} - s_k \leq h_{\max}, \\ & s_0 = 0,\ \tau_{\min} \leq \tau_k \leq \tau_{\max},\ \sum_{v=k-\overline{\delta}}^{k} m_v \leq \overline{\delta}, \forall k \in \mathbb{N} \Bigg\}, \end{aligned} \tag{7.6}$$

which includes variable sampling intervals, small and large delays, and packet dropouts. Note that in this case we allow for large delays in the sense that τ_k might be larger than h_k.

7.3.2 *Discrete-Time Modeling Approaches*

7.3.2.1 The Exact Discrete-Time NCS Model

To arrive at a discrete-time description of the NCS, the equation (7.4b) of the continuous-time control input $u^*(t)$ is reformulated to indicate explicitly which control inputs u_l are active in the sampling interval $[s_k, s_{k+1})$. Such a reformulation is needed to derive the discrete-time NCS model, which will ultimately be employed in the stability analysis and controller synthesis methods.

Lemma 7.1. *Consider the continuous-time NCS as defined in* (7.4) *and the admissible sequences of sampling instants, delays, and packet dropouts as defined in* (7.6). *Define* $\underline{d} := \lfloor \frac{\tau_{\min}}{h_{\max}} \rfloor$, *the largest integer smaller than or equal to* $\frac{\tau_{\min}}{h_{\max}}$ *and* $\overline{d} := \lceil \frac{\tau_{\max}}{h_{\min}} \rceil$, *the smallest integer larger than or equal to* $\frac{\tau_{\max}}{h_{\min}}$. *Then, the control action* $u^*(t)$ *in the sampling interval* $[s_k, s_{k+1})$ *is described by*

$$u^*(t) = u_{k+j-\overline{d}-\overline{\delta}} \ \text{for}\ t \in [s_k + t_j^k, s_k + t_{j+1}^k), \tag{7.7}$$

where the actuation update instants $t_j^k \in [0, h_k]$ *are defined as*

$$\begin{aligned} t_j^k = \min\Bigg\{ & \max\{0, \tau_{k+j-\overline{d}-\overline{\delta}} - \sum_{l=k+j-\overline{\delta}-\overline{d}}^{k-1} h_l\} + m_{k+j-\overline{d}-\overline{\delta}} h_{\max}, \\ & \max\{0, \tau_{k+j-\overline{d}-\overline{\delta}+1} - \sum_{l=k+j+1-\overline{\delta}-\overline{d}}^{k-1} h_l\} + m_{k+j-\overline{d}-\overline{\delta}+1} h_{\max}, \\ & \dots, \max\{0, \tau_{k-\underline{d}} - \sum_{l=k-\underline{d}}^{k-1} h_l\} + m_{k-\underline{d}} h_{\max}, h_k \Bigg\}, \end{aligned} \tag{7.8}$$

with $t_j^k \leq t_{j+1}^k$ *and* $j \in \{0, 1, \dots, \overline{d} + \overline{\delta} - \underline{d}\}$ *(see Figure 7.3). Moreover,* $0 = t_0^k \leq t_1^k \leq \dots \leq t_{\overline{d}+\overline{\delta}-\underline{d}}^k \leq t_{\overline{d}+\overline{\delta}-\underline{d}+1}^k := h_k$.

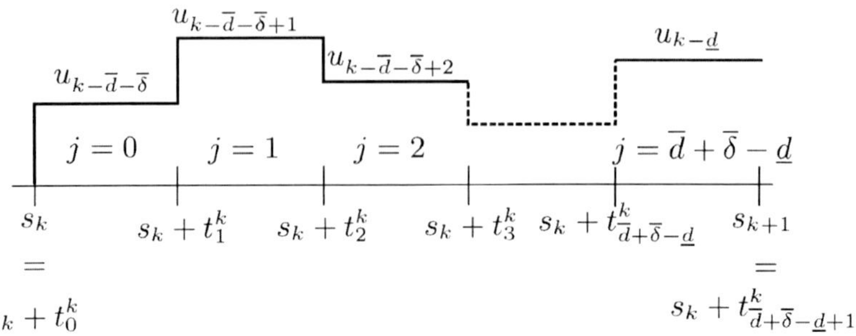

Fig. 7.3 Graphical interpretation of the actuation update instants t_j^k

Proof. The proof is given in [11], see also [14, 9]. ■

Note that the above lemma first of all indicates that the only control values that can be active in the interval $[s_k, s_{k+1}]$ are $u_{k-\overline{d}-\overline{\delta}}, \ldots, u_{k-\underline{d}}$. Secondly, (7.7) indicates that $u_{k+j-\overline{d}-\overline{\delta}}$ is active in $[s_k+t_j^k, s_k+t_{j+1}^k)$. Note that when $t_j^k = t_{j+1}^k$ this essentially means that the value $u_{k+j-\overline{d}-\overline{\delta}}$ is not active in the interval $[s_k, s_{k+1}]$ (*e.g.* due to a dropout or more recent information arriving earlier). The exact values of t_j^k are determined by the exact realization of the delays, sampling intervals and dropouts as present in the right-hand side of (7.8).

Based on Lemma 7.1, a discrete-time NCS model can be obtained now by exact integration of (7.4) leading to

$$x_{k+1} = e^{Ah_k}x_k + \sum_{j=0}^{\overline{d}+\overline{\delta}-\underline{d}} \int_{h_k-t_{j+1}^k}^{h_k-t_j^k} e^{As}ds\,B u_{k+j-\overline{d}-\overline{\delta}} \tag{7.9}$$

with t_j^k as defined in Lemma 7.1.

Let θ_k denote the vector of uncertain parameters consisting of the sampling interval and the actuation update instants

$$\theta_k := (h_k, t_1^k, \ldots, t^k_{\overline{d}+\overline{\delta}-\underline{d}}). \tag{7.10}$$

Using now the lifted state vector

$$\xi_k = \begin{pmatrix} x_k^T & u_{k-1}^T & \cdots & u_{k-\overline{d}-\overline{\delta}}^T \end{pmatrix}^T$$

that includes the current system state and past system inputs, we obtain the lifted model

$$\xi_{k+1} = \tilde{A}(\theta_k)\xi_k + \tilde{B}(\theta_k)u_k, \tag{7.11}$$

where

$$
\tilde{A}(\theta_k) = \begin{pmatrix} \Lambda(\theta_k) & M_{\overline{d}+\overline{\delta}-1}(\theta_k) & M_{\overline{d}+\overline{\delta}-2}(\theta_k) & \dots & M_1(\theta_k) & M_0(\theta_k) \\ 0 & 0 & 0 & \dots & 0 & 0 \\ 0 & I & 0 & \dots & 0 & 0 \\ \vdots & \ddots & \ddots & \ddots & \vdots & \vdots \\ \vdots & & \ddots & \ddots & 0 & 0 \\ 0 & \dots & \dots & 0 & I & 0 \end{pmatrix}
$$

and

$$
\tilde{B}(\theta_k) = \begin{pmatrix} M_{\overline{d}+\overline{\delta}}(\theta_k) \\ I \\ 0 \\ \vdots \\ 0 \end{pmatrix}
$$

with $\Lambda(\theta_k) = e^{Ah_k}$ and

$$
M_j(\theta_k) = \begin{cases} \int_{h_k - t_{j+1}^k}^{h_k - t_j^k} e^{As} ds B & \text{if } 0 \le j \le \overline{d} + \overline{\delta} - \underline{d}, \\ 0 & \text{if } \overline{d} + \overline{\delta} - \underline{d} < j \le \overline{d} + \overline{\delta}. \end{cases} \tag{7.12}
$$

Remark 7.1. Essentially, the uncertainty parameters $m_{k-\bar{d}-\bar{\delta}}, \dots, m_{k-\underline{\delta}}$ are included implicitly into the parameter θ_k using the expressions (7.8) for the actuation update times. When we will derive upper and lower bounds on t_j^k, this induces some conservatism if packet dropouts are present. However, the advantage of not including $m_{k-\bar{d}-\bar{\delta}}, \dots, m_{k-\underline{\delta}}$ explicitly in θ_k is that the number of uncertainty parameters is smaller thereby reducing the complexity of the stability analysis. Alternative models for dropouts are discussed and compared in [90] (see also Remark 7.18).

Remark 7.2. In the above model set-up a time-driven modeling paradigm (exact integration from sample instant to sample instant) was adopted. An alternative discrete-time modeling approach was proposed in [43], which uses an event-driven paradigm (integrating from event instant to event instant, where the events include sampling, updating of control values, *etc.*).

To illustrate the developments so far, let us consider the following example.

Example 7.1. The example consists of a second-order motion control example, obtained from the document printing domain. In particular, a single motor driving a roller-pair is considered, as depicted in Figure 7.4, which obeys the dynamics:

$$
\ddot{x}_s = \frac{q r_R}{J_M + q^2 J_R} u, \tag{7.13}
$$

with $J_M = 1.95 \cdot 10^{-5}\text{kgm}^2$ the inertia of the motor, $J_R = 6.5 \cdot 10^{-5}\ \text{kgm}^2$ the inertia of the roller-pair, $r_R = 14 \cdot 10^{-3}$ m the radius of the roller, $q = 0.2$ the transmission ratio between motor and upper roller, x_s the sheet position and u the motor torque.

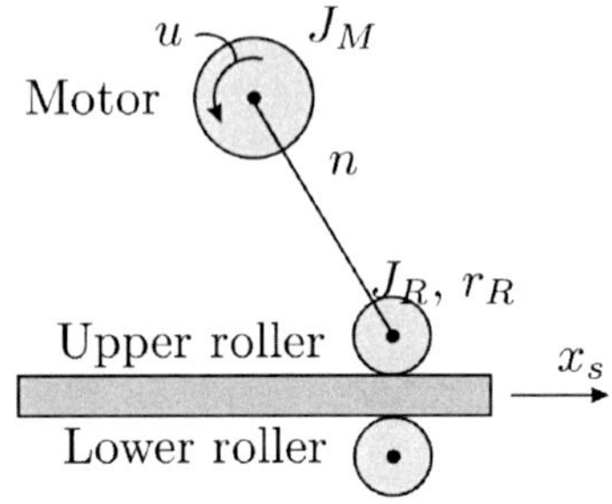

Fig. 7.4 Schematic overview of the motor-roller example

The continuous-time state-space representation of (7.13) is given by (7.1), with $A = \begin{bmatrix} 0 & 1 \\ 0 & 0 \end{bmatrix}$, $B = \begin{bmatrix} 0 \\ b \end{bmatrix}$, with $b := \frac{qr_R}{J_M + q^2 J_R} = 126.7\,(\text{kgm})^{-1}$, and $x(t) = \begin{bmatrix} x_s(t) \; \dot{x}_s(t) \end{bmatrix}^T$. For the sake of simplicity, consider the case that the sampling interval h is constant, the delays $\tau_k \in [\tau_{\min}, \tau_{\max}]$, $\forall k \in \mathbb{N}$, $\tau_{\min} = 0$ and $\tau_{\max} = h$ (*i.e.* the small delay case with $\overline{d} = 1$, $\underline{d} = 0$) and $\overline{\delta} = 0$ (no packet dropouts). The exact discrete-time model can be written in the form (7.11), where the extended discrete-time state consists of the state of the continuous-time model and the previous control action (due to the small delay): $\xi_k = \begin{bmatrix} x_k^T \; u_{k-1}^T \end{bmatrix}^T$. Moreover, the uncertain parameter $\theta_k = t_1^k = \tau_k$ (see (7.8)), since the small delay case, a constant sampling interval and no packet dropouts are considered. The (uncertain) matrices in (7.11) are given by

$$\begin{aligned} \Lambda = e^{Ah} = \begin{bmatrix} 1 & h \\ 0 & 1 \end{bmatrix}, \; M_0(\tau_k) &= \begin{bmatrix} \frac{b}{2}\left(h^2 - (h-\tau_k)^2\right) \\ b\,(h - (h-\tau_k)) \end{bmatrix}, \\ M_1(\tau_k) = \int_0^h e^{As} ds B - M_0(\tau_k) &= \begin{bmatrix} \frac{1}{2} b (h-\tau_k)^2 \\ b(h-\tau_k) \end{bmatrix} \end{aligned} \tag{7.14}$$

and the overall model (7.11) reduces to

$$\xi_{k+1} = \tilde{A}(\theta_k)\xi_k + \tilde{B}(\theta_k) u_k \tag{7.15a}$$

$$= \begin{pmatrix} \Lambda & M_0(\tau_k) \\ 0 & 0 \end{pmatrix} \xi_k + \begin{pmatrix} M_1(\tau_k) \\ 1 \end{pmatrix} u_k \tag{7.15b}$$

$$= \begin{pmatrix} 1 & h & \frac{b}{2}\left(h^2 - (h-\tau_k)^2\right) \\ 0 & 1 & b\,(h - (h-\tau_k)) \\ 0 & 0 & 0 \end{pmatrix} \xi_k + \begin{pmatrix} \frac{1}{2} b (h-\tau_k)^2 \\ b(h-\tau_k) \\ 1 \end{pmatrix} u_k. \tag{7.15c}$$

The latter equalities show clearly that we are dealing with a parameter-varying linear system, in which one can observe the basic terms $h - \tau_k$ and $(h - \tau_k)^2$ as uncertainty terms. We will now present a general procedure how to find and overapproximate these uncertainty terms such that the system becomes amendable for stability analysis and controller synthesis.

7.3.2.2 The Polytopic Overapproximation

A first step towards the stability analysis is transforming the bounds on the delays, sampling intervals and dropouts ($\tau_{\min}$, $\tau_{\max}$, $h_{\min}$, $h_{\max}$ and $\bar{\delta}$) to upper and lower bounds on t_j^k. These computations are done in [9, 11] and lead to bounds $t_{j,\min}, t_{j,\max}$, *i.e.*, $t_j^k \in [t_{j,\min}, t_{j,\max}]$ for all $k \in \mathbb{N}$, see [9, 11] for the exact expressions. Together with the fact that $h_k \in [h_{\min}, h_{\max}]$, one can define the uncertainty set

$$\Theta = \{\theta_k \in \mathbb{R}^{\overline{d}+\overline{\delta}-\underline{d}+1} \,|h_k \in [h_{\min}, h_{\max}], t_j^k \in [t_{j,\min}, t_{j,\max}],$$
$$1 \le j \le \overline{d}+\overline{\delta}-\underline{d}, 0 \le t_1^k \le \ldots \le t^k_{\overline{d}+\overline{\delta}-\underline{d}} \le h_k\}, \quad (7.16)$$

such that $\theta_k \in \Theta$ for all $k \in \mathbb{N}$.

The stability analysis for the uncertain system (7.11) with the uncertainty parameter $\theta_k \in \Theta$ (given a discrete-time controller such as a lifted state feedback $u_k = -K\xi_k$) is now essentially a *robust* stability analysis problem. The obstruction to apply various robust stability techniques directly is that the uncertainty appears in an *exponential* fashion as observed from the form of $M_j(\theta_k)$ and $\Lambda(\theta_k)$. To render the formulation (7.11) amendable for robust stability analysis, overapproximation techniques can be employed to embed the original model (as tight as possible) in a "larger" model that has useful structural properties such as discrete-time polytopic models with (or without) additional norm-bounded uncertainties. If robust stability (or other properties) can be proven for this polytopic overapproximation, then this also implies the robust stability of the original discrete-time NCS model. As these polytopic models are suitable for the application of available *robust* stability methods, this provides a means to tackle the NCS stability analysis problem.

In the literature, many different ways of constructing such polytopic embeddings of the uncertain system are proposed: overapproximation techniques are based on interval matrices [10], the real Jordan form [11, 14, 13, 12, 69], the Taylor series [42], gridding and norm-bounding [72, 26, 79, 77, 22], and the Cayley-Hamilton theorem [29, 30]. There are also some approaches that are related to gridding and norm-bounding such as [3, 23] in which essentially one grid point is taken corresponding to one nominal sampling interval or nominal delay and the variation of sampling intervals/delays is captured in the norm-bounded uncertainties. Typically, [72, 26, 79, 77, 22] use more grid points to reduce the size of the norm-bounded uncertainties and thereby the conservatism of the overapproximation and resulting stability conditions. For the sake of brevity, we will only discuss one of these overapproximation techniques to illustrate the main ideas. We opt here to use the real Jordan form approach as adopted in [11, 14, 13, 12]. For a comprehensive overview and comparison of these overapproximation techniques we refer the interested reader to [39].

Real Jordan Form

To derive the stability analysis and control synthesis conditions, the model (7.11) is rewritten using the real Jordan form [44] of the continuous-time system matrix A.

Basically, the state matrix is expressed as $A = TJT^{-1}$ with J the real Jordan form and T an invertible matrix. This leads to a generic model of the form

$$\xi_{k+1} = \left(F_0 + \sum_{i=1}^{\zeta} \alpha_i(\theta_k) F_i \right) \xi_k + \left(G_0 + \sum_{i=1}^{\zeta} \alpha_i(\theta_k) G_i \right) u_k \tag{7.17}$$

with θ_k defined in (7.10) and $\zeta = (\overline{d} + \overline{\delta} - \underline{d} + 1)\nu$ the number of time-varying functions α_i. Here, ν is the degree of the minimal polynomial $q_{\min}$ of A. Note that the minimal polynomial of A is the monic polynomial p of smallest degree that satisfies $p(A) = 0$. The minimal polynomial can be easily obtained [44] from the (complex) Jordan form. Actually, ν is equal to the sum of all the maximal dimensions of the *complex* Jordan blocks corresponding to all the *distinct* eigenvalues of A. Clearly, $\nu \leq n$, where n is the dimension of the state vector x. Note that $\nu = n$ when the geometric multiplicity of each distinct eigenvalue of A is equal to one and $\nu < n$ when the geometric multiplicity of an eigenvalue is larger than one. A typical function $\alpha_i(\theta_k)$ is of the form $(h_k - t_j^k)^l e^{\lambda(h_k - t_j^k)}$, when λ is a real eigenvalue of A, and of the form $(h_k - t_j^k)^l e^{a(h_k - t_j^k)} \cos(b(h_k - t_j^k))$ or $(h_k - t_j^k)^l e^{a(h_k - t_j^k)} \sin(b(h_k - t_j^k))$ when λ is a complex eigenvalue ($\lambda = a + bi$) of A with $l = 0, 1 \ldots, r_j$, where r_j is related to the size of the Jordan blocks corresponding to λ. For more details on the use of the real Jordan form to obtain the NCS model, the reader is referred to [39] or to Appendix B in [9].

Using bounds on the uncertain parameters $\theta_k = (h_k, t_1^k, \ldots, t_{\overline{d}+\overline{\delta}-\underline{d}}^k)$ described by the set Θ in (7.16) the set of matrix pairs

$$\mathscr{F}\mathscr{G} = \left\{ \left(F_0 + \sum_{i=1}^{\zeta} \alpha_i(\theta) F_i, G_0 + \sum_{i=1}^{\zeta} \alpha_i(\theta) G_i \right) \mid \theta \in \Theta \right\} \tag{7.18}$$

can be formulated that contains all possible matrix combinations in (7.17) and thus also in (7.11). Based on this infinite set $\mathscr{F}\mathscr{G}$ of matrices, stability analysis (for a given controller) and the design of stabilizing controllers can be carried out for the NCS (7.4). To overcome the infinite dimension of the set $\mathscr{F}\mathscr{G}$, a *polytopic overapproximation* of the set is used. Denote the maximum and minimum value of $\alpha_i(\theta_k)$, respectively, by

$$\overline{\alpha}_i = \max_{\theta_k \in \Theta} \alpha_i(\theta_k), \ \underline{\alpha}_i = \min_{\theta_k \in \Theta} \alpha_i(\theta_k) \tag{7.19}$$

with Θ defined in (7.16). Then the set of matrices $\mathscr{F}\mathscr{G}$, given in (7.18), is a subset of $co(\mathscr{H}_{FG})$, where 'co' denotes the convex hull and $\mathscr{H}_{FG}$ is the *finite* set of matrix pairs given by

$$\mathscr{H}_{FG} = \left\{ \left((F_0 + \sum_{i=1}^{\zeta} \alpha_i F_i), (G_0 + \sum_{i=1}^{\zeta} \alpha_i G_i) \right) : \alpha_i \in \{\underline{\alpha}_i, \overline{\alpha}_i\}, i = 1, 2, \ldots, \zeta \right\}. \tag{7.20}$$

The set of vertices $\mathscr{H}_{FG}$ is written as $\mathscr{H}_{FG} = \{(H_{F,j}, H_{G,j}) \mid j = 1,2,\ldots,2^{\zeta}\}$ for enumeration purposes later. Hence, we have that

$$\mathscr{F}\mathscr{G} \subseteq co(\mathscr{H}_{FG}) := \{\sum_{j=1}^{\zeta} \beta_j (H_{F,j}, H_{G,j}) \mid \beta = (\beta_1, \ldots, \beta_{\zeta})^T \in \mathbb{B}\}, \tag{7.21}$$

where

$$\mathbb{B} := \{\beta \in \mathbb{R}^{\zeta} \mid \sum_{j=1}^{\zeta} \beta_j = 1 \text{ and } \beta_j \geq 0 \text{ for all } j = 1, \ldots, \zeta\}. \tag{7.22}$$

Hence, we obtain the polytopic system

$$\xi_{k+1} = (F_0 + \sum_{j=1}^{\zeta} \beta_j^k F_j)\xi_k + (G_0 + \sum_{j=1}^{\zeta} \beta_j^k G_j)u_k \tag{7.23}$$

with $\beta^k \in \mathbb{B}$ for each $k \in \mathbb{N}$. Due to (7.21) any input/state trajectory generated by (7.11) for some sequence $\{\theta_k\}_{k\in\mathbb{N}}$ with $\theta_k \in \Theta$, $k \in \mathbb{N}$ is also an input/state trajectory of (7.23) for some sequence $\{\beta^k\}_{k\in\mathbb{N}}$ with $\beta^k \in \mathbb{B}$, $k \in \mathbb{N}$.

Example 7.2. Revisit the motion control system as in Example 7.1. We take $h = 0.001$ constant, $\tau_{\min} = 0$, $\tau_{\max} = h$ and $\overline{\delta} = 0$, which leads to the exact discrete-time representation as given by (7.15). Let us now illustrate the procedure for convex overapproximation based on the real Jordan form, as explicated above, using this example. The exact discrete-time model (7.15) can be written in the form (7.17), with the uncertain functions $\alpha_1(\tau_k) = h - \tau_k$ and $\alpha_2(\tau_k) = (h - \tau_k)^2$, and

$$\begin{aligned} F_0 &= \begin{bmatrix} 1 & h & \frac{b}{2}h^2 \\ 0 & 1 & bh \\ 0 & 0 & 0 \end{bmatrix}, \quad G_0 = \begin{bmatrix} 0 \\ 0 \\ 1 \end{bmatrix} \\ F_1 &= \begin{bmatrix} 0 & 0 & 0 \\ 0 & 0 & -b \\ 0 & 0 & 0 \end{bmatrix}, \quad G_1 = \begin{bmatrix} 0 \\ b \\ 0 \end{bmatrix} \\ F_2 &= \begin{bmatrix} 0 & 0 & -\frac{1}{2}b \\ 0 & 0 & 0 \\ 0 & 0 & 0 \end{bmatrix}, \quad G_2 = \begin{bmatrix} \frac{1}{2}b \\ 0 \\ 0 \end{bmatrix}. \end{aligned} \tag{7.24}$$

Hence, the number of uncertain functions α_i is $\zeta = 2$.

In order to illustrate the conservatism introduced by the overapproximation of the set of matrices $\mathscr{F}\mathscr{G}$ in (7.18) by the convex hull of the set of matrices $\mathscr{H}_{FG}$ in (7.20), it is important to realize that the only uncertain matrix in the discrete-time system (7.15) is the matrix (2×1)-matrix $M_0(\tau_k)$ given in (7.14). Namely, $M_1(\tau_k)$ in (7.15) can be written as $M_1(\tau_k) = \int_0^h e^{As} ds B - M_0(\tau_k)$, see (7.14). The matrix $M_0(\tau_k)$ can be written as follows in terms of the uncertain functions:

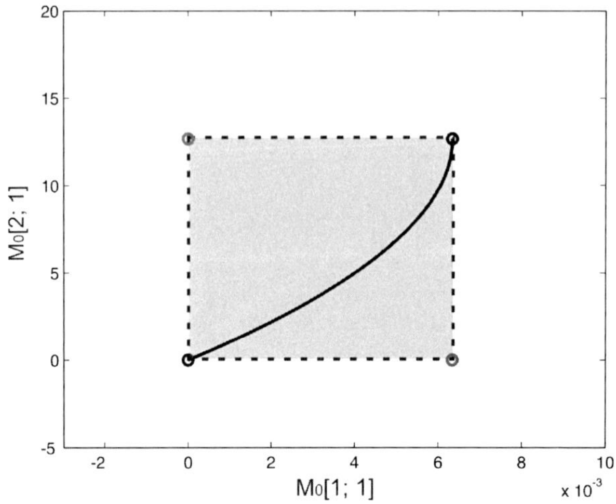

Fig. 7.5 Overapproximation of the set of 2×1-matrices $\{M_0(\tau) \mid \tau \in [\tau_{\min}, \tau_{\max}]\}$ ($M_0[1,1]$ on the horizontal axis and $M_0[2,1]$ on the vertical axis) for $h = 0.001$ and $\tau_k \in [0, h]$. The circles indicate the vertices $M_{0,i}$, $i = 1,2,3,4$, defined in (7.26) and the gray area the convex hull of these vertices

$$M_0(\tau_k) = \begin{bmatrix} \frac{1}{2}b(h^2 - \alpha_2(\tau_k)) \\ b(h - \alpha_1(\tau_k)) \end{bmatrix}. \tag{7.25}$$

Now, the overapproximation of $\mathscr{F}\mathscr{G}$ by the convex hull of $\mathscr{H}_{FG}$ basically means overapproximating $\{M_0(\tau) \mid \tau \in [\tau_{\min}, \tau_{\max}]\}$, with $\tau_{\min} = 0$ and $\tau_{\max} = h$, by the convex hull of the following set of four ($2^\zeta = 4$) generators:

$$\begin{aligned} M_{0,1} &= \begin{bmatrix} \frac{1}{2}b\left(h^2 - (h - \tau_{\max})^2\right) \\ b\left(h - (h - \tau_{\max})\right) \end{bmatrix} = \begin{bmatrix} \frac{1}{2}bh^2 \\ bh \end{bmatrix} \\ M_{0,2} &= \begin{bmatrix} \frac{1}{2}b\left(h^2 - (h - \tau_{\min})^2\right) \\ b\left(h - (h - \tau_{\max})\right) \end{bmatrix} = \begin{bmatrix} 0 \\ bh \end{bmatrix} \\ M_{0,3} &= \begin{bmatrix} \frac{1}{2}b\left(h^2 - (h - \tau_{\max})^2\right) \\ b\left(h - (h - \tau_{\min})\right) \end{bmatrix} = \begin{bmatrix} \frac{1}{2}bh^2 \\ 0 \end{bmatrix} \\ M_{0,4} &= \begin{bmatrix} \frac{1}{2}b\left(h^2 - (h - \tau_{\min})^2\right) \\ b\left(h - (h - \tau_{\min})\right) \end{bmatrix} = \begin{bmatrix} 0 \\ 0 \end{bmatrix}. \end{aligned} \tag{7.26}$$

The polytopic overapproximation is visualized in Figure 7.5.

7.3.2.3 Stability Analysis

In this section, we consider the stability analysis of the NCS (7.11) (or equivalently (7.17)) in closed loop with a state feedback controller. From a control design perspective, when dealing with a system such as (7.11), it is natural to design a state feedback controller using the full lifted state variable ξ_k of the model (7.11), *i.e.*,

$$u_k = -K\xi_k. \tag{7.27}$$

However, from the point of view of the NCS (7.4), this is equivalent to using a dynamical controller of the form

$$u_k = -K_0 x_k - K_1 u_{k-1} \ldots - K_{\bar{d}+\bar{\delta}} u_{k-\bar{d}-\bar{\delta}}. \tag{7.28}$$

The use of such a *dynamic* control law requires a reconsideration of the assumptions made earlier that allowed, amongst others, to lump all the delays τ_k^{sc}, τ_k^c and τ_k^{ca} in one parameter τ_k (see Section 7.3.1). Using a dynamic control law as in (7.28) actually leads to more restrictive assumptions on the network modeling setup as $y_k = x_k$ should always arrive at the controller after the moment that u_{k-1} is sent to the actuator, *i.e.* $s_k + \tau_k^{sc} > s_{k-1} + \tau_{k-1}^{sc} + \tau_{k-1}^c$ as otherwise special precautions are needed to handle out-of-order arrival of measured outputs resulting in longer delays. In addition, the adopted modeling setup and controller in (7.28) require that no packet dropouts occur between the sensors and the controller. Although modeling dropouts alternatively as prolongations of the sampling interval (see, *e.g.*, the comparison in [90]) might alleviate these issues to some extent, dropouts in the channel between the controller and the actuators introduce similar complications in this case.

The mentioned issues do not occur for genuine static state feedbacks of the form

$$u_k = -\bar{K}x_k = -\begin{bmatrix}\bar{K} & 0\end{bmatrix}\xi_k =: -K\xi_k \tag{7.29}$$

and the more restrictive assumptions, as mentioned above, are not needed. This enhances its applicability. For this reason, in the controller synthesis section we will focus on the design of a controller in the form (7.29), and we will provide references for the design of lifted state feedbacks as in (7.27), see Remark 7.7. However, the design of state feedbacks as in (7.29) requires the design of a *structured* feedback gain $K = \begin{bmatrix}\bar{K} & 0\end{bmatrix}$, which is known to be a notoriously difficult problem. A solution for this hard problem will be provided. However, we first derive stability conditions for the NCS given a state feedback as in (7.27) or (7.29). In the stability analysis it is assumed implicitly that in case the lifted state feedback controller (7.27) is used the more restrictive assumptions on the network setup, as mentioned above, are satisfied.

The closed-loop system resulting from interconnecting (7.11) and (7.27) can be formulated as follows:

$$\xi_{k+1} = \tilde{A}_{cl}(\theta_k)\xi_k \text{ with } \tilde{A}_{cl}(\theta_k) = \left(\tilde{A}(\theta_k) - \tilde{B}(\theta_k)K\right), \tag{7.30}$$

with $\theta_k \in \Theta$ for all $k \in \mathbb{N}$, or equivalently, after exploiting the real Jordan form as in (7.17), as

$$\xi_{k+1} = F_{cl}(\theta_k)\xi_k, \tag{7.31}$$

with

$$F_{cl}(\theta_k) = \left[(F_0 - G_0 K) + \sum_{i=1}^{\zeta} \alpha_i(\theta_k)(F_i - G_i K)\right]\xi_k. \tag{7.32}$$

Clearly, $F_{cl}(\theta_k) \in \mathscr{F}_{cl}$, $k \in \mathbb{N}$, where

$$\mathscr{F}_{cl} = \left\{ \left(F_0 - G_0 K \right) + \sum_{i=1}^{\zeta} \alpha_i(\theta) \left(F_i - G_i K \right) | \theta \in \Theta \right\}. \tag{7.33}$$

Given the fact that $\mathscr{F}\mathscr{G} \subseteq co(\mathscr{H}_{FG})$ with $\mathscr{F}\mathscr{G}$ as in (7.18) and $\mathscr{H}_{FG}$ as in (7.20), it follows that

$$\mathscr{F}_{cl} \subseteq co\left(\mathscr{H}_{F_{cl}}\right) \tag{7.34}$$

with

$$\mathscr{H}_{F_{cl}} = \left\{ \left(F_0 - G_0 K \right) + \sum_{i=1}^{\zeta} \alpha_i \left(F_i - G_i K \right) : \alpha_i \in \{\underline{\alpha}_i, \overline{\alpha}_i\}, i = 1, 2, \ldots, \zeta \right\}. \tag{7.35}$$

We will also write the set of vertices $\mathscr{H}_{F_{cl}}$ as $\mathscr{H}_{F_{cl}} = \{H_{F_{cl},j} \mid j = 1, 2, \ldots, 2^{\zeta}\}$ for enumeration purposes. Hence,

$$\mathscr{F}_{cl} \subseteq co\{H_{F_{cl},1}, \ldots, H_{F_{cl},2^{\zeta}}\}. \tag{7.36}$$

Using the finite set $\mathscr{H}_{F_{cl}}$ of 2^{ζ} vertices, a finite number of LMI-based stability conditions can be formulated using [15, 16]. The resulting stability characterization for the closed-loop system (7.30) using parameter-dependent Lyapunov functions is given in the following theorem in which we use the notation $\succ$ and $\prec$ to indicate positive definiteness and negative definiteness, respectively, of a matrix.

Theorem 7.1. *Consider the discrete-time NCS model* (7.11) *and the state feedback controller* (7.27)*, with the network-induced uncertainties* $\theta_k \in \Theta$, $\forall k \in \mathbb{N}$, *and* Θ *defined in* (7.16)*. If there exist matrices* $P_j = P_j^T \succ 0$, $j = 1, 2, \ldots, 2^{\zeta}$, *that satisfy*

$$H_{F_{cl},j}^T P_l H_{F_{cl},j} - P_j \prec 0, \text{ for all } j, l \in \{1, 2, \ldots, 2^{\zeta}\}, \tag{7.37}$$

with $H_{F_{cl},j} \in \mathscr{H}_{F_{cl}}$, $j = 1, 2, \ldots, 2^{\zeta}$, *and* $\mathscr{H}_{F_{cl}}$ *defined in* (7.35)*, then the origin of the closed-loop NCS system* (7.11), (7.27) *is a globally exponentially stable (GES) equilibrium point.*

Proof. The proof is a direct consequence of the results in [11, 14, 41]. ∎

Remark 7.3. It can be shown that under the conditions of Theorem 7.1 also the intersample behavior is bounded, see *e.g.* [10, 9, 11, 14]. Using the results in [64], this also implies that the equilibrium point $x = 0$ of the sampled-data NCS (7.4), (7.7), (7.8), (7.27) is GES.

Remark 7.4. This theorem exploits the following Lyapunov function

$$V(\xi_k, \beta^k) = \xi_k^T P(\beta^k) \xi_k \tag{7.38}$$

based on the inclusion (7.36), which guarantees that (7.31) can be overapproximated by the polytopic system

$$\xi_{k+1} = (\sum_{j=1}^{2^\zeta} \beta_j^k H_{F_{cl},j})\xi_k \tag{7.39}$$

with $\beta^k \in \mathbb{B}$, $k \in \mathbb{N}$, and $\mathbb{B}$ defined in (7.22). The parameter-dependent Lyapunov function $V(\xi_k, \beta^k)$ is then given by $\xi_k^T P(\beta^k)\xi_k = \xi_k^T \sum_{j=1}^{2^\zeta} \beta_j^k P_j \xi_k$. In [11, 41] it is shown that if the LMIs in the above theorem are satisfied then they imply the existence of a Lyapunov-Krasovskii functional (LKF) of the form

$$V(x_k, \ldots, x_{k-\overline{d}-\overline{\delta}}, \theta_k) = \sum_{i=0}^{\overline{d}+\overline{\delta}} \sum_{j=0}^{\overline{d}+\overline{\delta}} x_{k-i}^T Q^{i,j}(\theta_k) x_{k-j}, \tag{7.40}$$

which is the most general (discrete-time) LKF that can be obtained using quadratic forms. Notice that when using this approach based on parameter-dependent quadratic Lyapunov functions the conservative upper bounds in the difference of the LKF, which are usually encountered in the literature to arrive at LKF-based stability conditions in LMI form, are avoided.

Remark 7.5. The case of a common quadratic Lyapunov function (CQLF) $V(\xi_k) = \xi_k^T P \xi_k$ is a particular case of this theorem by taking $P_j = P$, $j = 1, \ldots, 2^\zeta$.

7.3.2.4 Design of Stabilizing Controllers

As already briefly mentioned, the main difficulty to synthesize a genuine state feedback (7.29) is that it results in a structured control synthesis problem, *i.e.*, a control law (7.27) needs to be designed with a specific structure, $K = \begin{pmatrix} \overline{K} & 0_{m\times(\overline{d}+\overline{\delta})m} \end{pmatrix}$. A solution to this structured controller synthesis problem is to apply the approach presented in [18]. Moreover, as was already exploited for the stability analysis problem above, such an approach allows for the use of a parameter-dependent Lyapunov function [15] that might result in less conservative controller synthesis results than the use of a common quadratic Lyapunov function. LMI conditions for synthesis of state feedback controllers as in (7.29) are given in the next theorem.

Theorem 7.2. *Consider the NCS model* (7.4), *and* (7.29), *and its discrete-time representation* (7.11), (7.29) *for sequences of sampling instants, delays, and packet dropouts* $\sigma \in \mathscr{S}$ *with* $\mathscr{S}$ *as in* (7.6). *Consider the equivalent representation* (7.17) *based on the Jordan form of A and the set of vertices* $\mathscr{H}_{FG}$ *defined in* (7.20).

If there exist symmetric positive definite matrices $Y_j \in \mathbb{R}^{(n+(\overline{d}+\overline{\delta})m)\times(n+(\overline{d}+\overline{\delta})m)}$, *a matrix* $\overline{Z} \in \mathbb{R}^{m\times n}$, *matrices* $X_j = \begin{pmatrix} \overline{X}_1 & 0 \\ X_{2,j} & X_{3,j} \end{pmatrix}$, *with* $\overline{X}_1 \in \mathbb{R}^{n\times n}$, $X_{2,j} \in \mathbb{R}^{(\overline{d}+\overline{\delta})m\times n}$, $X_{3,j} \in \mathbb{R}^{(\overline{d}+\overline{\delta})m\times(\overline{d}+\overline{\delta})m}$, $j = 1, 2, \ldots, 2^\zeta$, *that satisfy*

$$\begin{pmatrix} X_j + X_j^T - Y_j & X_j^T H_{F,j}^T - \left(\overline{Z}\ 0\right)^T H_{G,j}^T \\ H_{F,j} X_j - H_{G,j} \left(\overline{Z}\ 0\right) & Y_l \end{pmatrix} \succ 0, \tag{7.41}$$

for all $j, l \in \left\{1, 2, \ldots, 2^{\zeta}\right\}$, *then the closed-loop NCS* (7.4) *and* (7.29) *with* $\overline{K} = \overline{Z}\,\overline{X}_1^{-1}$ *is globally exponentially stable (GES) for sequences of sampling instants, delays, and packet dropouts* $\sigma \in \mathscr{S}$.

Proof. For the proof, see [11]. ∎

Note that in the above theorem the stability is directly formulated for the continuous-time NCS model (7.4) and (7.29) using the ideas in Remark 7.3.

Remark 7.6. The case of a common quadratic Lyapunov function (CQLF) $V(\xi) = \xi_k^T P \xi_k$ is a particular case of this theorem by taking $Y_j = Y$, for all $j = 1, \ldots, 2^{\zeta}$, with $P = Y^{-1}$.

Remark 7.7. If one is still interested in using a lifted state feedback (7.27) despite the mentioned disadvantages, then Theorem 7.2 can be modified by replacing the matrices X_j $j = 1, 2, \ldots, \zeta$ by a constant matrix X without a specific structure and replacing $\left(\overline{Z}\ 0\right)$ by Z. The extended state feedback controller is obtained then by $K = ZX^{-1}$.

Remark 7.8. The derived discrete-time models based on polytopic overapproximations as in (7.23) are suitable for control design using model predictive control (MPC) as well. For instance, the MPC techniques in [49] can be used for this purpose as was indicated in [29, 30].

Remark 7.9. The design of output-based dynamic discrete-time controllers that result in stable closed-loop NCSs is at present an unsolved problem. Due to the adopted polytopic overapproximations, the problem is basically a design problem for a robustly stabilizing output-based dynamic controller for a polytopic system, which is considered to be hard problem in the literature. The stability analysis for these type of controllers (under small delay assumptions) is solved, even in the presence of communication constraints, see Section 7.4.2 below.

Remark 7.10. Here, we only presented results on the stability and stabilization of NCSs. However, extensions exist that provide constructive LMI conditions guaranteeing input-to-state stability, see [86, 85]. In [86, 85] the input-to-state stability property is exploited to solve the (approximate) tracking problem for linear NCS with time-varying (small) delays and time-varying sampling intervals.

Example 7.3. Consider again Example 7.1. As a first instance, assume that the sensor sampling interval $h = 0.001$ is constant and that the controller is given by (7.29) with $\bar{K} = \left(50\ K_2\right)$. The controller gains K_2 that stabilize the system with time-varying delays $\tau_k \in [0, \tau_{\max}]$, with $\tau_{\max} \leq 2h$, are determined using Theorem 7.1 for the case of a common quadratic Lyapunov function, resulting in the gray area in Figure 7.6. In other words for a fixed value of K_2 the NCS is stable for all $\tau_k \in [0, \tau_{\max}]$

as long as $\tau_{\max}$ lies in the gray area for the corresponding value of K_2. Clearly, the large delay case is considered here. To assess the conservatism of the computed stability region, the stability region for *constant* time-delays equal to $\tau_{\max}$ is depicted by the dash-dotted line in Figure 7.6. This comparison reveals the fact that the stability bound is hardly conservative for this example, as the stability region for *time-varying* delays should always lie within the stability region for *constant* delays. In Figure 7.6, also a periodic delay sequence $\tau_1, \tau_2, \tau_1, \tau_2, \ldots$, with $\tau_1 = 0.2h$ and $\tau_2 = 0.6h$, is depicted which has been shown in [10, 14] to induce instability. The latter observation is another indicator for the fact that the stability boundary for uncertain, time-varying delays as in Figure 7.6 is hardly conservative.

For more examples, illustrating both Theorems 7.1 and 7.2 including the case including packet dropouts and the exploitation of parameter-dependent Lyapunov functions, we refer the interested reader to [11, 90, 12, 9].

7.3.3 Sampled-Data Modeling Approaches

In this section, we discuss a modeling and analysis approach for NCS with delays, time-varying sampling intervals and packet dropouts as developed in [58, 60, 59]. Herein, the sampled-data NCS model is formulated in terms of so-called impulsive delay-differential equations (DDEs). Before going into details, we would like to make the following observations:

- This approach studies the stability of the sampled-data NCS without exploiting any form of discretization of a continuous-time plant model as in the discrete-time approach;

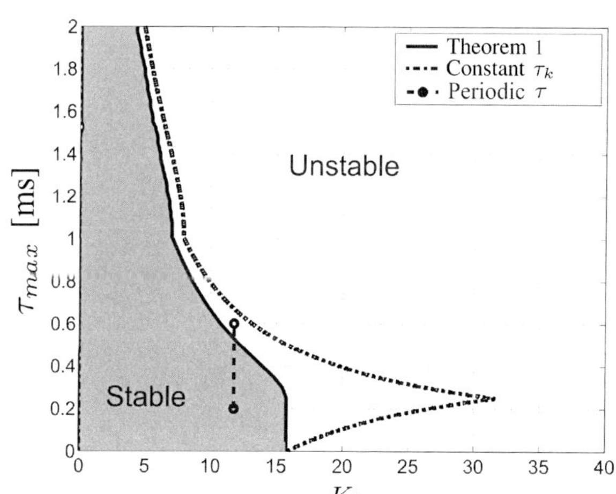

Fig. 7.6 Stability region in terms of K_2 and time-varying delays $\tau_k \in [0, \tau_{\max}]$ (for $h = 0.001$, $K_1 = 50$) for Theorem 7.1, with a common quadratic Lyapunov function, and for constant delays equal to $\tau_{\max}$

- The model in terms of impulsive DDEs shows great similarity with the modeling of the sampled-data NCS using the hybrid systems formalism, see, *e.g.*, [65, 66, 6, 37, 38, 35], as will be discussed in Section 7.4.1. However, the continuous-time approach described in Section 7.4.1 is an emulation-type approach, where controllers are designed in continuous-time, whereas here *discrete-time* (state feedback) controllers are considered and included directly in the sampled-data NCS model. Also the sampled-data approach using impulsive DDEs has not incorporated the presence of communication constraints and the resulting scheduling protocols as has been done in the continuous-time approach, see [65, 66, 6, 37, 38, 35] and Section 7.4.1.
- The modeling framework of impulsive DDEs in principle allows to consider nonlinear systems for which stability results for nonlinear impulsive DDEs have been presented, *e.g.*, in [60, 59]. However, only for the case of *linear* NCS constructive LMI-based stability conditions have been formulated. The continuous-time approach in Section 7.4.1 (see [65, 66, 6, 37, 38, 35]) has stability conditions for nonlinear NCSs as well.

Consider the linear continuous-time plant (7.1) and a discrete-time static state feedback controller as in (7.29), *i.e.* $u_k = -\bar{K}x_k$. The state measurements $x_k := x(s_k)$ are sampled at the sampling instants s_k satisfying (7.2), which are non-equidistantly spaced in time due to the time-varying sampling intervals $h_k > 0$, with $h_k \in [h_{\min}, h_{\max}]$ for all $k \in \mathbb{N}$. The sequence of sampling instants $s_0, s_1, s_2, \ldots$ is strictly increasing in the sense that $s_{k+1} > s_k$, for all $k \in \mathbb{N}$. As in Section 7.3.1, it is assumed that the sensor-to-controller delay, the computational delay and the controller-to-actuator delay can be lumped into a single delay τ_k, with $\tau_k \in [\tau_{\min}, \tau_{\max}]$ for all $k \in \mathbb{N}$. Summarizing, $\{(s_k, \tau_k)\}_{k\in\mathbb{N}} \in \bar{\mathscr{S}}$, where

$$\bar{\mathscr{S}} := \left\{ \{(s_k, \tau_k)\}_{k\in\mathbb{N}} \mid h_{\min} \leq s_{k+1} - s_k \leq h_{\max}, s_0 = 0, \tau_k \in [\tau_{\min}, \tau_{\max}] \text{ for all } k \in \mathbb{N} \right\} \tag{7.42}$$

represents the admissible sequences of sampling times and delays. Note that packet dropouts are not considered explicitly in this approach, but can be accounted for by considering packet drops as an elongation of the effective sampling interval, see also Remark 7.18 below. Let us denote by $r_k = s_k + \tau_k$ the k-th control update instant with $r_0 = \tau_0$. Both the small and large delay cases can be considered, where we allow the delays τ_k to be larger than the sampling intervals h_k with the understanding that the sequence of input update times $\{r_0, r_1, r_2, \ldots\}$ remains strictly increasing. In essence, this means that if a sample arrives at the destination in an out-of-order fashion (*i.e.*, an old sample arrives the destination after the most recent one), it should be rejected (and is effectively deleted from the sequence s_k).

Now, the sampled-data NCS system can be formulated as

$$\begin{cases} \dot{x} &= Ax + Bu^*(t), \;\; x(0) = x_0 \\ u^*(t) &= u_k, \qquad\qquad r_k \leq t \leq r_{k+1}, \\ u_k &= -\bar{K}x_k, \end{cases} . \tag{7.43}$$

Alternatively, the sampled-data NCS system can more compactly be formulated as

$$\dot{x} = Ax - B\bar{K}x(s_k), \quad r_k \leq t \leq r_{k+1}, \tag{7.44}$$

with initial condition given by x_0 and $x(s_{-1})$.

Let us introduce the definition $v_1(t) := x(s_k)$ for $t \in [r_k, r_{k+1})$, where $v_1(t)$ represents a piece-wise constant signal reflecting a delayed version of the most recently sampled state (that is not rejected). Moreover, when we introduce $\zeta(t) := \left[x^T(t)\ v_1^T(t)\right]^T$, we can write the dynamics of the NCS (7.43) (or (7.44)) as an impulsive DDE of the form

$$\dot{\zeta}(t) = F\zeta(t), \qquad t \in [r_k, r_{k+1}) \tag{7.45a}$$

$$\zeta(r_{k+1}) = \begin{bmatrix} x(r_{k+1}) \\ x(s_{k+1}) \end{bmatrix}, \qquad k \in \mathbb{N} \tag{7.45b}$$

with $\zeta(t)$ right-continuous, the initial condition $\zeta(0) := \left[x^T(0)\ x^T(s_{-1})\right]^T$, and

$$F := \begin{bmatrix} A & -B\bar{K} \\ 0 & 0 \end{bmatrix}.$$

Consider the following positive-definite candidate Lyapunov functional

$$\begin{aligned} V :=& x^T Px + \int_{t-\rho_1}^{t} (\rho_{1max} - t + s)\dot{x}^T(s)R_1\dot{x}(s)ds \\ &+ \int_{t-\rho_2}^{t} (\rho_{2max} - t + s)\dot{x}^T(s)R_2\dot{x}(s)ds + \int_{t-\tau_{\min}}^{t} (\tau_{\min} - t + s)\dot{x}^T(s)R_3\dot{x}(s)ds \\ &+ \int_{t-\rho_1}^{t-\tau_{\min}} (\rho_{1max} - t + s)\dot{x}^T(s)R_4\dot{x}(s)ds + (\rho_{1max} - \tau_{\min}) \int_{t-\tau_{\min}}^{t} \dot{x}^T(s)R_4\dot{x}(s)ds \\ &+ \int_{t-\tau_{\min}}^{t} x^T(s)Zx(s)ds + (\rho_{1max} - \rho_1)(x - v_2)^T X(x - v_2) \end{aligned} \tag{7.46}$$

with P, X, Z, R_i, $i = 1, \ldots, 4$, positive definite matrices,

$$v_2(t) := x(r_k),\ \rho_1(t) := t - s_k,\ \rho_2(t) := t - r_k, \quad \text{for} \quad r_k \leq t < r_{k+1},$$

and

$$\rho_{1max} := \sup_{t \geq 0} \rho_1(t), \quad \rho_{2max} := \sup_{t \geq 0} \rho_2(t).$$

Note that $\rho_1(t)$ and $\rho_2(t)$ are sawtooth-like functions of time representing, within a control update interval, the elapsed time since the last (not rejected) sampling instant and the elapsed time since the last (not rejected) control update, respectively. The evolution of this Lyapunov functional is discontinuous at the control update times r_k, due to the jump in ζ in (7.45b), but a decrease of V over the jump is guaranteed by construction.

The next theorem formulates LMI-based conditions for global exponential stability of the NCS (7.45) for any sequence of sampling instants and delays taken from the class $\mathscr{S}$ as in (7.42).

Theorem 7.3. *[58, 59] If there exist positive definite matrices P, X, Z, R_i, $i = 1,\ldots,4$, and not necessarily symmetric matrices N_i, $i = 1,\ldots,4$, satisfying the LMIs*

$$\begin{bmatrix} M_1 + (\beta - \tau_{\min})(M_2 + M_3) & \tau_{\max} N_1 & \tau_{\min} N_3 \\ * & -\tau_{\max} R_1 & 0 \\ * & * & -\tau_{\max} R_3 \end{bmatrix} \prec 0, \tag{7.47a}$$

$$\begin{bmatrix} M_1 + (\beta - \tau_{\min}) M_2 & \tau_{\max} N_1 & \tau_{\min} N_3 & (\beta - \tau_{\min})(N_1 + N_2) & (\beta - \tau_{\min}) N_4 \\ * & -\tau_{\max} R_1 & 0 & 0 & 0 \\ * & * & -\tau_{\min} R_3 & 0 & 0 \\ * & * & * & -(\beta - \tau_{\min})(R_1 + R_2) & 0 \\ * & * & * & * & -(\beta - \tau_{\min}) R_4 \end{bmatrix} \prec 0, \tag{7.47b}$$

where $\beta := h_{\max} + \tau_{\max}$, $\bar{F} := \begin{bmatrix} A & -B\bar{K} & 0 & 0 \end{bmatrix}$,

$$\begin{aligned} M_1 := {} & \bar{F}^T \begin{bmatrix} P & 0 & 0 & 0 \end{bmatrix} + \begin{bmatrix} P \\ 0 \\ 0 \\ 0 \end{bmatrix} \bar{F} + \tau_{\min} F^T (R_1 + R_3) F - \begin{bmatrix} I \\ 0 \\ -I \\ 0 \end{bmatrix} X \begin{bmatrix} I \\ 0 \\ -I \\ 0 \end{bmatrix}^T + \begin{bmatrix} I \\ 0 \\ 0 \\ 0 \end{bmatrix} Z \begin{bmatrix} I \\ 0 \\ 0 \\ 0 \end{bmatrix}^T \\ & - \begin{bmatrix} 0 \\ 0 \\ 0 \\ I \end{bmatrix} Z \begin{bmatrix} 0 \\ 0 \\ 0 \\ I \end{bmatrix}^T - N_1 \begin{bmatrix} I & -I & 0 & 0 \end{bmatrix} - \begin{bmatrix} I \\ -I \\ 0 \\ 0 \end{bmatrix} N_1^T - N_2 \begin{bmatrix} I & 0 & -I & 0 \end{bmatrix} - \begin{bmatrix} I \\ 0 \\ -I \\ 0 \end{bmatrix} N_2^T \\ & - N_3 \begin{bmatrix} I & 0 & 0 & -I \end{bmatrix} - \begin{bmatrix} I \\ 0 \\ 0 \\ -I \end{bmatrix} N_3^T - N_4 \begin{bmatrix} 0 & -I & 0 & I \end{bmatrix} - \begin{bmatrix} 0 \\ -I \\ 0 \\ I \end{bmatrix} N_4^T, \end{aligned}$$

$$M_2 := \bar{F}^T (R_1 + R_2 + R_4) \bar{F},$$

$$M_3 := \begin{bmatrix} I \\ 0 \\ -I \\ 0 \end{bmatrix} X \bar{F} + \bar{F}^T X \begin{bmatrix} I & 0 & -I & 0 \end{bmatrix},$$

then, system (7.45) is globally exponentially stable for any sequence of delays and sampling instants taken from the class $\mathscr{S}$ as in (7.42).

Proof. For the proof, see [59]. ∎

Remark 7.11. The proof of Theorem 7.3 exploits stability results for nonlinear impulsive DDEs as presented in [60, 59].

Remark 7.12. The conditions in Theorem 7.3 do not explicitly depend on the values of $h_{\min}$. Consequently, this approach towards modeling NCSs may result in more conservative conditions in comparison to those obtained using the discrete-time approach discussed in Section 7.3.2, when $0 \ll h_{\min} \simeq h_{\max}$. The reason is that the discrete-time approach can actually exploit the knowledge that $h_{\min} > 0$.

Remark 7.13. When considering the control synthesis problem, *i.e.* when the control gain $\bar{K}$ is considered unknown, the LMIs in Theorem 7.3 generally become bilinear matrix inequalities (BMIs). However, for the case without delays in [60, 61] LMI-based control synthesis conditions for static state feedback controllers have been proposed.

Remark 7.14. In [58], results on the stability analysis of linear NCS with continuous-time, dynamic output feedback controllers are presented. Herein, it is assumed that these continuous-time controllers can be evaluated exactly on the sampling instants (by exact discretization and some form of time-stamping of the sampled measurements).

Example 7.4. We now reconsider the motion control example of Example 7.1 and use it to compare the discrete-time and sampled-data approach.

First consider the case of a constant sampling interval $h = 0.005$, but with time-varying and uncertain delays in the set $[0, \tau_{\max}]$. Applying Theorem 7.3 in a slightly modified form (for the special case in which a Lyapunov functional (7.46) with $Z = R_i = 0$, $i = 1, \ldots, 4$, is exploited) leads to stability guarantee for the NCS for a maximal delay up to $\tau_{\max} = 0.33h$, see [86, 85]. Comparing this with the discrete-time approach using Theorem 7.1 (for the special case of a common quadratic Lyapunov function) shows that stability can be guaranteed up to a maximal delay of $\tau_{\max} = 0.94h$.

Next, we consider the case in which the sampling interval is variable, *i.e.*, $h_k \in [h_{\min}, h_{\max}]$, $k \in \mathbb{N}$ and the delay is zero. More specifically, we take $h_{\min} = h_{\max}/1.5$, so $h_{\min} \neq 0$. Using the discrete-time approach in Theorem 7.1 (for the special case of a common quadratic Lyapunov function), stability can be assured almost up to $h_{\max} = 1.34 \times 10^{-2}$, which is the sampling interval for which the system with a constant sampling interval (and no delay) becomes unstable. This fact shows that the proposed discrete-time stability conditions as in Theorem 7.1 are not conservative in this example. Using the impulsive DDE approach, stability can only be guaranteed up to $h_{\max} = 9 \times 10^{-3}$. Hence, for this motion control example the discrete-time approach clearly outperforms the sampled-data approach as far as the characterisation of stability is concerned.

Remark 7.15. In [86, 85], an extension of Theorem 7.3 (for the special case that $Z = R_i = 0$, $i = 1, \ldots, 4$), guarantees input-to-state stability in the face of perturbations. This extension is exploited to solve the (approximate) tracking problem for NCS with time-varying delays and sampling intervals. It is important to note that the input-to-state stability gains from additive perturbations to the states of the NCS provided by the impulsive DDE modeling approach are much tighter than those obtained using the discrete-time modeling and analysis approach as shown in

[86, 85]. The conservatism in the estimates of the input-to-state stability gain using the discrete-time approach are mainly due to the conservative upperbounding of the intersample behavior. In this respect it seems that the impulsive DDE approach is beneficial in studying such performance related issues.

7.4 NCS Including Communication Constraints

In this section we will discuss stability analysis approaches that incorporate communication constraints. Specifically, communication is constrained in the sense that the number of control inputs and measured outputs that can be transmitted over a network is limited. At each transmission time only one of the nodes consisting of particular actuators and/or sensors will obtain access to the network to communicate its data. Which node obtains access is determined by a scheduling protocol. As we will see this complicates the description and the analysis of the NCS considerably. The communication constraints and protocol will actually introduce (additional) discrete effects in the problem, which will require modeling and stability analysis techniques from the hybrid systems domain [89, 31].

We will present a continuous-time/emulation approach in Section 7.4.1 and a discrete-time approach in Section 7.4.2. Both approaches have their own advantages and disadvantages as we will conclude at the end.

7.4.1 Continuous-Time (Emulation) Approaches

In this section, we introduce the continuous-time model that will be used to describe NCSs including communication constraints as well as varying transmission intervals and transmission delays. Dropouts and quantization effects can be included as discussed in [35] and in Remark 7.18, but for the ease of exposition we will not consider them below. The model that we discuss in this section was derived in [37, 38] and forms an extension of the NCS models used before in [65] that were motivated by the work in [92]. The emulation approach is characterized by the design procedure that, first, a stabilizing continuous-time controller for the continuous-time plant is designed (ignoring any network effects). Next, we study under which network effects (level of delays, size of sampling intervals, type of protocol used for the communication scheduling) the NCS inherits the stability properties from the network-free continuous-time closed-loop system.

7.4.1.1 Description of the NCS

Consider the continuous-time plant

$$\dot{x}_p = f_p(x_p, \hat{u}), \qquad y = g_p(x_p) \tag{7.48}$$

that is sampled. Here, $x_p \in \mathbb{R}^{n_p}$ denotes the state of the plant, $\hat{u} \in \mathbb{R}^{n_u}$ denotes the most recent control values available at the plant and $y \in \mathbb{R}^{n_y}$ is the output of the plant. The controller is given by

$$\dot{x}_c = f_c(x_c, \hat{y}), \qquad u = g_c(x_c), \tag{7.49}$$

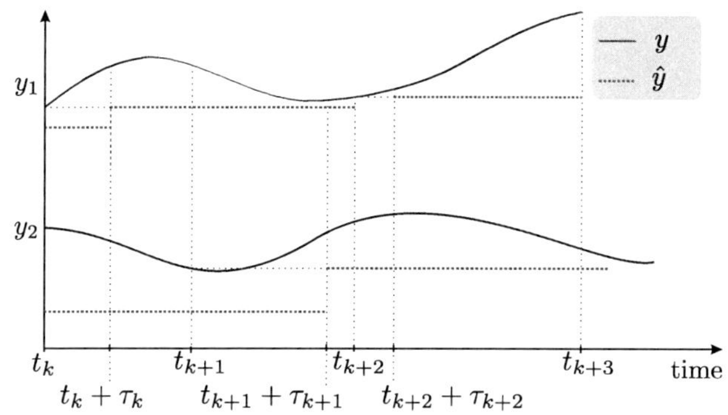

Fig. 7.7 Illustration of a typical evolution of y and $\hat{y}$ for 2 nodes

where the variable $x_c \in \mathbb{R}^{n_c}$ is the state of the controller, $\hat{y} \in \mathbb{R}^{n_y}$ is the most recent output measurement of the plant that is available at the controller and $u \in \mathbb{R}^{n_u}$ denotes the control input. At times t_k, $k \in \mathbb{N}$, (parts of) the input u at the controller and/or the output y at the plant are sampled and transmitted over the network. The transmission times satisfy $0 \leq t_0 < t_1 < t_2 < \ldots$. Even stronger, we assume that there exists a $\delta > 0$ (which can be arbitrarily small) such that the transmission intervals $t_{k+1} - t_k$ satisfy $\delta \leq t_{k+1} - t_k \leq h_{mati}$ for all $k \in \mathbb{N}$, where h_{mati} denotes the maximally allowable transmission interval (MATI). At each transmission time t_k, $k \in \mathbb{N}$, the protocol determines which of the nodes $j \in \{1, 2, \ldots, N\}$ is granted access to the network. Each node corresponds to a collection of sensors or actuators. The sensors/actuators corresponding to the node, which is granted access, collect their values in $y(t_k)$ or $u(t_k)$ that will be sent over the communication channel. They will arrive after a transmission delay of τ_k time units at the controller or actuator, respectively. This results in updates of the corresponding entries in $\hat{y}$ or $\hat{u}$ at the arrival times $t_k + \tau_k$, $k \in \mathbb{N}$, which were denoted by r_k in the previous section. The situation described above is illustrated for y and $\hat{y}$ in Figure 7.7 for the situation that there are two nodes and the nodes get access to the network in an alternating fashion.

It is assumed that there are bounds on the maximal delay in the sense that $\tau_k \in [0, \tau_{mad}]$, $k \in \mathbb{N}$, where $0 \leq \tau_{mad} \leq h_{mati}$ is the maximally allowable delay (MAD). In particular, we will use the following standing assumption in the sequel.

Assumption 7.4. *The transmission times satisfy* $\delta \leq t_{k+1} - t_k < h_{mati}$, $k \in \mathbb{N}$ *and the delays satisfy* $0 \leq \tau_k \leq \min\{\tau_{mad}, t_{k+1} - t_k\}$, $k \in \mathbb{N}$, *where* $\delta \in (0, h_{mati}]$ *is arbitrary.*

This assumption implies that each transmitted packet arrives before the next sample is taken meaning that only the small delay case is considered here[1]. In various situations $\tau_k \leq h_k := t_{k+1} - t_k$, $k \in \mathbb{N}$ is a realistic assumption. Indeed, if two or more

[1] Extensions of this continuous-time approach including large delays do not exist to this date.

nodes are sharing one communication channel and one of the nodes is transmitting its data, the channel is busy and hence other nodes cannot access the network, which guarantees $\tau_k \leq h_k = t_{k+1} - t_k$, $k \in \mathbb{N}$.

Remark 7.16. Compared to the notation in the previous section we have that $h_{\max} = h_{mati}$ and $h_{\min} = \delta$, which can actually be chosen arbitrarily close to 0. For the delays we have that $\tau_{\min} = 0$ and $\tau_{\max} = \tau_{mad}$. We used here the terms MATI and MAD (h_{mati} and τ_{mad}) as done in the literature [65, 66, 6, 37, 38, 35] on the continuous-time approach.

The updates of $\hat{y}$ and $\hat{u}$ satisfy

$$\hat{y}((t_k + \tau_k)^+) = y(t_k) + h_y(k, e(t_k)) \tag{7.50a}$$

$$\hat{u}((t_k + \tau_k)^+) = u(t_k) + h_u(k, e(t_k)) \tag{7.50b}$$

at $t_k + \tau_k$, where e denotes the vector $(e_y, e_u) := (e_y^T, e_u^T)^T$ with $e_y := \hat{y} - y$, $e_u := \hat{u} - u$ and h_u and h_y are update functions related to the protocol. Hence, $e \in \mathbb{R}^{n_e}$ with $n_e = n_y + n_u$. If the NCS has N nodes, then the error vector e can be partitioned as $e = (e_1^T, e_2^T, \ldots, e_N^T)^T$. The update functions h_y and h_u are related to the protocol of which we give two well-known examples below. Typically when the j-th node gets access to the network at some transmission time t_k the corresponding values in $\hat{y}$ and $\hat{u}$ have a jump at $t_k + \tau_k$ to the corresponding transmitted values in $y(t_k)$ and $u(t_k)$, since the quantization effects are assumed to be negligible. For instance, when y_j is transmitted at time t_k, it holds that $h_{y,j}(k, e(t_k)) = 0$ meaning that $\hat{y}_j((t_k + \tau_k)^+) = y_j(t_k)$. However, for reasons of generality, more freedom is allowed in the protocols given by $h = (h_y, h_u) := (h_y^T, h_u^T)^T$. Two well-known examples are the Round Robin (RR) protocol the Try-Once-Discard (TOD) protocol (sometimes also called the maximum-error-first protocol). For 2 nodes the RR protocol is given by

$$h(k,e) = \begin{cases} \begin{pmatrix} 0 \\ e_2 \end{pmatrix}, & \text{if } k = 0, 2, 4, 6, \ldots \\ \begin{pmatrix} e_1 \\ 0 \end{pmatrix}, & \text{if } k = 1, 3, 5, 7, \ldots \end{cases}$$

Hence, the two nodes get access to the network in an alternating fashion: When the transmission counter is even the first node gets access, when the counter is odd the second node can send its data. As such, the RR protocol is a *static* protocol in the sense that the order in which the nodes get access to the network is fixed. In contrast, the TOD protocol is a *dynamic* scheduling protocol, which is given for two nodes by

$$h(k,e) = \begin{cases} \begin{pmatrix} 0 \\ e_2 \end{pmatrix}, & \text{if } |e_1| \geq |e_2| \\ \begin{pmatrix} e_1 \\ 0 \end{pmatrix}, & \text{if } |e_2| > |e_1|, \end{cases}$$

Here $|\cdot|$ denotes the Euclidean norm in $\mathbb{R}^n$ and later we will also use $\langle\cdot,\cdot\rangle$ as the corresponding inner product. Hence, the TOD protocol gives access to the node with the largest difference between the latest transmitted value of the corresponding inputs/outputs and the current value of these inputs/outputs. Indeed, the node with the largest network-induced error e_i is allowed to transmit its signal values. Extensions of these protocols to more than 2 nodes are straightforward.

In between the updates of the values of $\hat{y}$ and $\hat{u}$, the network is assumed to operate in a zero-order-hold (ZOH) fashion, meaning that the values of $\hat{y}$ and $\hat{u}$ remain constant in between the updating times $t_k+\tau_k$ and $t_{k+1}+\tau_{k+1}$:

$$\dot{\hat{y}} = 0, \quad \dot{\hat{u}} = 0. \tag{7.51}$$

To compute the resets of e at the update or arrival times $\{t_i+\tau_k\}_{k\in\mathbb{N}}$, we proceed as follows:

$$\begin{aligned} e_y((t_k+\tau_k)^+) &= \hat{y}((t_k+\tau_k)^+) - y(t_k+\tau_k) = y(t_k) + h_y(k,e(t_k)) - y(t_k+\tau_k) \\ &= h_y(k,e(t_k)) + \underbrace{y(t_k)-\hat{y}(t_k)}_{-e(t_k)} + \underbrace{\hat{y}(t_k+\tau_k)-y(t_k+\tau_k)}_{e(t_k+\tau_k)} \\ &= h_y(k,e(t_k)) - e(t_k) + e(t_k+\tau_k). \end{aligned}$$

In the third equality it is used that $\hat{y}(t_{s_i}) = \hat{y}(t_{s_i}+\tau_k)$, which holds due to the ZOH character of the network.

A similar derivation holds for e_u, leading to the following model for the NCS:

$$\left.\begin{aligned} \dot{x}(t) &= f(x(t),e(t)) \\ \dot{e}(t) &= g(x(t),e(t)) \end{aligned}\right\} \quad t\in[t_k,t_k+\tau_k) \tag{7.52a}$$

$$e((t_k+\tau_k)^+) = h(k,e(t_k)) - e(t_k) + e(t_k+\tau_k), \tag{7.52b}$$

where $x=(x_p,x_c)\in\mathbb{R}^{n_x}$ with $n_x = n_p+n_c$, f, g are appropriately defined functions depending on f_p, g_p, f_c and g_c and $h=(h_y,h_u)$. See [65] for the explicit expressions of f and g.

Remark 7.17. The model (7.52) reduces to the model used in [65, 66] in absence of delays, *i.e.* $\tau_k=0$ for all $k\in\mathbb{N}$. Indeed, then (7.52) becomes

$$\left.\begin{aligned} \dot{x}(t) &= f(x(t),e(t)) \\ \dot{e}(t) &= g(x(t),e(t)) \end{aligned}\right\} \quad t\in[t_k,t_{k+1}) \tag{7.53a}$$

$$e(t_k^+) = h(k,e(t_k)). \tag{7.53b}$$

Assumption 7.5. *f and g are continuous and h is locally bounded.* ∎

Observe that the system $\dot{x}=f(x,0)$ is the closed-loop system (7.48)-(7.49) without the network ($e=0$).

The stability problem that is considered is formulated as follows.

Problem 7.1. Suppose that the controller (7.49) was designed for the plant (7.48) rendering the continuous-time closed-loop system (7.48)-(7.49) (or equivalently, $\dot{x} = f(x,0)$) stable in some sense. Determine the value of h_{mati} and τ_{mad} so that the NCS given by (7.52) is stable as well when the transmission intervals and delays satisfy Assumption 7.4. ■

Remark 7.18. Of course, there are certain extensions that can be made to the above setup. The inclusion of packet dropouts is relatively easy, if one models them as prolongations of the transmission interval. Indeed, if we assume that there is a bound $\bar{\delta} \in \mathbb{N}$ on the maximum number of successive dropouts, the stability bounds derived below are still valid for the MATI given by $h'_{mati} := \frac{h_{mati}}{\bar{\delta}+1}$, where h_{mati} is the obtained MATI for the dropout-free case.

Remark 7.19. In case $h(k,e) = 0$ for all $k \in \mathbb{N}$ and $e \in \mathbb{R}^{n_e}$, the above model essentially reduces to a sampled-data systems (without communication constraints) with a continuous-time controller. In this particular case the impulsive DDE in the sampled-data approach of Section 7.3.3 (see Remark 7.14) and the continuous-time NCS model as presented here are related.

7.4.1.2 Reformulation in a Hybrid System Framework

To facilitate the stability analysis, the above NCS model is transformed into the hybrid system framework as developed in [31]. To do so, the auxiliary variables $s \in \mathbb{R}^n$, $\kappa \in \mathbb{N}$, $\tau \in \mathbb{R}_{\geq 0}$ and $\ell \in \{0,1\}$ are introduced to reformulate the model in terms of so-called flow equations and reset equations. The variable s is an auxiliary variable containing the memory in (7.52b) storing the value $h(k,e(t_k)) - e(t_k)$ for the update of e at the update instant $t_k + \tau_k$, κ is a counter keeping track of the number of the transmission, τ is a timer to constrain both the transmission interval as well as the transmission delay and ℓ is a Boolean keeping track whether the next event is a transmission event or an update event. To be precise, when $\ell = 0$ the next event will be related to transmission and when $\ell = 1$ the next event will be an update. Note that here use is made of the fact that only small delays are considered as this implies that transmission and update events occur in an alternating manner.

The hybrid system Σ_{NCS} is given by the flow equations

$$\left.\begin{aligned} \dot{x} &= f(x,e) \\ \dot{e} &= g(x,e) \\ \dot{s} &= 0 \\ \dot{\kappa} &= 0 \\ \dot{\tau} &= 1 \\ \dot{\ell} &= 0 \end{aligned}\right\} (\ell = 0 \wedge \tau \in [0, h_{mati}]) \vee (\ell = 1 \wedge \tau \in [0, \tau_{mad}]) \tag{7.54}$$

and the reset equations are obtained by combining the "transmission reset relations," active at the transmission instants $\{t_k\}_{k\in\mathbb{N}}$, and the "update reset relations", active at the update instants $\{t_k + \tau_k\}_{k\in\mathbb{N}}$, given by

$$(x^+, e^+, s^+, \tau^+, \kappa^+, \ell^+) = G(x, e, s, \tau, \kappa, \ell), \text{ when}$$
$$(\ell = 0 \wedge \tau \in [\delta, h_{mati}]) \vee (\ell = 1 \wedge \tau \in [0, \tau_{mad}]) \quad (7.55)$$

with G given by the transmission resets (when $\ell = 0$)

$$G(x, e, s, \tau, \kappa, 0) = (x, e, h(\kappa, e) - e, 0, \kappa + 1, 1) \quad (7.56)$$

and the update resets (when $\ell = 1$)

$$G(x, e, s, \tau, \kappa, 1) = (x, s + e, -s - e, \tau, \kappa, 0). \quad (7.57)$$

7.4.1.3 Lyapunov-Based Stability Analysis

A Lyapunov function for Σ_{NCS} will be constructed based on the following conditions for the reset part (the protocol) and the flow part of the system.

Conditions on the Reset Part

Condition 7.6. *The protocol given by h is UGES (uniformly globally exponentially stable), meaning that there exists a function $W : \mathbb{N} \times \mathbb{R}^{n_e} \to \mathbb{R}_{\geq 0}$ that is locally Lipschitz in its second argument such that*

$$\underline{\alpha}_W |e| \leq W(\kappa, e) \leq \overline{\alpha}_W |e| \quad (7.58a)$$
$$W(\kappa + 1, h(\kappa, e)) \leq \lambda W(\kappa, e) \quad (7.58b)$$

for constants $0 < \underline{\alpha}_W \leq \overline{\alpha}_W$ and $0 < \lambda < 1$. ■

Additionally it is assumed here that

$$W(\kappa + 1, e) \leq \lambda_W W(\kappa, e) \quad (7.59)$$

for some constant $\lambda_W \geq 1$ and that for almost all $e \in \mathbb{R}^{n_e}$ and all $\kappa \in \mathbb{N}$

$$\left| \frac{\partial W}{\partial e}(\kappa, e) \right| \leq M_1 \quad (7.60)$$

for some constant $M_1 > 0$. For all protocols discussed in [92, 91, 65, 6] such Lyapunov functions and corresponding constants exist. For instance, if N is the number of nodes in the network, for the RR protocol $\lambda_{RR} = \sqrt{\frac{N-1}{N}}$, $\underline{\alpha}_{W_{RR}} = 1$, $\overline{\alpha}_{W_{RR}} = \sqrt{N}$, $\lambda_{W_{RR}} = \sqrt{N}$, $M_{1,RR} = \sqrt{N}$ and for the TOD protocol $\lambda_{TOD} = \sqrt{\frac{N-1}{N}}$, $\underline{\alpha}_{W_{TOD}} = \overline{\alpha}_{W_{TOD}} = 1$, $\lambda_{W_{TOD}} = 1$, $M_{1,TOD} = 1$. In particular $W_{TOD}(i, e) = |e|$. See [38, 65] for the proofs.

Conditions on the Flow Part

The following growth condition on the flow of the NCS model (7.52) is used:

$$|g(x,e)| \leq m_x(x) + M_e|e|, \tag{7.61}$$

where $m_x : \mathbb{R}^{n_x} \to \mathbb{R}_{\geq 0}$ and $M_e \geq 0$ is a constant. Moreover, the following is additionally used.

Condition 7.7. *There exists a locally Lipschitz continuous function* $V : \mathbb{R}^{n_x} \to \mathbb{R}_{\geq 0}$ *satisfying the bounds*

$$\underline{\alpha}_V(|x|) \leq V(x) \leq \overline{\alpha}_V(|x|) \tag{7.62}$$

for some $\mathscr{K}_\infty$*-functions*[2] $\underline{\alpha}_V$ *and* $\overline{\alpha}_V$, *and the condition*

$$\langle \nabla V(x), f(x,e) \rangle \leq -m_x^2(x) - \rho(|x|) + (\gamma^2 - \varepsilon) W^2(\kappa, e) \tag{7.63}$$

for almost all $x \in \mathbb{R}^{n_x}$ *and all* $e \in \mathbb{R}^{n_e}$ *with* $\rho \in \mathscr{K}_\infty$, *for some* $\gamma > 0$ *and* $0 < \varepsilon < \max\{\gamma^2, 1\}$.

Essentially, the condition above is a Lyapunov-based formulation for the system $\dot{x} = f(x,e)$ to have an $\mathscr{L}_2$ gain [88] from $W^2(\kappa,e)$ to $m_x^2(x)$ strictly smaller than γ together with global asymptotic stability in case $e = 0$.

Stability Result

Lumping the above parameters into four new ones given by

$$L_0 = \frac{M_1 M_e}{\underline{\alpha}_W}; \; L_1 = \frac{M_1 M_e \lambda_W}{\lambda \underline{\alpha}_W}; \; \gamma_0 = M_1 \gamma; \; \gamma_1 = \frac{M_1 \gamma \lambda_W}{\lambda} \tag{7.64}$$

we can provide the following conditions on MAD and MATI to guarantee stability of Σ_{NCS}. Indeed, consider now the differential equations

$$\dot{\phi}_0 = -2L_0\phi_0 - \gamma_0(\phi_0^2 + 1) \tag{7.65a}$$

$$\dot{\phi}_1 = -2L_1\phi_1 - \gamma_0(\phi_1^2 + \frac{\gamma_1^2}{\gamma_0^2}). \tag{7.65b}$$

Observe that the solutions to these differential equations are strictly decreasing as long as $\phi_\ell(\tau) \geq 0$, $\ell = 0, 1$. Define the equilibrium set as

$$\mathscr{E} := \{(x, e, s, \kappa, \tau, \ell) \mid x = 0, \; e = s = 0\}.$$

Theorem 7.8. *Consider the system* Σ_{NCS} *such that Assumptions 7.4 and 7.5 are satisfied. Let Condition 7.6 with* (7.59) *and* (7.60) *and Condition 7.7 with* (7.61) *hold. Suppose* $h_{mati} \geq \tau_{mad} \geq 0$ *satisfy*

$$\phi_0(\tau) \geq \lambda^2 \phi_1(0) \textit{ for all } 0 \leq \tau \leq h_{mati} \tag{7.66a}$$

$$\phi_1(\tau) \geq \phi_0(\tau) \textit{ for all } 0 \leq \tau \leq \tau_{mad} \tag{7.66b}$$

[2] A function $\alpha : \mathbb{R}_+ \to \mathbb{R}_+$ is called a $\mathscr{K}$-function, if it is continuous, strictly increasing and $\alpha(0) = 0$. A $\mathscr{K}$-function α is called a $\mathscr{K}_\infty$-function if $\alpha(s) \to \infty$ if $s \to \infty$. Examples of $\mathscr{K}_\infty$-functions are $\alpha(s) = cs^\lambda$ for some $c > 0$ and $\lambda > 0$.

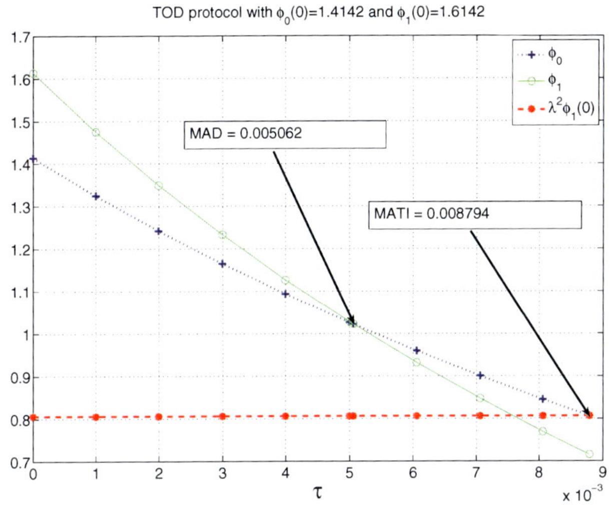

Fig. 7.8 Illustration of how the solutions ϕ_ℓ, $\ell = 0, 1$, lead to MAD and MATI

for solutions ϕ_0 *and* ϕ_1 *of* (7.65) *corresponding to certain chosen initial conditions* $\phi_\ell(0) > 0$, $\ell = 0, 1$, *with* $\phi_1(0) \geq \phi_0(0) \geq \lambda^2\phi_1(0) \geq 0$ *and* $\phi_0(h_{mati}) > 0$. *Then for the system* Σ_{NCS} *the set* $\mathscr{E}$ *is uniformly globally asymptotically stable (UGAS).* ■

The proof is based on constructing Lyapunov functions $U(\xi)$ for Σ_{NCS}, using the solutions ϕ_0 and ϕ_1 to (7.65), that satisfy $U(\xi^+) \leq U(\xi)$ at reset times and $\dot{U}(\xi) < 0$ during flow. See [38] for the proof and the exact definition of UGAS of $\mathscr{E}$, which implies (next to Lyapunov stability of $\mathscr{E}$) that $x(t) \to 0$, $e(t) \to 0$ and $s(t) \to 0$, when $t \to \infty$.

From the above theorem quantitative numbers for h_{mati} and τ_{mad} can be obtained by constructing the solutions to (7.65) for certain initial conditions. Computing the τ value of the intersection of ϕ_0 and the constant line $\lambda^2\phi_1(0)$ provides h_{mati} according to (7.66a), while the intersection of ϕ_0 and ϕ_1 gives a value for τ_{mad} due to (7.66b). In Figure 7.8 this is illustrated. Different values of the initial conditions $\phi_0(0)$ and $\phi_1(0)$ lead, of course, to different solutions ϕ_0 and ϕ_1 of the differential equations (7.65) and thus different h_{mati} and τ_{mad}. As a result, tradeoff curves between h_{mati} and τ_{mad} can be obtained that indicate when stability of the NCS is still guaranteed. This will be illustrated below for the benchmark example of the batch reactor. Before showing the example, a systematic procedure to determine these tradeoff curves is provided.

Systematic Procedure for the Determination of MATI and MAD

The main steps in the procedure to compute the tradeoff curves between MATI and MAD are given as follows:

Procedure 7.9. *Given Σ_{NCS} apply the following steps:*

1. *Construct a Lyapunov function W for the UGES protocol as in Condition 7.6 with the constants $\underline{\alpha}_W$, $\overline{\alpha}_W$, λ, λ_W and M_1 as in* (7.58), (7.59) *and* (7.60). *Suitable Lyapunov functions and the corresponding constants are available for many protocols in the literature [66, 65, 38];*
2. *Compute the function m_x and the constant M_e as in* (7.61) *bounding g as in* (7.52)*;*
3. *Compute for $\dot{x} = f(x,e)$ in the NCS model* (7.52) *the $\mathcal{L}_2$ gain from $W(\kappa,e)$ to $m_x(x)$ in the sense that* (7.62)-(7.63) *is satisfied for a (storage) function V for some small $0 < \varepsilon < \max\{\gamma^2, 1\}$ and $\rho \in \mathcal{K}_\infty$. When f is linear, this can be done using LMIs. Of course, here the 'emulated' controller should guarantee that such a property is satisfied;*
4. *Use now* (7.64) *to obtain L_0, L_1, γ_0 and γ_1;*
5. *For initial conditions $\phi_0(0)$ and $\phi_1(0)$ with $\lambda^2\phi_1(0) \leq \phi_0(0) < \phi_1(0)$ compute (numerically) the solutions ϕ_0 and ϕ_1 to* (7.65) *and find (the largest values of) h_{mati} and τ_{mad} such that* (7.66) *are satisfied. The largest values can be found by determining the intersection of ϕ_0 and ϕ_1 (giving τ_{mad}) and the intersection of ϕ_0 with $\lambda^2\phi_1(0)$ (giving h_{mati}). Repeat this step for various values of the initial conditions thereby obtaining various combinations of h_{mati} and τ_{mad} leading to tradeoff curves.*

This procedure is systematic in nature and can consequently be applied in a straightforward manner.

Delay-Free Results

In the above setting taken from [38, 37] both varying h_k and τ_k are allowed. The case without delays ($\tau_{mad} = 0$) has been treated in the earlier works [92, 91, 66, 65, 6]. Basically, the least conservative of them, being [6], uses slightly weaker versions of Condition 7.6, Condition 7.7 and

$$\left\langle \frac{\partial W}{\partial e}(\kappa,e), g(x,e) \right\rangle \leq LW(\kappa,e) + m_x(x) \tag{7.67}$$

for all $\kappa \in \mathbb{N}$ and almost all $e \in \mathbb{R}^{n_e}$. Instead of four parameters as in (7.64), in [6] only the parameters γ and L are used next to λ to determine h_{mati} (as $\tau_{mad} = 0$). Also the two differential equations that are formulated in (7.65) reduce to only one differential equation given by

$$\dot{\phi} = -2L\phi - \gamma(\phi^2 + 1) \tag{7.68}$$

and the initial condition is chosen as $\phi(0) = \lambda^{-1}$. The conditions (7.66) reduce to $\phi(\tau) \geq \lambda$ for all $0 \leq \tau \leq h_{mati}$ to guarantee stability of Σ_{NCS}. Hence, the value of τ for which $\phi(\tau) = \lambda$ determines the h_{mati} that can be guaranteed. Interestingly, due to the fact that there is only one differential equation, h_{mati} can be analytically computed and results in

$$h_{mati} = \begin{cases} \frac{1}{Lr}\arctan\left(\frac{r(1-\lambda)}{2\frac{\lambda}{1+\lambda}(\frac{\gamma}{L})+1+\lambda}\right), & \gamma > L \\ \frac{1-\lambda}{L(1+\lambda)}, & \gamma = L \\ \frac{1}{Lr}\text{arctanh}\left(\frac{r(1-\lambda)}{2\frac{\lambda}{1+\lambda}(\frac{\gamma}{L})+1+\lambda}\right), & \gamma < L, \end{cases} \tag{7.69}$$

where $r = \sqrt{|(\frac{\gamma}{L})^2 - 1|}$.

Application to the Benchmark Example of the Batch Reactor

In this part the discussed results are applied to the case study of the batch reactor, which has developed over the years as a benchmark example in NCSs [6, 92, 65]. The functions in the NCS (7.52) for the batch reactor are given by the linear functions $f(x,e,w) = A_{11}x + A_{12}e$ and $g(x,e,w) = A_{21}x + A_{22}e$ in which the numerical values for A_{ij}, $i,j = 1,2$, are provided in [65, 92] and given by

$$A_{11} = \begin{pmatrix} 1.3800 & -0.2077 & 6.7150 & -5.6760 & 0 & 0 \\ -0.5814 & -15.6480 & 0 & 0.6750 & -11.3580 & 0 \\ -14.6630 & 2.0010 & -22.3840 & 21.6230 & -2.2720 & -25.1680 \\ 0.0480 & 2.0010 & 1.3430 & -2.1040 & -2.2720 & 0 \\ 0 & 1.0000 & 0 & 0 & 0 & 0 \\ 1.0000 & 0 & 1.0000 & -1.0000 & 0 & 0 \end{pmatrix};$$

$$A_{12} = \begin{pmatrix} 0 & 0 \\ 0 & -11.3580 \\ -15.7300 & -2.2720 \\ 0 & -2.2720 \\ 0 & 1.0000 \\ 1.0000 & 0 \end{pmatrix};$$

$$A_{21} = \begin{pmatrix} 13.3310 & 0.2077 & 17.0120 & -18.0510 & 0 & 25.1680 \\ 0.5814 & 15.6480 & 0 & -0.6750 & 11.3580 & 0 \end{pmatrix};$$

$$A_{22} = \begin{pmatrix} 15.7300 & 0 \\ 0 & 11.3580 \end{pmatrix}.$$

The batch reactor, which is open-loop unstable, has $n_u = 2$ inputs, $n_y = 2$ outputs, $n_p = 4$ plant states and $n_c = 2$ controller states and $N = 2$ nodes (only the outputs are assumed to be sent over the network). See [65, 92] for more details on this example.

For all the technical details of the application of Procedure 7.9 to this benchmark example the reader is referred to [37, 38]. Here we show only the outcomes. Figure 7.9 shows the stability regions in terms of MAD and MATI for the TOD and the RR protocols for the batch reactor as can be proven on the basis of the above results. Interestingly, this shows tradeoff curves between MAD and MATI: a larger MAD requires a smaller MATI in order to guarantee stability. In addition, the delay-free results as obtained in [6], which improved the earlier bounds in [65], are exactly recovered. These delay-free results amount for the TOD protocol to $\tau_{mad} = 0$ and $h_{mati} = 0.0108$ and for the RR protocol to $\tau_{mad} = 0$ and $h_{mati} = 0.0090$. Next to

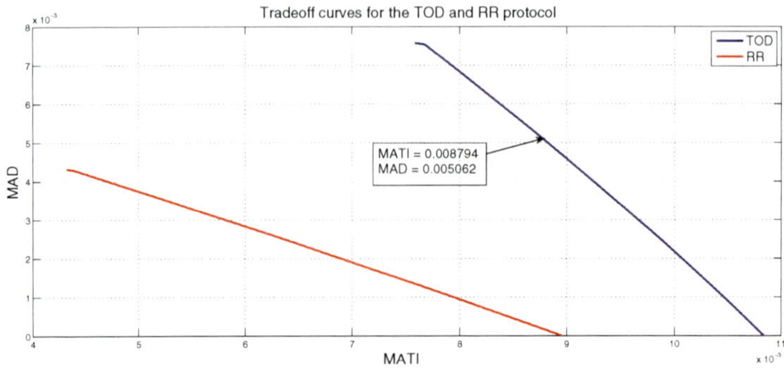

Fig. 7.9 Tradeoff curves between MATI and MAD

finding tradeoffs between MAD and MATI, different protocols can be compared with respect to each other. In Figure 7.9, it is apparent that for the task of stabilization of the unstable batch reactor the TOD protocol outperforms the RR protocol in the sense that it can allow for larger delays and larger transmission intervals. Note that in Figure 7.9 the particular combination $\tau_{mati} = 0.008794$ and $\tau_{mad} = 0.005062$ corresponding to Figure 7.8 is highlighted.

Extension of these results to include guarantees on disturbance attenuation properties in the sense of $\mathscr{L}_p$ gains from certain disturbance inputs to to-be-controlled outputs are reported as well in [38, 37]. In case of the batch reactor this would yield results as depicted in Figure 7.10 for the $\mathscr{L}_2$ gain. This picture shows tradeoffs between the network properties MAD and MATI on the one hand and control performance in terms of $\mathscr{L}_2$ gain from a specific disturbance input to a controlled output variable. These tradeoff curves are very useful for control and network designers to make well founded design decisions.

7.4.2 Discrete-Time Approach

The continuous-time (emulation) approach as presented in Section 7.4.1 applies to general *continuous-time* nonlinear plants and controllers. However, it does not include the possibility of allowing the controller to be formulated in discrete time. The case of discrete-time controllers has been considered in [17], where however, a fixed transmission interval and no delay are assumed. Another feature of the continuous-time approach is that the lower bounds on the transmission intervals h_k and delays τ_k are always equal to zero (*i.e.*, $h_k \in (\delta, h_{\text{mati}}]$, $\tau_k \in [0, \tau_{\text{mad}}]$, where δ could be chosen arbitrarily close to 0). The ability to handle discrete-time controllers and nonzero lower bounds on the transmission intervals and delays is highly relevant from a practical point of view, because controllers are typically implemented in a digital and, thus, discrete-time form. Furthermore, finite communication bandwidth introduces nonzero lower bounds on the transmission intervals and transmission delays. The discrete-time approach surveyed here (see [22, 21]) studies these highly relevant situations as well, although in a *linear* context. The linearity property is exploited

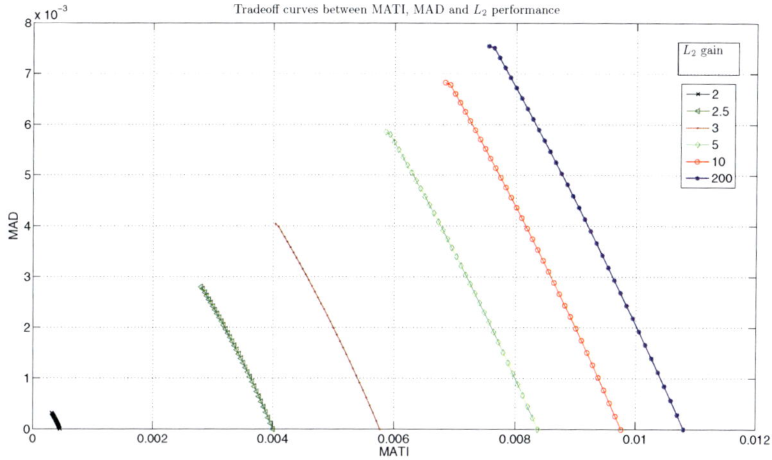

Fig. 7.10 Tradeoff curves between MATI and MAD for various levels of the $\mathscr{L}_2$ gain of the NCS with the TOD protocol

in the stability analysis and leads to less conservative results than the continuous-time approach. However, note that the continuous-time approach can accommodate for NCSs based on nonlinear plants and controllers and general (UGES) protocols, features that the discrete-time approach does not offer.

7.4.2.1 The Exact Discrete-Time NCS Model

As mentioned, the discrete-time approach applies in a *linear* context, which means that (7.48) is replaced by the linear time-invariant (LTI) continuous-time plant given by

$$\begin{aligned} \dot{x}^p(t) &= A^p x^p(t) + B^p \hat{u}(t) \\ y(t) &= C^p x^p(t), \end{aligned} \tag{7.70}$$

where $x^p \in \mathbb{R}^{n_p}$ denotes the state of the plant, $\hat{u} \in \mathbb{R}^{n_u}$ the most recently received control variable, $y \in \mathbb{R}^{n_y}$ the (measured) output of the plant and $t \in \mathbb{R}^+$ the time. The controller, also an LTI system, is assumed to be given in either continuous time by

$$\begin{aligned} \dot{x}^c(t) &= A^c x^c(t) + B^c \hat{y}(t) \\ u(t) &= C^c x^c(t) + D^c \hat{y}(t), \end{aligned} \tag{7.71a}$$

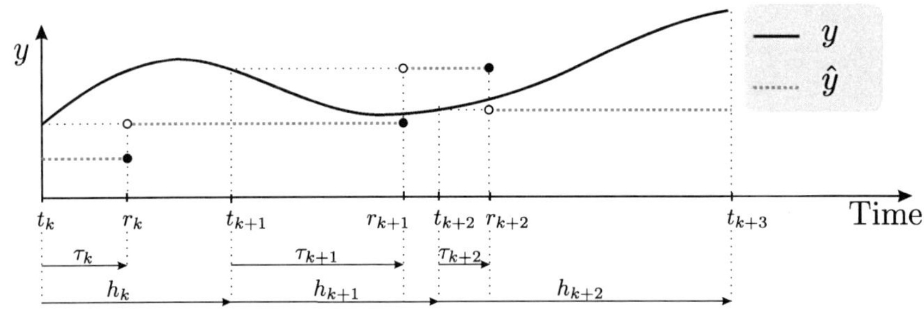

Fig. 7.11 Illustration of a typical evolution of y and $\hat{y}$

or in discrete time by

$$\begin{aligned} x^c_{k+1} &= A^c x^c_k + B^c \hat{y}_k \\ u(t_k) &= C^c x^c_k + D^c \hat{y}(t_k). \end{aligned} \tag{7.71b}$$

In parallel with Section 7.4.1 (only subscripts becoming superscripts, as subscripts are used to indicate the counter k of the discrete-time step), $x^c \in \mathbb{R}^{n_c}$ denotes the state of the controller, $\hat{y} \in \mathbb{R}^{n_y}$ the most recently received output of the plant and $u \in \mathbb{R}^{n_u}$ denotes the controller output. At transmission instant $t_k, k \in \mathbb{N}$, (parts of) the outputs of the plant $y(t_k)$ and controller $u(t_k)$ are sampled and are transmitted over the network. It is assumed that they arrive at instant $r_k = t_k + \tau_k$, called the arrival instant, where τ_k denotes the communication delay. The situation described above is illustrated in Figure 7.11. In the case of a discrete-time controller (7.71b), the states of the controller x^c_{k+1} are updated using $\hat{y}_k := \lim_{t \downarrow r_k} \hat{y}(t)$, directly after $\hat{y}$ is updated. Note that in this case, the update of x^c_{k+1} in (7.71b) has to be performed in the time interval $(r_k, t_{k+1}]$.

The functioning of the network will now be explained in more detail by defining these 'most recently received' $\hat{y}$ and $\hat{u}$ exactly. As in the continuous-time (emulation) approach in Section 7.4.1, the plant is equipped with sensors and actuators that are grouped into N nodes. At each transmission instant t_k, $k \in \mathbb{N}$, one node, denoted by $\sigma_k \in \{1, \ldots, N\}$, obtains access to the network and transmits its corresponding values. These transmitted values are received and implemented on the controller or the plant at arrival instant r_k. As was assumed in Section 7.4.1, a transmission only occurs after the previous transmission has arrived, *i.e.*, $t_{k+1} > r_k \geqslant t_k$, for all $k \in \mathbb{N}$. In other words, also here the small delay case is treated in the sense that the delay is smaller than the transmission interval $\tau_k \leq h_k := t_{k+1} - t_k$. After each transmission and reception, the values in $\hat{y}$ and $\hat{u}$ are updated using the newly received values, while the other values in $\hat{y}$ and $\hat{u}$ remain the same, as no additional information has been received for them. This leads to the constrained data exchange expressed as

$$\begin{cases} \hat{y}(t) = \Gamma^y_{\sigma_k} y(t_k) + (I - \Gamma^y_{\sigma_k}) \hat{y}(t_k) \\ \hat{u}(t) = \Gamma^u_{\sigma_k} u(t_k) + (I - \Gamma^u_{\sigma_k}) \hat{u}(t_k) \end{cases} \tag{7.72}$$

for all $t \in (r_k, r_{k+1}]$, where $\Gamma_{\sigma_k} := \operatorname{diag}\left(\Gamma^y_{\sigma_k}, \Gamma^u_{\sigma_k}\right)$ is the diagonal matrix given by

$$\Gamma_i = \operatorname{diag}\left(\gamma_{i,1}, \ldots, \gamma_{i,n_y+n_u}\right), \tag{7.73}$$

where $\sigma_k = i$ and the elements $\gamma_{i,j}$, $j \in \{1, \ldots, n_y\}$, are equal to one, if plant output y^j is in node i, elements $\gamma_{i,j+n_y}$, $j \in \{1, \ldots, n_u\}$, are equal to one, if controller output u^j is in node i, and are zero elsewhere. Note that (7.72) is directly related to (7.50) in the continuous-time approach with $h_y(k, e_y(t_k)) = (I - \Gamma^y_{\sigma_k})e(t_k)$, $h_u(k, e(t_k)) = (I - \Gamma^u_{\sigma_k})e_u(t_k)$, $e_y(t) = \hat{y}(t) - y(t)$ and $e_u(t) = \hat{u}(t) - u(t)$.

The value of $\sigma_k \in \{1, \ldots, N\}$ in (7.72) indicates which node is given access to the network at transmission instant t_k, $k \in \mathbb{N}$. Indeed, (7.72) reflects that the values in $\hat{u}$ and $\hat{y}$ corresponding to node σ_k are updated just after r_k, with the corresponding transmitted values at time t_k, while the others remain the same. A scheduling protocol determines the sequence $(\sigma_0, \sigma_1, \ldots)$ such as the Round Robin and Try-Once-Discard protocols discussed earlier.

The transmission instants t_k, as well as the arrival instants r_k, $k \in \mathbb{N}$ are not necessarily distributed equidistantly in time. Hence, both the transmission intervals $h_k := t_{k+1} - t_k$ and the transmission delays $\tau_k := r_k - t_k$ are varying in time, as is also illustrated in Figure 7.11. It is assumed that the variations in the transmission interval and delays are bounded and are contained in the sets $[h_{\min}, h_{\max}]$ and $[\tau_{\min}, \tau_{\max}]$, respectively, with $h_{\max} \geq h_{\min} \geq 0$ and $\tau_{\max} \geq \tau_{\min} \geq 0$. Since it is assumed that each transmission delay τ_k is smaller than the corresponding transmission interval h_k, it holds that $(h_k, \tau_k) \in \Psi$, for all $k \in \mathbb{N}$, where

$$\Psi := \left\{(h, \tau) \in \mathbb{R}^2 \mid h \in [h_{\min}, h_{\max}], \tau \in [\tau_{\min}, \min\{h, \tau_{\max}\}]\right\}. \tag{7.74}$$

Note that in comparison with Section 7.4.1, h_{mati} would correspond to $h_{\max}$ and τ_{mad} to $\tau_{\max}$. However, in Section 7.4.1 it was assumed that $\tau_{\min} = 0$ and $h_{\min} = \delta$, where δ could be chosen arbitrarily small, due to the emulation type of approach, while that is not the case here. Therefore, here the different notation using $h_{\min}$, $h_{\max}$, $\tau_{\min}$ and $\tau_{\max}$ is used.

To analyse stability of the NCS described above, it is transformed into a discrete-time model. In this framework, a discrete-time equivalent of (7.70) is needed. Additionally, when a continuous-time controller is used, also a discretization of (7.71a) is needed. To arrive at this description, define the network-induced error as

$$\begin{cases} e^y(t) := \hat{y}(t) - y(t) \\ e^u(t) := \hat{u}(t) - u(t). \end{cases} \tag{7.75}$$

By exact discretization of (7.70) and/or (7.71a) a discrete-time switched uncertain system can be obtained that describes the evolution of the states between t_k and $t_{k+1} = t_k + h_k$. In order to do so, define $x^p_k := x^p(t_k)$, $u_k := u(t_k)$, $\hat{u}_k := \lim_{t \downarrow r_k} \hat{u}(t)$ and $e^u_k := e^u(t_k)$. This results in three different models each describing a particular NCS. The first and the second model cover the situation where both the plant and the controller outputs are transmitted over the network, differing by the fact that the controller is given by (7.71a) and (7.71b), respectively. In the third model, it is assumed that the controller is given by (7.71a) and that only the plant outputs y are transmitted over the network and u are sent continuously via an ideal nonnetworked connection. This particular case is included, because it is often used in examples in NCS literature, *e.g.*, for the benchmark example of the batch reactor as discussed before in Section 7.4.1.

The NCS Model with Continuous-Time Controller (7.71a)

For an NCS having continuous-time controller (7.71a), the complete NCS model is obtained by combining (7.72), (7.75) with exact discretizations of plant (7.70) and controller (7.71a) and defining

$$\bar{x}_k := \begin{bmatrix} x_k^{p\top} & x_k^{c\top} & e_k^{y\top} & e_k^{u\top} \end{bmatrix}^\top. \tag{7.76}$$

This results in the discrete-time model

$$\bar{x}_{k+1} = \underbrace{\begin{bmatrix} A_{h_k} + E_{h_k} BDC & E_{h_k} BD - E_{h_k - \tau_k} B \Gamma_{\sigma_k} \\ C(I - A_{h_k} - E_{h_k} BDC) & I - D^{-1}\Gamma_{\sigma_k} + C(E_{h_k - \tau_k} B \Gamma_{\sigma_k} - E_{h_k} BD) \end{bmatrix}}_{=:\tilde{A}_{\sigma_k, h_k, \tau_k}} \bar{x}_k \tag{7.77}$$

in which $\tilde{A}_{\sigma_k, h_k, \tau_k} \in \mathbb{R}^{n \times n}$, with $n = n_p + n_c + n_y + n_u$, and

$$A_{h_k} := \text{diag}(e^{A^p h_k}, e^{A^c h_k}), \qquad B := \begin{bmatrix} 0 & B^p \\ B^c & 0 \end{bmatrix}, C \qquad := \text{diag}(C^p, C^c), \tag{7.79a}$$

$$D := \begin{bmatrix} I & 0 \\ D^c & I \end{bmatrix}, \; E_\rho := \text{diag}(\int_0^\rho e^{A^p s} ds, \int_0^\rho e^{A^c s} ds), \quad \rho \in \mathbb{R}. \tag{7.79b}$$

The NCS Model with Discrete-Time Controller (7.71b)

For an NCS having controller (7.71b), the complete NCS model is obtained by combining (7.71b), (7.72), (7.75), and an exact discretization of the continuous-time plant (7.70), also resulting in (7.77), in which now

$$A_{h_k} := \text{diag}(e^{A^p h_k}, A^c), \qquad B := \begin{bmatrix} 0 & B^p \\ B^c & 0 \end{bmatrix}, C \qquad := \text{diag}(C^p, C^c), \tag{7.80a}$$

$$D := \begin{bmatrix} I & 0 \\ D^c & I \end{bmatrix}, \; E_\rho := \text{diag}(\int_0^\rho e^{A^p s} ds, I), \quad \rho \in \mathbb{R}. \tag{7.80b}$$

The NCS Model if Only y is Transmitted over the Network

In this case it is assumed that only the outputs of the plant are transmitted over the network and the controller communicates its values continuously and without delay.

Therefore it holds that $u(t) = \hat{u}(t)$, for all $t \in \mathbb{R}^+$, which allows the combination of (7.70) and (7.71a) into

$$\begin{bmatrix} \dot{x}^p(t) \\ \dot{x}^c(t) \end{bmatrix} = \begin{bmatrix} A^p & B^pC^c \\ 0 & A^c \end{bmatrix} \begin{bmatrix} x^p(t) \\ x^c(t) \end{bmatrix} + \begin{bmatrix} B^pD^c \\ B^c \end{bmatrix} \hat{y}(t). \tag{7.81}$$

Since $\hat{y}$ is still updated according to (7.72), the evolution of the states between t_k and $t_{k+1} = t_k + h_k$ can also be described by exact discretization. In this case, (7.76) reduces to

$$\bar{x}_k := \begin{bmatrix} x_k^{p\top} & x_k^{c\top} & e_k^{y\top} \end{bmatrix}^\top, \tag{7.82}$$

resulting in (7.77), in which

$$A_{h_k} := e^{\begin{bmatrix} A^p & B^pC^c \\ 0 & A^c \end{bmatrix} h_k}, \qquad B := \begin{bmatrix} B^pD^c \\ B^c \end{bmatrix}, \qquad C := \begin{bmatrix} C^p & 0 \end{bmatrix}, \tag{7.83a}$$

$$D := I,\ E_\rho := \int_0^\rho e^{\begin{bmatrix} A^p & B^pC^c \\ 0 & A^c \end{bmatrix} s} ds, \quad \rho \in \mathbb{R}. \tag{7.83b}$$

Protocols as a Switching Function

Based on the previous modeling steps, the NCS is reformulated as the discrete-time switched uncertain system (7.77). In this framework, protocols are considered as the switching function determining σ_k. The two protocols mentioned before, namely the Try-Once-Discard (TOD) and the Round-Robin (RR) protocol, are considered and generalized into the classes of 'quadratic' and 'periodic' protocols, respectively.

A *quadratic* protocol is a protocol, for which the switching function can be written as

$$\sigma_k = \arg \min_{i=1,\dots,N} \bar{x}_k^\top P_i \bar{x}_k, \tag{7.84}$$

where P_i, $i \in \{1,\dots,N\}$, are certain given matrices. In fact, the TOD protocol belongs to this class of protocols, see [22, 21].

A *periodic* protocol is a protocol that satisfies for some $\tilde{N} \in \mathbb{N}$

$$\sigma_{k+\tilde{N}} = \sigma_k \tag{7.85}$$

for all $k \in \mathbb{N}$. $\tilde{N}$ is then called the period of the protocol. Clearly, the RR protocol belongs to this class.

The above modeling approach now provides a description of the NCS system in the form of a *discrete-time switched linear uncertain system* given by (7.77) and one of the protocols, characterized by (7.84) or (7.85). The system switches between N linear uncertain systems and the switching is due to the fact that only one node accesses the network at each transmission instant. The uncertainty is caused by the fact that the transmission intervals and the transmission delays $(h_k, \tau_k) \in \Psi$ are varying over time.

Remark 7.20. If there is only one node $N = 1$ and $\Gamma_i = I$, the setting of Section 7.3.2 for the case of small delays is recovered. If in addition, $\tau_{max} = 0$ and $h_{min} = h_{max}$, then the NCS reduces to a standard sampled-data system (in case of a continuous-time plant and discrete-time controller), see, *e.g.*, [24, 8].

7.4.2.2 The Polytopic Overapproximation

As in the case without communication constraints (see Section 7.3.2), also in the NCS models derived in the previous section the uncertainty appears in an exponential manner (see the terms A_{h_k}, $A_{h_k-\tau_k}$, E_{h_k} and $E_{h_k-\tau_k}$ in (7.77)). To convert these descriptions into a suitable form for robust stability analysis, a polytopic overapproximation method is exploited. Essentially any of the available overapproximation methods (*e.g.*, the one based on the real Jordan form discussed before) can be applied. However, in [21], in which the above modeling is presented, a combination of gridding and norm-bounding is combined into a new and efficient method. The gridding method in [21] has the advantage that if the exact discrete-time model (so before the overapproximation) is exponentially stable proven by a parameter-dependent quadratic Lyapunov function [15], then the overapproximated polytopic system with a sufficiently dense set of grid points in Ψ also has a parameter-dependent quadratic Lyapunov function. The Lyapunov function for the polytopic system can then be found by LMIs. In other words, if the original NCS system is "quadratically stable," then the LMIs derived in [21] will prove this for a sufficiently dense gridding. The reader is referred to [22, 21] for full details. Actually, in a recent paper [39] a comparison was made between the various polytopic overapproximation methods and it was argued that the gridding method of [22, 21] has the most favorable properties of the studied methods.

7.4.2.3 Application to the Batch Reactor

The exact same setup will be analyzed as for the continuous-time approach in Subsection 7.4.1.3 and focus on the TOD and RR protocol and assume that the controller is directly connected to the actuator, *i.e.*, only the (two) outputs are transmitted via the network. Using the LMIs as in [21] the aim is to construct combinations of h_{max} and τ_{max} for which the NCS is stable, and it is assumed that $\tau_{min} = 0$ and $h_{min} = 10^{-4}$. This results in tradeoff curves, as shown in Figure 7.12. These tradeoff curves can be used to impose or select a suitable network with certain communication delay and bandwidth requirements. Note that bandwidth is inversely proportional with the transmission interval.

Moreover, in Fig 7.12, also the tradeoff curves as obtained for the continuous-time approach (Subsection 7.4.1.3) are given. It can be observed that the discrete-time approach provides less conservative results than the continuous-time approach (at least for this example). More interestingly, in case there is no delay, *i.e.*, $\tau_{min} = \tau_{max} = 0$, the maximum allowable transmission interval h_{max} obtained in [6], which provide the least conservative results known in literature so far, was $h_{max} = 0.0108$, while the discrete-time approach results in $h_{max} = 0.066$. In [92],

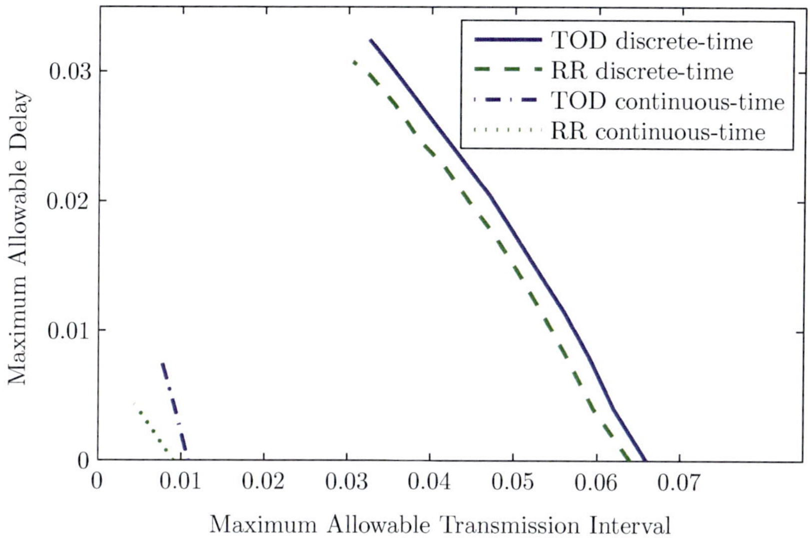

Fig. 7.12 Tradeoff curves between allowable transmission intervals and transmission delays for two different protocols, where "continuous-time" refers to the approach in Section 7.4.1, while the other curves refer to the discrete-time approach as discussed in Section 7.4.2

the largest $h_{\max}$ for which stability can be guaranteed was estimated (using simulations) to be approximately 0.08 for the TOD protocol, so the result of $h_{\max} = 0.066$ as found here already approaches this value accurately. Furthermore, for the RR protocol, [6] provides the bound $h_{\max} = 0.009$ in the delay-free case, while the discrete-time approach yields stability for $h_{\max} = 0.064$. Also in [92], for a constant transmission interval, *i.e.* $h_{\min} = h_{\max}$, the bound 0.0657 was obtained for the RR protocol. Obviously, the case where the transmission interval is constant, provides an upper bound on the true MATI. Therefore, one can conclude that for this example, the discrete-time methodology reduces conservatism significantly in comparison to existing approaches including the continuous-time approach discussed in Section 7.4.1. Furthermore, the discrete-time approach even approximates known estimates of the true MATI ($h_{mati} = h_{\max}$) closely. In addition, the discrete-time approach applies to situations (non-zero lower bounds and discrete-time controllers, see [21] for examples) that cannot be handled by the continuous-time methodologies.

7.4.3 *Comparison of Discrete-Time and Continuous-Time Approaches*

Interestingly, both the discrete-time and the continuous-time approaches exploit a NCS model that is intrinsically of a hybrid nature. The continuous-time (emulation) approach results in *hybrid inclusions* with flows and resets [31], while the discrete-time approach uses *uncertain switched linear systems* that are overapproximated by *uncertain switched polytopic systems*.

There are some clear (dis)advantages of both methods. The continuous-time (emulation) approach as presented in Subsection 7.4.1 applies to general *continuous-time* nonlinear plants and controllers, and general (UGES) protocols. In addition, in the continuous-time approach $\mathscr{L}_p$ gain analysis of the original continuous-time NCS can be done in a straightforward manner (see [38]), while this not true for the discrete-time approach. However, the continuous-time approach does not allow for discrete-time controllers and cannot handle nonzero lower bounds on the transmission intervals h_k and delays τ_k. The discrete-time approach as discussed in Subsection 7.4.2 can allow for both *continuous-time* and *discrete-time* controllers and *non-zero* lowerbounds on delays and transmission intervals. However, it applies to the case of *linear* plants and controllers and specific protocols (periodic and quadratic protocols) only, although it can do this in a significantly less conservative manner than the "general-purpose" continuous-time approach.

7.5 Conclusions

In this overview we discussed various approaches to the modeling, stability analysis and stabilizing controller synthesis of NCSs with varying delays, varying transmission intervals, packet dropouts and communication constraints. The methods discussed in this chapter all assume hard bounds on the size of the varying delays, transmission intervals and the maximal number of subsequent dropouts. Roughly speaking, three main lines can be distinguished, as summarized below.

(i) The discrete-time approach is based on a discrete-time NCS model, which can be used for both discrete-time and continuous-time linear controllers and linear plants. LMI-based stability conditions are derived by using common or parameter-dependent quadratic Lyapunov functions for an overapproximated polytopic model. Different methods are available for performing the polytopic overapproximation of the discretized NCS model in which the delay and sampling interval uncertainties appear in an exponential fashion, see [39]. Within this research line the largest number of network-induced imperfections are treated in [21] that includes varying sampling intervals and delays, dropouts and communication constraints. Both discrete-time and continuous-time controllers can be handled. However, in [21] only delays smaller than the sampling interval are considered. For the most comprehensive discrete-time approach including large delays, but without communication constraints, we refer to [14, 11].

(ii) The sampled-data approach uses (impulsive) delay-differential equations describing continuous-time sampled-data NCS models with discrete-time controllers. Extensions of the classical Lyapunov-Krasovskii functionals are exploited to give stability guarantees for *linear* plants and controllers. The conditions also result in LMIs. Communication constraints have not (yet) been treated using (impulsive) delay-differential equations.

(iii) The continuous-time (emulation) approach studies continuous-time sampled-data NCS models with continuous-time controllers and transforms these into hybrid models. Continuous-time Lyapunov functions are constructed by combining individual Lyapunov functions of the network-free closed-loop system and the network protocol (or adopting directly small gain arguments) to assess stability of the NCS. Typically only continuous-time controllers are treated at present, however both controller and plant can be nonlinear.

Although already three main research lines of NCSs exist for stability analysis (where we did not even discuss the stochastic approaches) and many papers are available, at present there are almost no complete frameworks that can handle all the 5 mentioned network-induced imperfections: (i) Variable sampling/transmission intervals; (ii) variable communication delays; (iii) Packet dropouts (iv) Communication constraints; (v) Quantization errors. Only recently one approach was presented in [35] that includes all 5 imperfections under quite restricted conditions (small delay case, particular quantizers, continuous-time controllers, *etc.*) The availability of a complete analysis and design frameworks based on either one of the main lines as discussed in this chapter or possibly on a completely new line, would be desirable. At present such a framework is not available, certainly not with the analysis and design techniques implemented in suitable software tools. One of the goals of this chapter was to survey the main techniques and discuss the open problems, which will hopefully stimulate further research in this direction in order to develop this envisioned framework and toolset in the near future.

Particular attention should be given to controller design methods, considering that for the three main lines these methods are still rather limited and extensions are needed. Indeed, most of the design techniques lead to static feedback controllers, while in industrial practice there is a strong need for output-based dynamic controllers. In the discrete-time approach ordinary and lifted state feedback controllers could be designed using LMI conditions, while in the emulation approach, continuous-time controllers are synthesized based on the network-free nonlinear system using any available method for the design of stabilizing controllers for nonlinear systems, which is in general a nontrivial task. As the emulation design does not incorporate any information on the network, it is hard to design controllers that are stabilizing and performing for sufficiently long delays and transmission intervals, although one can aim at obtaining favorable properties through the presented stability conditions. In addition, one might wonder if continuous-time controllers are useful in practical problems as most NCS setups will require digital discrete-time controllers that are tailored towards non-zero lower bounds on delays and transmission intervals. Of course, one option is to implement the continuous-time controller using numerical integration schemes, however it is unclear if these controllers will have the desired properties in the end. Also for the sampled-data approach constructive design methods seem to be missing in the literature, certainly in an efficient form. Hence, as a final conclusion one can state that the controller design and controller/protocol co-design techniques for NCSs are still in their infancy and deserve a lot of attention in the years to come.

Acknowledgements. The authors are grateful to João Hespanha, Nick Bauer and Tijs Donkers for giving detailed comments on an earlier version of this manuscript.

References

1. Antunes, D., Hespanha, J., Silvestre, C.: Control of impulsive renewal systems: Application to direct design in networked control. In: IEEE Conference on Decision and Control, pp. 6882–6887 (2009)
2. Antunes, D., Hespanha, J., Silvestre, C.: Stability of impulsive systems driven by renewal processes. In: Proc. Amer. Contr. Conf., pp. 4032–4037 (2009)
3. Balluchi, A., Murrieri, P., Sangiovanni-Vincentelli, A.L.: Controller synthesis on non-uniform and uncertain discrete-time domains. In: Morari, M., Thiele, L. (eds.) HSCC 2005. LNCS, vol. 3414, pp. 118–133. Springer, Heidelberg (2005)
4. Brockett, R.: Stabilization of motor networks. In: Proc. of the 34th IEEE Conf. on Decision and Control, vol. 2, pp. 1484–1488 (1995)
5. Brockett, R.W., Liberzon, D.: Quantized feedback stabilization of linear systems. IEEE Trans. Autom. Control 45, 1279–1289 (2000)
6. Carnevale, D., Teel, A.R., Nešić, D.: A Lyapunov proof of improved maximum allowable transfer interval for networked control systems. IEEE Trans. Aut. Control 52, 892–897 (2007)
7. Chaillet, A., Bicchi, A.: Delay compensation in packet-switching networked controlled sytems. In: IEEE Conf. Decision and Control, pp. 3620–3625 (2008)
8. Chen, T., Francis, B.A.: Optimal Sampled-data Control Systems. Springer, London (1995)
9. Cloosterman, M.: Control over Communication Networks: Modeling, Analysis, and Synthesis. PhD thesis Eindhoven University of Technology (2008)
10. Cloosterman, M., van de Wouw, N., Heemels, W.P.M.H., Nijmeijer, H.: Robust stability of networked control systems with time-varying network-induced delays. In: Proc. IEEE Conf. on Decision and Control, San Diego, USA, December 2006, pp. 4980–4985 (2006)
11. Cloosterman, M.B.G., Hetel, L., van de Wouw, N., Heemels, W.P.M.H., Daafouz, J., Nijmeijer, H.: Controller synthesis for networked control systems. Automatica 46, 1584–1594 (2010)
12. Cloosterman, M.B.G., van de Wouw, N., Heemels, W.P.M., Nijmeijer, H.: Stabilization of networked control systems with large delays and packet dropouts. In: American Control Conference, pp. 4991–4996 (2008)
13. Cloosterman, M.B.G., van de Wouw, N., Heemels, W.P.M.H., Nijmeijer, H.: Stability of networked control systems with large delays. In: 46th IEEE Conference on Decision and Control, pp. 5017–5022 (2007)
14. Cloosterman, M.B.G., van de Wouw, N., Heemels, W.P.M.H., Nijmeijer, H.: Stability of networked control systems with uncertain time-varying delays. IEEE Trans. Autom. Control 54(7), 1575–1580 (2009)
15. Daafouz, J., Bernussou, J.: Parameter dependent Lyapunov functions for discrete time systems with time varying parametric uncertainties. Systems & Control Letters 43, 355–359 (2001)
16. Daafouz, J., Riedinger, P., Iung, C.: Stability analysis and control synthesis for switched systems: A switched Lyapunov function approach. IEEE Transactions on Automatic Control 47, 1883–1887 (2002)

17. Dačić, D.B., Nešić, D.: Quadratic stabilization of linear networked control systems via simultaneous protocol and controller design. Automatica 43, 1145–1155 (2007)
18. de Oliveira, M.C., Bernussou, J., Geromel, J.C.: A new discrete-time robust stability condition. Systems & Control Letters 37, 261–265 (1999)
19. Delchamps, D.F.: Stabilizing a linear system with quantized state feedback. IEEE Trans. Autom. Control 35, 916–924 (1990)
20. Donkers, M.C.F., Heemels, W.P.M.H., Bernardini, D., Bemporad, A., Shneer, V.: Stability analysis of stochastic networked control systems. In: Proc. Amer. Contr. Conf., pp. 3684–3689 (2010)
21. Donkers, M.C.F., Heemels, W.P.M.H., van de Wouw, N., Hetel, L.: Stability analysis of networked control systems using a switched linear systems approach. Submitted for journal publication
22. Donkers, M.C.F., Hetel, L., Heemels, W.P.M.H., van de Wouw, N., Steinbuch, M.: Stability analysis of networked control systems using a switched linear systems approach. In: Majumdar, R., Tabuada, P. (eds.) HSCC 2009. LNCS, vol. 5469, pp. 150–164. Springer, Heidelberg (2009)
23. Dritsas, L., Tzes, A.: Robust stability analysis of networked systems with varying delays. International Journal of Control (2009)
24. Franklin, G.F., Powell, J.D., Workman, M.L.: Digital control of dynamic systems. Addison-Wesley Pub. Co. Inc., Reading (1990)
25. Fridman, E., Seuret, A., Richard, J.P.: Robust sampled-data stabilization of linear systems: an input delay approach. Automatica 40, 1441–1446 (2004)
26. Fujioka, H.: Stability analysis for a class of networked/embedded control systems: A discrete-time approach. In: Proc. of the American Control Conf., pp. 4997–5002 (2008)
27. Gao, H., Chen, T., Lam, J.: A new delay system approach to network-based control. Automatica 44(1), 39–52 (2008)
28. Garcia-Rivera, M., Barreiro, A.: Analysis of networked control systems with drops and variable delays. Automatica 43, 2054–2059 (2007)
29. Gielen, R., Olaru, S., Lazar, M.: On polytopic embeddings as a modeling framework for networked control systems. In: Proc. 3rd Int. Workshop on Assessment and Future Directions of Nonlinear Model Predictive Control, Pavia, Italy (2008)
30. Gielen, R.H., Olaru, S., Lazar, M., Heemels, W.P.M.H., van de Wouw, N., Niculescu, S.-I.: On polytopic inclusions as a modeling framework for systems with time-varying delays. Automatica 46(3), 615–619 (2010)
31. Goebel, R., Sanfelice, R., Teel, A.R.: Hybrid dynamical systems. IEEE Control Systems Magazine 29(2), 28–93 (2009)
32. Gupta, V., Chung, T.H., Hassibi, B., Murray, R.M.: On a stochastic sensor selection algorithm with applications in sensor scheduling and sensor coverage. Automatica 42(2), 251–260 (2006)
33. Hea, Y., Wang, Q.G., Wua, M., Lin, C.: Delay-range-dependent stability for systems with time-varying delay. Automatica 43, 371–376 (2007)
34. Heemels, W.P.M.H., Gorter, R.J.A., van Zijl, A., Bosch, P.P.J.v.d., Weiland, S., Hendrix, W.H.A., Vonder, M.R.: Asynchronous measurement and control: a case study on motor synchronisation. Control Engineering Practice 7(12), 1467–1482 (1999)
35. Heemels, W.P.M.H., Nešić, D., Teel, A.R., van de Wouw, N.: Networked and quantized control systems with communication delays. In: Proc. Joint 48th IEEE Conference on Decision and Control (CDC) and 28th Chinese Control Conference, Shanghai, China, pp. 7929–7935 (2009)
36. Heemels, W.P.M.H., Siahaan, H., Juloski, A., Weiland, S.: Control of quantized linear systems: an l_1-optimal control approach. In: Proc. American Control Conference, Denver, USA, pp. 3502–3507 (2003)

37. Heemels, W.P.M.H., Teel, A.R., van de Wouw, N., Nešić, D.: Networked control systems with communication constraints: tradeoffs between sampling intervals and delays. In: Proc. European Control Conference in Budapest, Hungary (2009)
38. Heemels, W.P.M.H., Teel, A.R., van de Wouw, N., Nešić, D.: Networked control systems with communication constraints: Tradeoffs betweens transmission intervals, delays and performance. IEEE Trans. Autom. Control 55(8), 1781–1796 (2010)
39. Heemels, W.P.M.H., van de Wouw, N., Gielen, R.H., Donkers, M.C.F., Hetel, L., Olaru, S., Lazar, M., Daafouz, J., Niculescu, S.: Comparison of overapproximation methods for stability analysis of networked control systems. In: 13th International Workshop on Hybrid Systems: Computation and Control, Stockholm, Sweden, pp. 181–190 (2010)
40. Hespanha, J.P., Naghshtabrizi, P., Xu, Y.: A survey of recent results in networked control systems. Proc. of the IEEE, 138–162 (2007)
41. Hetel, L., Cloosterman, M.B.G., van de Wouw, N., Heemels, W.P.M.H., Daafouz, J., Nijmeijer, H.: Comparison of stability characterisations for networked control systems. In: Proc. Joint 48th IEEE Conference on Decision and Control (CDC) and 28th Chinese Control Conference, Shanghai, China, pp. 7911–7916 (2009)
42. Hetel, L., Daafouz, J., Iung, C.: Stabilization of arbitrary switched linear systems with unknown time-varying delays. IEEE Trans. Autom. Control 51(10), 1668–1674 (2006)
43. Hetel, L., Daafouz, J., Iung, C.: Analysis and control of LTI and switched systems in digital loops via an event-based modeling. International Journal of Control 81(7), 1125–1138 (2008)
44. Horn, R., Johnson, C.R.: Matrix Analysis. Cambridge University Press, Cambridge (1985)
45. Hristu, D., Morgansen, K.: Limited communication control. Systems & Control Letters 37(4), 193–205 (1999)
46. Ishii, H.: H-infty control with limited communication and message losses. Systems and Control Letters 57, 322–331 (2008)
47. Ishii, H., Francis, B.A.: Stabilization with control networks. Automatica 38, 1745–1751 (2002)
48. Kao, C.-Y., Lincoln, B.: Simple stability criteria for systems with time-varying delays. Automatica 40, 1429–1434 (2004)
49. Kothare, M., Balakrishnan, V., Morari, M.: Robust constrained model predictive control using linear matrix inequalities. Automatica 32, 1361–1379 (1996)
50. Krtolica, R., Ozguner, U., Chan, H., Goktas, H., Winkelman, J., Liubakka, M.: Stability of linear feedback systems with random communication delays. In: Proc. American Control Conf, pp. 2648–2653 (1991)
51. Liberzon, D.: On stabilization of linear systems with limited information. IEEE Trans. Autom. Control 48(2), 304–307 (2003)
52. Liberzon, D.: Quantization, time delays, and nonlinear stabilization. IEEE Trans. Autom. Control 51(7), 1190–1195 (2006)
53. Matveev, A.S., Savkin, A.V.: The problem of state estimation via asynchronous communication channels with irregular transmission times. IEEE Trans. Autom. Control 48(4), 670–676 (2003)
54. Mirkin, L.: Some remarks on the use of time-varying delay to model sample-and-hold circuits. IEEE Transactions on Automatic Control 52(6), 1109–1112 (2007)
55. Montestruque, L.A., Antsaklis, P.: Stochastic stability for model-based networked control systems. In: Proc. American Control Conf., pp. 4119–4124 (2003)
56. Montestruque, L.A., Antsaklis, P.: Stability of model-based networked control systems with time-varying transmission times. IEEE Trans. Autom. Control 49(9), 1562–1572 (2004)

57. Naghshtabrizi, P., Hespanha, J.P.: Designing an observer-based controller for a network control system. In: Proc. of the 44th Conference on Decision and Control, and the European Control Conference 2005, Séville, Spain, pp. 848–853 (December 2005)
58. Naghshtabrizi, P., Hespanha, J.P.: Stability of network control systems with variable sampling and delays. In: Proc. of the Forty-Fourth Annual Allerton Conf. on Communication, Control, and Computing (2006)
59. Naghshtabrizi, P., Hespanha, J.P., Teel, A.R.: Stability of delay impulsive systems with application to networked control systems. In: Proc. American Control Conference, New York, USA, pp. 4899–4904 (2007)
60. Naghshtabrizi, P., Hespanha, J.P., Teel, A.R.: Exponential stability of impulsive systems with application to uncertain sampled-data systems. Systems & Control Letters 57(5), 378–385 (2008)
61. Naghshtabrizi, P., Hespanha, J.P., Teel, A.R.: On the robust stability and stabilization of sampled-data systems: A hybrid system approach. In: Proc. of the 45th Conf. on Decision and Contr. (December 2006)
62. Nair, G.N., Evans, R.J.: Stabilizability of stochastic linear systems with finite feedback data rates. SIAM J. Control Optim. 43, 413–436 (2004)
63. Nešić, D., Liberzon, D.: A unified framework for design and analysis of networked and quantized control systems. IEEE Trans. Autom. Control 54(4), 732–747 (2009)
64. Nešić, D., Teel, A.R., Sontag, E.D.: Formulas relating KL-stability estimates of discrete-time and sampled-data nonlinear systems. Systems & Control Letters 38(1), 49–60 (1999)
65. Nešić, D., Teel, A.R.: Input-output stability properties of networked control systems. IEEE Trans. Autom. Control 49(10), 1650–1667 (2004)
66. Nešić, D., Teel, A.R.: Input-to-state stability of networked control systems. Automatica 40, 2121–2128 (2004)
67. Nilsson, J., Bernhardsson, B., Wittenmark, B.: Stochastic analysis and control of real-time systems with random time delays. Automatica 34, 57–64 (1998)
68. Nilsson, J.: Real-Time Control Systems with Delays. PhD thesis, Dept. of Automatic Control, Lund Inst. of Techn., Lund, Sweden (1998)
69. Olaru, S., Niculescu, S.-I.: Predictive Control for Linear Systems with Delayed Input Subject to Constraints. In: Proceedings of 17th IFAC World Congress 2008 17th IFAC World Congress 2008, CD-ROM, Seul, Korea (2008)
70. Pan, Y.-J., Marquez, H.J., Chen, T.: Stabilization of remote control systems with unknown time varying delays by LMI techniques. Int. Journal of Control 79(7), 752–763 (2006)
71. Rehbinder, H., Sanfridson, M.: Scheduling of a limited communication channel for optimal control. Automatica 40(3), 491–500 (2004)
72. Sala, A.: Computer control under time-varying sampling period: An LMI gridding approach. Automatica 41(12), 2077–2082 (2005)
73. Schenato, L.: To zero or to hold control inputs with lossy links? IEEE Trans. Autom. Control 54, 1093–1099 (2009)
74. Seiler, P., Sengupta, R.: An $\mathscr{H}_\infty$ approach to networked control. IEEE Trans. Autom. Control 50, 356–364 (2005)
75. Shi, Y., Yu, B.: Output feedback stabilization of networked control systems with random delays modeled by markov chains. IEEE Trans. Autom. Control 54, 1668–1674 (2009)
76. Sinopoli, B., Schenato, L., Franceschetti, M., Poolla, K., Jordan, M.I., Sastry, S.S.: Kalman filtering with intermittent observations. IEEE Trans. Autom. Control 49(9), 1453–1464 (2004)

77. Skaf, J., Boyd, S.: Analysis and synthesis of state-feedback controllers with timing jitter. IEEE Transactions on Automatic Control 54(3) (2009)
78. Smith, S.C., Seiler, P.: Estimation with lossy measurements: jump estimators for jump systems. IEEE Trans. Autom. Control 48(12), 2163–2171 (2003)
79. Suh, Y.S.: Stability and stabilization of nonuniform sampling systems. Automatica 44, 3222–3226 (2008)
80. Tabbara, M., Nešić, D.: Input-output stability of networked control systems with stochastic protocols and channels. IEEE Trans. Autom. Control 53, 1160–1175 (2008)
81. Tabbara, M., Nešić, D., Teel, A.R.: Stability of wireless and wireline networked control systems. IEEE Trans. Autom. Control 52(9), 1615–1630 (2007)
82. Tatikonda, S., Mitter, S.K.: Control under communication constraints. IEEE Trans. Autom. Control 49, 1056–1068 (2004)
83. Tipsuwan, Y., Chow, M.-Y.: Control methodologies in networked control systems. Control Engineering Practice 11, 1099–1111 (2003)
84. Tsumura, K., Ishii, H., Hoshina, H.: Tradeoffs between quantization and packet loss in networked control of linear systems. Automatica 45, 2963–2970 (2009)
85. van de Wouw, N., Naghshtabrizi, P., Cloosterman, M., Hespanha, J.P.: Tracking control for networked control systems. In: Proc. of the 46th IEEE Conference on Decision and Control, New Orleans, LA, USA, pp. 4441–4446 (December 2007)
86. van de Wouw, N., Naghshtabrizi, P., Cloosterman, M.B.G., Hespanha, J.P.: Tracking control for sampled-data systems with uncertain sampling intervals and delays. Intern. Journ. Robust and Nonlinear Control 20(4), 387–411 (2010)
87. van de Wouw, N., Nešić, D., Heemels, W.P.M.H.: A discrete-time framework for stability analysis of nonlinear networked control systems. (submitted, 2010)
88. van der Schaft, A.J.: L2-Gain and Passivity in Nonlinear Control, 2nd edn. Springer, Heidelberg (1999)
89. van der Schaft, A.J., Schumacher, J.M.: An Introduction to Hybrid Dynamical Systems. LNCIS, vol. 251. Springer, London (2000)
90. van Schendel, J.J.C., Donkers, M.C.F., Heemels, W.P.M.H., van de Wouw, N.: On dropout modelling for stability analysis of networked control systems. In: Proc. American Control Conf. (2010)
91. Walsh, G.C., Belidman, O., Bushnell, L.G.: Asymptotic behavior of nonlinear networked control systems. IEEE Trans. Automat. Contr. 46, 1093–1097 (2001)
92. Walsh, G.C., Ye, H., Bushnell, L.G.: Stability analysis of networked control systems. IEEE Trans. on Control Systems Technology 10(3), 438–446 (2002)
93. Wang, Y., Yang, G.: Multiple communication channels-based packet dropout compensation for networked control systems. IET Control Theory and Applications 2, 717–727 (2008)
94. Wei, G., Wang, Z., He, X., Shu, H.: Filtering for networked stochastic time-delay systems with sector nonlinearity. IEEE Trans. Circuits and Systems II: Express Briefs 56, 71–75 (2009)
95. Wittenmark, B., Nilsson, J., Törngren, M.: Timing problems in real-time control systems. In: Proc. of the Amer. Control Conf., Seattle, USA, pp. 2000–2004 (1995)
96. Xie, G., Wang, L.: Stabilization of networked control systems with time-varying network-induced delay. In: Proc. of the 43rd IEEE Conference on Decision and Control, Atlantis, Paradise Island, Bahamas, pp. 3551–3556 (December 2004)
97. Yang, F., Wang, Z., Hung, Y.S., Gani, M.: $\mathscr{H}_\infty$ control for networked systems with random communication delays. IEEE Trans. Autom. Control 51, 511–518 (2006)
98. Yang, T.C.: Networked control system: a brief survey. IEE Proc.-Control Theory Appl. 153(4), 403–412 (2006)

99. Yu, M., Wang, L., Chu, T.: Sampled-data stabilization of networked control systems with nonlinearity. IEE Proc.-Control Theory Appl. 152(6), 609–614 (2005)
100. Yue, D., Han, Q.-L., Peng, C.: State feedback controller design of networked control systems. IEEE Trans. on circuits and systems 51(11), 640–644 (2004)
101. Zhang, L., Shi, Y., Chen, T., Huang, B.: A new method for stabilization of networked control systems with random delays. IEEE Trans. Autom. Control 50(8), 1177–1181 (2005)
102. Zhang, W., Branicky, M.S., Phillips, S.M.: Stability of networked control systems. IEEE Control Systems Magazine 21(1), 84–99 (2001)

Chapter 8
Feedback Control over Limited Capacity Channels

Hideaki Ishii

Abstract. In this chapter, we present recent approaches towards networked control systems (NCSs) motivated by capacity constraints in communication channels. Consideration of such constraints is important since the channels are shared by various system components, and thus the communication rate necessary for control-related signals must be explicitly taken into account. The chapter consists of two main parts. In the first part, we discuss control problems involving quantization effects. While such problems have a long history in the control field, the characteristic aspect here is that the problems have strong ties to the theoretical question on how much information is necessary for the purpose of feedback control. The second part provides an information theoretic approach to the well-known Bode's integral formulae, where properties of sensitivity functions are characterized by the plant poles and zeros. We will observe the usefulness of notions such as entropy in deriving such formulae. The results in both parts exhibit fundamental limitations that arise due to the presence of capacity limited channels and are hence unique to networked control. Moreover, we will describe the close interplay between the two fields of control and communication.

8.1 Introduction

In recent years, the importance of the role played by communication in control systems has continuously grown. Indeed, in various control applications, we can find systems employing real-time remote controllers using the Internet or large-scale plants with numerous embedded components connected by shared channels. Such developments in applications have been enabled by the vast progress in communication technologies in the last twenty years.

The impact of networked control systems (NCSs) has also brought influences to the theoretical side of the control field. Consequently, new approaches for networked control have been obtained, incorporating the characteristics and effects of communication in the analysis and design of control systems; general references on

A. Bemporad, M. Heemels, and M. Johansson: Networked Control Systems, LNCIS 406, pp. 255–291.
springerlink.com

this topic include [2, 1, 26]. It must be noted that different from applications in the communications field, real-time requirements are much more strict in NCSs. This is because communication channels are employed for the purpose of feedback control, where the overall system performance is very sensitive to any delay in signal transmission and/or processing.

One of the main challenges in networked control is the constraint on capacity of the communication channels. When shared channels are used by different system components, the data rate of each signal must be counted to ensure that the total is less than the capacity of the channel. Otherwise, the performance of the overall system would degrade and even worse, the system may lose stability. Effects due to capacity constraints include time delays, losses of data, scheduling of transmissions, and encoding/quantization of signals.

In this chapter, we mainly focus on quantization, by highlighting the interplay between the two theories on control and communication. By quantization, we mean the conversion of real-valued signals to those taking discrete values, which is necessary to send data over digital channels. The new problems that we must address are (i) to introduce models of the capacity limited channels into control systems and then (ii) to find how much data rate is required for each control signal. A fundamental question unique to these problems comes down to the following: How much capacity is needed for the communication in control systems to guarantee certain stability and/or performance properties? This question will be addressed in different forms including the celebrated minimum data rate theorem.

We motivate the issue of channel capacity in NCSs through two examples.

Example 8.1. A common example of NCSs is the automotive local area networks (LANs). In-vehicle electronic devices have dramatically increased for control of various components including the engine, active suspensions, and cruise control. To reduce the amount of cables, networks are used to connect sensors, actuators, and controllers. At the plant side, electronic control units (ECUs) are placed, capable to communicate to each other and to remotely located controllers.

Automotive LANs are usually composed of several networks, each of which is responsible for a specific task and thus may employ a different protocol with appropriate capacity; see, *e.g.*, [47]. For example, a slow network with data rate 125 kbps can be used for body control such as the air conditioner, door locks, and meters. For such purposes, CAN is a useful protocol. On the other hand, for the powertrain including the engine, transmission, and steering, a much faster and more reliable network would be necessary; a network of 5 Mbps with the FlexRay protocol may be used.

In practice, the required data rate for each signal would be determined by engineers, who find the appropriate data size and the transmission frequency. To coordinate the transmissions by the many components, scheduling of the timings is necessary as well. Then, those rates are summed up. It is critical that the total is below the capacity of the channel. We would like to know how control theory can take part in this type of design procedures. $\triangledown$

Example 8.2. We next consider the networked control of the familiar inverted pendulum system (*e.g.*, [17]). The objective is to keep a pendulum on top of a cart upright by applying force to the cart and by measuring the cart position and the pendulum angle. Here, we pay attention only to the angle measurement, and suppose that this is done by an encoder with sampling period 1 ms. The encoder output goes through an AD converter using, say, 8 bits. In this case, the transmission requires at least 8000 bits per second (bps). Through experience (*e.g.*, in undergraduate labs), we know that if the controller is properly designed, such an encoder is sufficient to stabilize the system.

Now, at the same sampling period, it may still be fine to reduce the number to 7 or 6 bits with some loss in performance. However, in the extreme case of only 2 bits or even 1 bit, we do not have a clear answer whether a stabilizing control strategy exists. With 1 bit, we can communicate only very coarse information such as the pendulum being on the left or the right of the upright position, but this would clearly be difficult. Hence, it seems that there is a threshold number of bits somewhere in between. Is it possible to explicitly characterize this threshold? If such a threshold exists, it is plausible that it depends on the level of instability of the plant, but how shall we measure this? ▽

An interesting aspect of NCSs is that two fields become relevant, not only control but also communication. New applications in NCSs have provided the need to explore problems lying in the intersection of them. Though it has been long acknowledged that the the two fields have much in common, there are certain differences that can be summarized as follows (see, *e.g.*, [41]). Communication or information theory deals with general problems of reliable transmission of data and the capacity of channels required for that purpose. However, this theory is usually not concerned with the content of the data, as long as the stochastic properties of the data are known in terms of the underlying probability distributions. On the other hand, in control, the signals within systems are specifically used for feedback control; the data rate is conventionally assumed to be infinite, enabling the signals to take real values.

In this chapter, we present results exhibiting the interplay of control and communication. It consists of two parts. The first part is the main one and deals with capacity constraints and, in particular, quantization issues in NCSs. The second part focuses on the so-called Bode's integral formula, where information theoretic tools such as entropy are shown to be useful for control problems. Results in both parts show fundamental limitations arising in NCSs. These limitations set certain theoretical hard bounds on what can be achieved and cannot be overcome. In this respect, these results are analogous in nature to Shannon's theory, which shows that given a stochastic model of a channel, the amount of information that can be reliably transmitted over it by any coding scheme is bounded.

This chapter is organized as follows: For the first part on control under capacity constraints, in Section 2, we provide an introduction to quantized control; the general problem setup is given with some details on the background. Then, two approaches for quantized control are presented. In Section 3, the minimum data rate

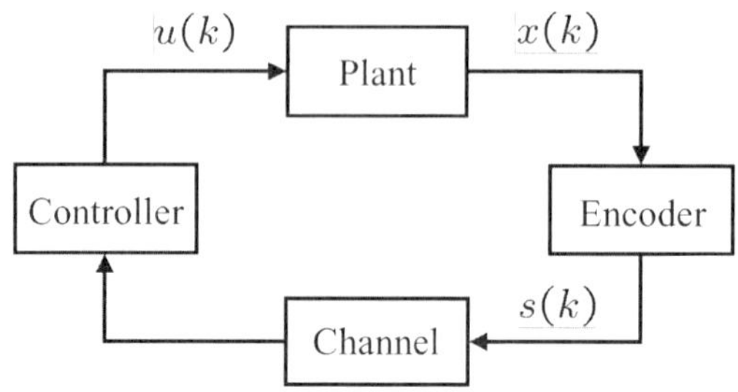

Fig. 8.1 Control under capacity constraints

problem is addressed, while in Section 4, we discuss another approach based on the notion of coarseness in quantization. Section 5 is devoted to the second part on an information theoretic approach to Bode's integral formula. Finally, in Section 6, we provide concluding remarks.

8.2 Control under Capacity Constraints: System Setup and Background

This section provides a general introduction for the first part of this chapter on quantized control.

In the field of systems control, it has been the convention to assume that signals within systems take real values. This assumption is fully justifiable under the situations where each signal is transmitted over its own dedicated channel, in which case the available bandwidth is large. The fact is that in the past, most control systems were designed in such a way. Hence, there was no need to question this assumption.

However, once the signals are to be sent over shared channels, where the capacity can no longer be assumed infinite, the situation can change significantly. Before transmission, on the sender side, signals must be converted so that they take only discrete values, which can then be represented in bits. This conversion is referred to as quantization. After quantization, some encoding schemes can be applied to the quantized data. This step adds extra information to the raw data to ensure that even if noise and error are introduced during the transmission, the original data can be recovered at the receiver side by applying proper decoding.

The focus of this section is the effect of quantization in control systems and especially very coarse quantization. The quantized control problems discussed in this chapter are mainly for the system setup in Figure 8.1. Here, the plant to be controlled has an output that goes to the encoder. There, possibly some preprocessing takes place before quantization so that the signal can be sent over the channel. For the most part of this chapter, it is assumed that the channel is noiseless and there is no communication delay. On the receiver side is the controller that generates the

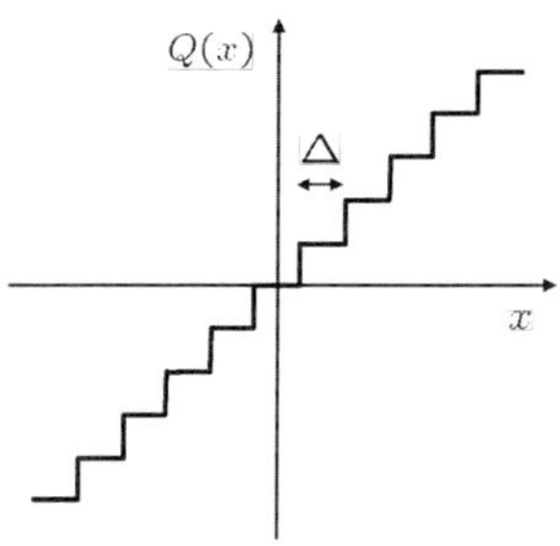

Fig. 8.2 Uniform quantizer

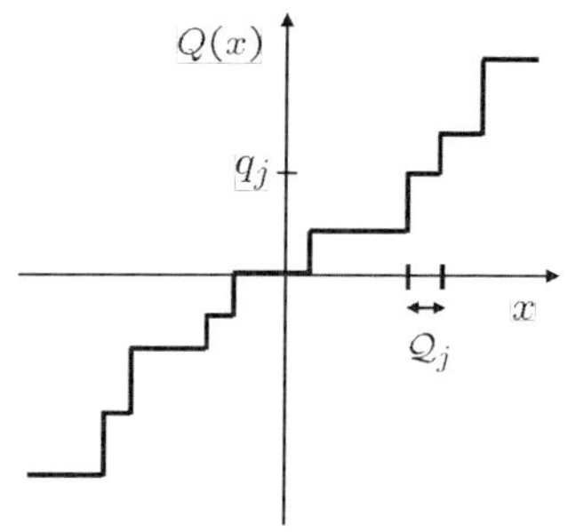

Fig. 8.3 General quantizer

control input. This setup is simple having only one channel in the feedback loop. It may represent a large-scale system whose sensors and actuators are separately located. Nevertheless, we will see that this is a good starting point in obtaining interesting and important theoretical results.

More specifically, in Figure 8.1, the plant is a discrete-time linear time-invariant (LTI) system whose state-space equation is given by

$$x(k+1) = Ax(k) + Bu(k), \tag{8.1}$$

where $x(k) \in \mathbb{R}^n$ is the state and $u(k) \in \mathbb{R}^m$ is the control input. We make the following assumptions: (i) The full state $x(k)$ is measured by sensors, (ii) the pair (A,B) is controllable (or, more generally, stabilizable), and (iii) the matrix A is unstable, having at least one eigenvalue with magnitude larger than or equal to 1[1]. Denote by λ_i^u, $i = 1,\dots,n_u$, the unstable eigenvalues of A.

By a quantizer, we mean a piecewise constant function. A typical one is the uniform quantizer Q_Δ. For the scalar case, this is defined by

$$Q_\Delta(x) = \Delta \left\lceil \frac{x - \frac{\Delta}{2}}{\Delta} \right\rceil, \quad x \in \mathbb{R}, \tag{8.2}$$

where $\Delta > 0$ is the step size and $\lceil \cdot \rceil : \mathbb{R} \to \mathbb{Z}$ is the ceiling function; see Figure 8.2. This quantizer essentially rounds off a number x to the closest multiple of Δ. Hence, the error caused by quantization is bounded as $|x - Q(x)| \le \Delta/2$ for every x. When the input x is a vector, one way is to quantize componentwise. For example, each entry x_i of x can be uniformly quantized.

It is important to note that for the purpose of systems control, uniform and/or componentwise quantization may not necessarily be the best choice to obtain good control performance. Hence, we do not have to limit ourselves to these types and introduce a more general notion of quantization as follows.

[1] In this chapter, the systems are in discrete time. Hence, we say a matrix is stable if all of its eigenvalues have magnitude less than 1.

Definition 8.1. Let $\mathscr{X}$ be a subset of $\mathbb{R}^n$. Then, let $\{\mathscr{Q}_j\}_{j\in\mathscr{S}}$ be a partition of $\mathscr{X}$ where $\mathscr{S}$ is the index set. That is, $\mathscr{Q}_i \cap \mathscr{Q}_j = \emptyset$ for $i \neq j$, and $\cup_{j\in\mathscr{S}} \mathscr{Q}_j = \mathscr{X}$. Each set $\mathscr{Q}_j$ is called a cell, and $\mathscr{X}$ is referred to as the region of the quantizer. Also, let $\{q_j\}_{j\in\mathscr{S}} \subset \mathbb{R}^n$ be the set of output values. The quantizer $Q : \mathscr{X} \to \{q_j\}_{j\in\mathscr{S}}$ is given by

$$Q(x) = q_j \quad \text{if } x \in \mathscr{Q}_j. \tag{8.3}$$

An example of this function for the scalar case is depicted in Figure 8.3. In this chapter, we will consider cases where the index set $\mathscr{S}$ is either finite or countable; its cardinality is denoted by N. What is transmitted over the channel is the index of the quantizer output, which can be recovered at the receiving end.

The objective of the quantized control problems in this chapter is stabilization of the unstable plant via quantized signals sent over the channel. This amounts to the joint design of the encoder-controller pair in Figure 8.1 and in particular the nonlinear function, the quantizer, as in (8.3). To keep the discussion simple, issues related to the performance of the overall system will not be addressed.

Notice that the operation of quantization can reduce the amount of information available for control since the real value (uncountable) is converted to a discrete one (finite or countable). So the quantized control problem may be interpreted as finding the amount of information required for control. This is a fundamental problem that has not been addressed until networks became acknowledged as a viable component for control systems.

Historical Remarks

We have some remarks on the history of quantized control.

First, we emphasize that this problem is not new. Classical works can be found in, *e.g.*, [4, 10]. These studies were motivated mainly by the use of digital controllers when computers became more accessible for control engineers. However, the quantizers studied there are limited to uniform quantizers. More important, the so-called additive noise model is employed. This model assumes that a quantizer can be approximated as $Q(x) = x + v$ for $x \in \mathbb{R}$, where v is the quantization error treated as uniformly bounded noise with $|v| \leq \Delta/2$. This approximation is known to be acceptable as long as quantization resolution is high so that Δ is sufficiently small compared to the size of the input x; however, in turn, this requires high data rate and thus may not be suitable for our purpose.

In his well-known work [12], Delchamps first pointed out some of the nonlinear effects that quantizers can cause within feedback control systems. The problem setup considered there is that of Figure 8.1. The control input is $u(k) = KQ(x(k))$, where Q is the componentwise uniform quantizer and K is a stabilizing state feedback gain when no quantization is present (*i.e.*, the matrix $A + BK$ is stable). It is fairly obvious that if the quantization is sufficiently fine, the state can be brought close to the origin. However, from the quantized information, it is impossible to know the exact location of the state, and thus the state will not go to the origin, but only stay around the origin. What Delchamps discovered is that the behavior there is in fact chaotic in a rigorous sense. Such a nonlinear behavior will never arise from the additive noise model. More recent work along this line can be found in [16].

Another result in Delchamps' work [12] is that if the plant is not so unstable (specifically, $|\lambda_i^u| < 2$ for all i), there exists a dynamic quantized control strategy of the form $u(k) = f_k(Q(x(k),\ldots,Q(x(0)))$ such that the state trajectory $x(k)$ goes arbitrarily close to the origin. In other words, this result shows that quantized information is sufficient to make precise control. Hence, after all, quantization may not be such an undesirable operation.

In what follows, we present two approaches on quantized control that have been recently studied by various researchers. Section 8.3 is devoted to the so-called minimum data rate problem and then in Section 8.4, the approach based on the notion of quantization coarseness is discussed.

8.3 The Minimum Data Rate for Stabilization

In this section, we provide an overview on the celebrated minimum data rate theorem. The material here is tutorial in nature, and we describe the results following their development that took place since the end of the 1990s.

8.3.1 Problem Formulation and Initial Results

First, we introduce the problem formulation and some initial results for the minimum data rate theorem. This problem was proposed by Wong and Brockett in their influential work [56].

Consider the quantized control system in Figure 8.1. We would like to find a stabilizing control law that utilizes a channel having a finite capacity. We employ a simple definition of data rate. Recall that the number of discrete values in the output of the quantizer is denoted by N. Here, we assume N to be finite. Then, the data rate for this transmission is given by

$$R := \lceil \log_2 N \rceil \text{ [bits/sample]}. \tag{8.4}$$

This represents the number of bits needed at each sampling time. (The sampling period is fixed in the discrete-time plant (8.1)).

Let's consider the simple case with a scalar plant ($n = 1$). The plant in (8.1) is unstable, so A has the magnitude larger than 1. Assume that the initial state $x(0)$ is within the interval $[-1,1]$. Here, we say that the system is stable if the state $x(k)$ remains within this interval for all times. The quantized strategy is as follows: The quantizer Q (as given in (8.3)) partitions this interval into N subintervals $\mathcal{Q}_j$. When the state $x(k)$ is in the subinterval $\mathcal{Q}_j$, the corresponding control q_j is applied at that time, that is,

$$u(k) = q_j \quad \text{if } x(k) \in \mathcal{Q}_j. \tag{8.5}$$

We would like to find how large the number N of subintervals should be to keep the state within the interval $[-1,1]$.

The next result due to Wong and Beckett [56] shows a necessary and sufficient condition for this problem.

Theorem 8.1. There exists a quantized control law in (8.5) such that the closed-loop system is stable in the sense that for each initial state $x(0)$, it holds $x(k) \in [-1,1]$, $\forall k \in \mathbb{Z}_+$, if and only if $N \geq |A|$, or equivalently,

$$R \geq \log_2 |A|. \tag{8.6}$$

This theorem highlights that, to achieve stabilization via quantized control, the minimum data rate depends only on the magnitude of the unstable pole of the plant. This implies that more unstable systems require more data rate. This is an intuitive and elegant result showing the presence of a fundamental limitation in networked control systems.

The proof of this theorem is fairly simple and can be given as follows:

Proof: For the necessity part, consider an interval $\mathcal{Q}_j$, $j \in \mathcal{S}$. At time k, if the state $x(k)$ is within this interval, then at the next time $k+1$, it should lie in another interval given by $\mathcal{R}_j := A\mathcal{Q}_j + q_j = \{Ax + q_j : x \in \mathcal{Q}_j\}$. Note that, compared to the original interval $\mathcal{Q}_j$, the width of this $\mathcal{R}_j$, which we denote by $|\mathcal{R}_j|$, expands by the ratio $|A|$ as

$$|\mathcal{R}_j| = |A| \cdot |\mathcal{Q}_j|, \quad j \in \mathcal{S}. \tag{8.7}$$

However, stability of the closed-loop system implies that this interval $\mathcal{R}_j$ must still be contained in $[-1,1]$. Thus, its width must be no greater than 2:

$$|\mathcal{R}_j| \leq 2, \quad j \in \mathcal{S}. \tag{8.8}$$

On the other hand, by definition, the intervals $\mathcal{Q}_j$, $j \in \mathcal{S}$, form a partition of $[-1,1]$. So the total of their widths is 2:

$$\sum_{j \in \mathcal{S}} |\mathcal{Q}_j| = 2. \tag{8.9}$$

Now, adding both sides of (8.7) for all j and then applying (8.9), we observe that

$$\sum_{j \in \mathcal{S}} |\mathcal{R}_j| = |A| \sum_{j \in \mathcal{S}} |\mathcal{Q}_j| = 2|A|. \tag{8.10}$$

By (8.8), it is immediate that $\sum_{j \in \mathcal{S}} |\mathcal{R}_j|$ can be bounded from above as

$$\sum_{j \in \mathcal{S}} |\mathcal{R}_j| \leq 2N. \tag{8.11}$$

Therefore, from (8.10) and (8.11), we arrive at the inequality that N must be no smaller than the magnitude of $|A|$. In terms of bits, this scheme requires at least $\log_2 |A|$ bits per time step.

The sufficiency part can be shown by employing a uniform quantizer that partitions $[-1,1]$ to $N \geq |A|$ subintervals of same widths; the output value q_j for each subinterval should be the mid-point multiplied by $-A/B$. □

By limiting the discussion to the scalar system case, we have seen the solution to the minimum data rate problem, stated as Theorem 8.1. The next question of interest is whether this result can be extended to the general n-dimensional case. It has been shown also in [56] that by following a similar argument, the necessity part can be generalized. This will be outlined next.

Consider the plant in (8.1), where the state $x(k)$ is measured. Let $\mathscr{X} \subset \mathbb{R}^n$ be a bounded set in the state space containing the origin. The control objective here is to keep the state within this set at all times: $x(k) \in \mathscr{X}$ for $k \in \mathbb{Z}_+$. The control strategy is to employ a static quantizer as given in Definition 8.1. In particular, the given set $\mathscr{X}$ is used as the quantizer region, and this set $\mathscr{X}$ is partitioned into N cells $\mathscr{Q}_j$, $j \in \mathscr{S}$. Similarly to the scalar case, the quantized control law is of the form (8.5).

The following theorem is also due to [56].

Theorem 8.2. Under the quantized control law (8.5), the closed-loop system is stable in the sense that for each initial state $x(0) \in \mathscr{X}$, it holds $x(k) \in \mathscr{X}$, $\forall k \in \mathbb{Z}_+$, only if $N \geq |\det A|$, or equivalently,

$$R \geq \sum_i \log_2 |\lambda_i^u|, \tag{8.12}$$

where λ_i^u denote the unstable eigenvalues of the matrix A.

The result shows that to keep the state within a prespecified bounded set for all times, the number N of partition sets must be no smaller than the product of the unstable eigenvalues of A, that is, the unstable poles of the plant. The inequality (8.12) is clearly a generalization of (8.6) in Theorem 8.1. The proof is very similar to the scalar case. Instead of measuring the widths of intervals, we must consider the volumes of the corresponding sets.

What was left unanswered in the work of [56] is whether there exists a control strategy for the general n-dimensional plant under which stabilization can be achieved at a data rate arbitrarily close to the lower bound shown in Theorem 8.2. This problem had led many researchers to look into this field as we will see in the remaining of this section.

8.3.2 Dynamic Quantizers

Before going into the general data rate problem, we introduce the technique of dynamic quantization, which will play an important role later. This was proposed by Brockett and Liberzon in [6].

In the discussion thus far, the quantizers given by Definition 8.1 have been assumed to be static functions. This means that the resolution of the quantized information remains the same throughout the time of the control operation. Clearly, from

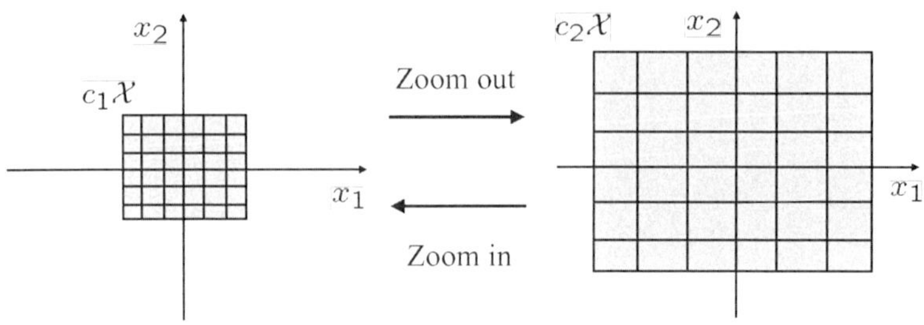

Fig. 8.4 Dynamic quantizer: The quantization region $\mathscr{X}$ is scaled with $c_1 < c_2$

the viewpoint of efficiently using the data rate, this may not be a good strategy. Furthermore, if noise enters the system or if the system lacks robustness, the state x may go beyond the region $\mathscr{X}$ of the quantizer. In such cases, the information regarding the location of x cannot be obtained.

The basic idea in the dynamic quantizer is to introduce scaling for Q according to the location of the state. If the uniform quantizer Q_Δ in (8.2) is employed for a scalar state x, this can be accomplished by making the step size Δ a time-varying parameter. More generally, given the quantizer Q in (8.3), we introduce a new parameter c and define the quantizer $Q_c : c\mathscr{X} \to \{cq_j\}$ with scaling by

$$Q_c(x) = cq_j \quad \text{if } x \in c\mathscr{Q}_j \tag{8.13}$$

or, equivalently, $Q_c(x) = cQ\big(x/c\big)$. Then, by tuning the parameter c in real time, we may obtain the state information with resolution required at each moment.

The mechanism is analogous to the zooming in digital cameras and can be described as follows (see Figure 8.4): If the state is outside its region, the quantizer "zooms out" so that the state can be captured within the region. This can be achieved by increasing the size of c. On the other hand, while the state is inside the region, we should "hold" the current region and apply the control. Once the state comes close to the origin, we can "zoom in" by reducing the size of c so that the quantization resolution becomes finer while the region becomes smaller. Repeating this zooming in, we can obtain asymptotic stabilization.

In the networked control system in Figure 8.1, the encoder and the controller should be initialized with the same parameter c for quantization. Then, for both the encoder and the controller to know which quantizer is being used, the information to be sent over the channel at each time is the following: The index of the cell in which the state lies and the state of the quantizer, which is either zoom in, zoom out, or hold. Note that since the quantizer region is only being scaled, the number N of the partition cells remains the same (similar to digital cameras, where the number of pixels does not change). This aspect can further be relaxed, as we will see next.

8.3.3 The Solution to the Minimum Data Rate Problem

For the minimum data rate problem, we are now ready to consider the general setup, where the dynamic quantizers just introduced will become useful. The description here is mainly based on [53] by Tatikonda and Mitter.

In the system in Figure 8.1, the plant (8.1) is n dimensional and its state x is measured. For the encoder and the controller, we consider more general classes as follows. The encoder is a function that takes as inputs, the past and current states x and the past quantized signals s that were sent over the channel. That is,

$$s(k) = E_k(x(k), \dots, x(0), s(k-1), \dots, s(0)). \tag{8.14}$$

We denote the number of quantized values at time k by $N(k)$, and thus the number of possible discrete values is a function of time. As a consequence, we modify the notion of data rate to that in the asymptotic average sense. This is denoted by $\overline{R}$ and is given by

$$\overline{R} := \limsup_{k\to\infty} \frac{1}{k} \sum_{\ell=0}^{k-1} \log_2 N(\ell) \text{ [bits/sample]}. \tag{8.15}$$

This rate $\overline{R}$ is the time average of the rates used over the time and hence may be any real number. It is different from the previous data rate notion since the rate R in (8.4) is fixed at all times and moreover must be an integer.

On the receiver side is the controller that generates the control input based on only the past quantized information as

$$u(k) = K_k(s(k-1), \dots, s(0)). \tag{8.16}$$

This function can be viewed as a decoding operation.

Under this setup, the data rate results in previous subsections can be generalized. The following theorem is due to [53] and provides a general solution to the minimum data rate problem. In particular, it gives a necessary and sufficient condition for achieving global asymptotic stabilization of the system in Figure 8.1. The condition is stated in terms of a lower bound on the average data rate $\overline{R}$.

Theorem 8.3. There exists a quantized control strategy in the form of (8.14) and (8.16) such that for each initial state $x(0) \in \mathbb{R}^n$, the state $x(k)$ converges to zero asymptotically if and only if the average data rate $\overline{R}$ in (8.15) satisfies

$$\overline{R} > \sum_i \log_2 |\lambda_i^u|. \tag{8.17}$$

This is a very general result considering that the classes of encoders in (8.14) and controllers in (8.16) are so broad. In particular, it shows that there is no control strategy that can stabilize the plant if the available data rate does not meet this bound.

This result in the deterministic form is due to Tatikonda and Mitter [53]; related results from the early stage can also be found in, *e.g.*, [46, 3, 36]. Around the same

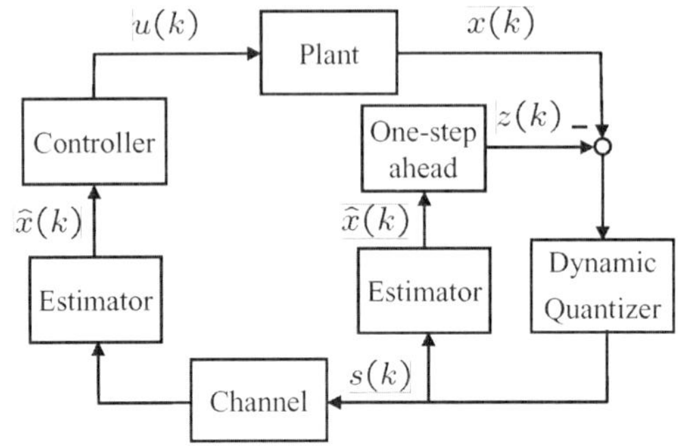

Fig. 8.5 Controller structure

time, in [42], Nair and Evans developed a stochastic counterpart, where the results take a similar form as (8.17); there will be more discussion on this later.

The necessity part of the proof essentially follows that of Theorem 8.2. The sufficiency part is proven by construction. That is, given an average data rate $\overline{R}$ satisfying the bound in (8.17), an explicit design procedure of a stabilizing control law including the quantizer, the transmission scheme, and the controller is given.

We briefly discuss the controller structure proposed in [53] for stabilization under the data rate condition (8.17). A schematic diagram is shown in Figure 8.5. Through the channel is sent the quantized signal $s(k)$, the output of the encoder side. This signal is used to reconstruct the state and in particular to generate its estimate denoted by $\hat{x}(k)$. Since the channel is noiseless, this estimate can be obtained on both sides of the channel. At the controller side, this is used to obtain the control input as $u(k) = K\hat{x}(k)$, where the feedback gain K is chosen such that $A+BK$ is a stable matrix.

On the encoder side, the estimate $\hat{x}(k)$ is utilized to find the state estimate $z(k)$ of one-step ahead via $z(k) = A\hat{x}(k) + Bu(k) = (A+BK)\hat{x}(k)$. Then, the difference between the estimate $z(k)$ and the new measurement $x(k+1)$ is computed and quantized. The quantizer here is a dynamic one. Its structure is however more involved than that in (8.13) given earlier. A characteristic aspect is that the quantization region is designed to be time varying but not just via scaling. This can be realized without much extra information because its dynamics relies on the model of the plant; see [53] for details.

We finally remark that in the case of output feedback where the state is not measured, we can place an observer at the sensor. In the encoding and quantization, the estimated state that the observer produces should be used instead of the state $x(k)$. It can be shown that global asymptotic stability can still be achieved under the same data rate condition given in (8.17).

We have further comments related to the minimum data rate theorem. First, it is important to note that the minimum data rate can be achieved only in the sense of average data rate as given in (8.15). Obviously, this is because the bound in (8.17) may not take an integer value. Hence, to use a data rate arbitrarily close to the bound, the original definition of data rate in (8.4) does not suffice.

Second, the stochastic control counterpart has been studied by a number of authors as well [34, 42, 54, 57, 40]. An interesting aspect in the work by Nair and Evans [42] is that the class of noises and disturbances considered there is more general than Gaussian. To achieve mean-square stabilization, exactly the same bound as in (8.17) has been obtained in terms of the unstable poles of the plant. This means that the noise statistics do not have any effect as long as stabilization is concerned. The situation is of course different if we consider the performance of the closed-loop system.

Finally, we note that the presentation here has been limited to the simple case with only one channel and a linear plant. In networked systems, we often employ multiple sensors and controllers that are physically distributed and thus connected by limited data-rate channels. Results for such cases can be found in, *e.g.*, [52, 58]. Also, quantized control of nonlinear systems has been considered from the networked control viewpoint; such studies include [11, 37].

8.4 The Coarsest Quantization for Stabilization

In this section, we describe another approach for quantization for networked control systems. The interest here also centers around control using limited information. The viewpoint is however different from that in the previous section for the minimum data rate theorem. We will find similar implications concerned with the necessary amount of information for stabilization problems.

In the previous section, we have seen that dynamic quantizers can be powerful and are indeed an essential tool for deriving the minimum data rate theorem. By incorporating the plant model into the encoder, the number of quantized values has been significantly reduced. Nevertheless, as shown in Figure 8.5, the controller structure there is cumbersome, requiring heavy computation for state estimation both on the sender side as well as on the receiver side.

In certain applications, resources for such computation may be costly or even unavailable. Hence, from the implementation side, it is of interest to employ simpler schemes. Obvious candidates are the uniform quantizers, which are memoryless. As mentioned earlier, however, such quantization may require a large data rate because the resolution is the same for any input value and thus may be unnecessarily fine. This raises the following question: What is a better way to distribute the quantization resolution for control purposes?

An interesting approach for the design of memoryless quantizers was proposed by Elia and Mitter in [15]. It is shown there that to achieve stability via a Lyapunov approach, a quantizer having the "coarsest" resolution can be characterized. Such quantization is fine around the origin, but coarse outside. Moreover, a certain parameter representing the coarseness can be determined by the unstable poles of the plant. Hence, this result shows another form of fundamental limitation involving control and quantization. This approach has been extended to incorporate

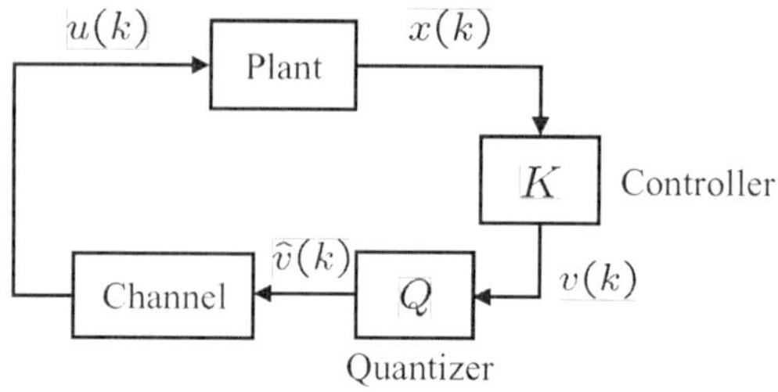

Fig. 8.6 Stabilization via coarsely quantized signals

performance issues based on H^2 and H^∞ norms in [19], and sampled-data control results are given in [33, 31, 32].

In this section, we first present the initial results of Elia and Mitter based on [15]. Then, we discuss two recent extensions for networked control with uncertainties: One result deals with unreliable communication from [55], and the other is for the case with unknown plants based on adaptive control from [24].

8.4.1 *The Coarsest Quantizers*

The system setup for this problem is shown in Figure 8.6. The system looks similar to the one before in Figure 8.1. Again, the plant is discrete-time LTI as in (8.1), and the state $x(k)$ is measured. The control input $u(k)$ is one dimensional. One difference is that in the current setting, the controller is on the sensor side and its output is given by $v(k) = Kx(k)$, where $K \in \mathbb{R}^{1\times n}$ is the state feedback gain. Hence, the signal to be quantized is the control input as

$$\hat{v}(k) = Q(v(k)) = Q(Kx(k)). \tag{8.18}$$

Here, we impose the restriction on the quantizer Q to be memoryless, but it is allowed to be any piecewise constant function of the form (8.3). In the following discussion, we will look only at the case where Q has the domain $\mathbb{R}$, resulting in a countable but infinite number N of quantization levels. Again, the channel is noiseless with no error or delay (*i.e.*, $u(k) = \hat{v}(k)$). Thus, the control input is expressed as

$$u(k) = Q(Kx(k)). \tag{8.19}$$

We consider stabilization of this system under the notion of quadratic stability based on a Lyapunov argument. We begin with a given quadratic function $V(x) = x^T Px$ for $x \in \mathbb{R}^n$, where the matrix $P \in \mathbb{R}^{n\times n}$ is positive definite. Under the control law in (8.19), the closed-loop system is said to be quadratically stable with respect to $V(x)$ if $V(x)$ becomes a Lyapunov function for the overall system, *i.e.*, it decreases along the state trajectory as

$$V(x(k+1)) - V(x(k)) < 0, \ \forall k: \ x(k) \neq 0.$$

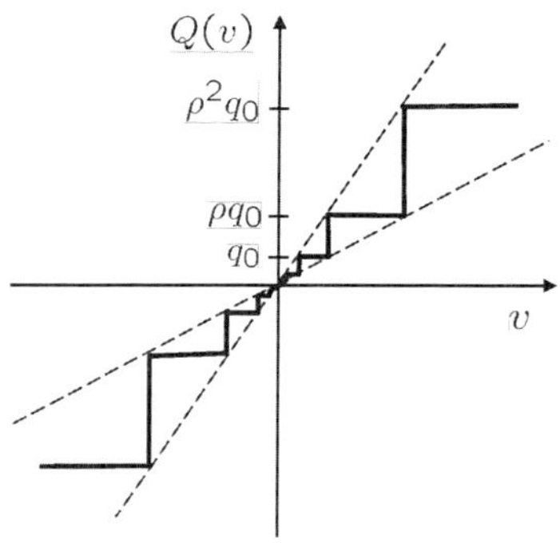

Fig. 8.7 Logarithmic quantizer

The inequality above implies that the energy in the system decreases at each time instant and will eventually become 0. The control law design based on a given $V(x)$ is a standard technique in nonlinear control using Lyapunov theory (*e.g.*, [35]). We note that in the linear plant case, for $V(x)$ to be a Lyapunov function, it is necessary that the matrix $P > 0$ satisfies the inequality

$$R := P - A^T PA + A^T PB(B^T PB)^{-1} B^T PA > 0 \tag{8.20}$$

In the interest of reducing the communication rate, we should look for a quantizer in (8.19) that is coarse. The coarseness of a quantizer Q is measured by how densely the quantized values are distributed. More specifically, define the density of the quantizer Q by

$$d_Q := \limsup_{\varepsilon \to 0} \frac{\sharp u[\varepsilon]}{-\ln \varepsilon}, \tag{8.21}$$

where $\sharp u[\varepsilon]$ denotes the number of quantized values q_j in the interval $[\varepsilon, 1/\varepsilon]$. This density d_Q can be viewed as the average number of quantization levels that Q has in a certain logarithmic sense.

The problem of finding the coarsest quantizer consists of two steps, which are formulated as follows: The first step is to find the quantizer Q with the minimum density d_Q subject to quadratic stability of the closed-loop system with respect to the given quadratic function $V(x) = x^T Px$. The second step is to find the quantizer that minimizes the density over all quadratic functions.

In this problem, the class of quantizers having the coarsest structure is shown to be that of the so-called logarithmic quantizers. These have the characteristic that the width of the partition cell becomes larger away from the origin, and finer closer to the origin. For the given quadratic function $V(x) = x^T Px$, let the state feedback gain be $K := -B^T PA/(B^T PB)$ and let

$$\rho := \frac{1+\delta}{1-\delta} \quad \text{with} \quad \delta := \sqrt{\frac{B^T PB}{B^T PAR^{-1} A^T PB}}. \tag{8.22}$$

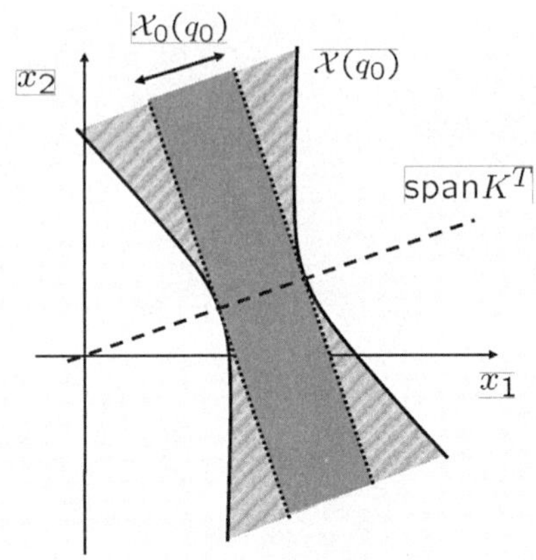

Fig. 8.8 The sets $\mathscr{X}(q_0)$ and $\mathscr{X}_0(q_0)$ in the state space

Then, under the control in (8.19), the coarsest quantizer is given by

$$Q(v) = \begin{cases} q_i & \text{if } v \in \left(\frac{\rho+1}{2}q_{i-1}, \frac{\rho+1}{2}q_i\right], \\ -q_i & \text{if } v \in \left[-\frac{\rho+1}{2}q_i, -\frac{\rho+1}{2}q_{i-1}\right), \\ 0 & \text{if } v = 0, \end{cases} \tag{8.23}$$

where the quantized values are $q_i = \rho^i q_0$ for $i \in \mathbb{Z}$ with $q_0 > 0$. This function is presented in Figure 8.7. In logarithmic quantizers, the key parameter $\rho > 1$ represents the expansion ratio, and a large ρ means a coarse quantizer.

The idea in finding the logarithmic quantizer above can be roughly explained as follows. For simplicity, assume $R = I$ in (8.20). First, take a real value q_0 and consider using this particular value for the control as $u = q_0$. Then, we can characterize the states where the function $V(x)$ will decrease in the next time step. The set of all such states is given by

$$\mathscr{X}(q_0) := \left\{x \in \mathbb{R}^n : V(Ax + Bq_0) - V(x) < 0\right\}.$$

It can be shown that this set $\mathscr{X}(q_0)$ is symmetric about the line spanned by the gain vector K^T; see Figure 8.8. Now, recall that it is Kx that is quantized. Thus, we must assign this control value q_0 to states contained in the set $\mathscr{X}(q_0)$ with the form $\alpha K^T + \beta$, where $\alpha \in \mathbb{R}$ and $\beta \in \mathrm{Ker}K$. The largest set $\mathscr{X}_0(q_0) \subset \mathscr{X}(q_0)$ containing all such states can be found to be a region between two hyperplanes orthogonal to K^T. In fact, it has the structure

$$\mathscr{X}_0(q_0) = \left\{\alpha K^T + \beta : \alpha \in [c_0 q_0, c_0 \rho q_0], \beta \in \mathrm{Ker}K\right\}$$

for some $c_0 > 0$. Clearly, the expansion ration ρ appears here. This discussion leads us to a logarithmic partitioning in the state space by the sets $\mathscr{X}(\pm q_i)$, $i \in \mathbb{Z}$, which essentially corresponds to the quantization function that is shown in (8.23).

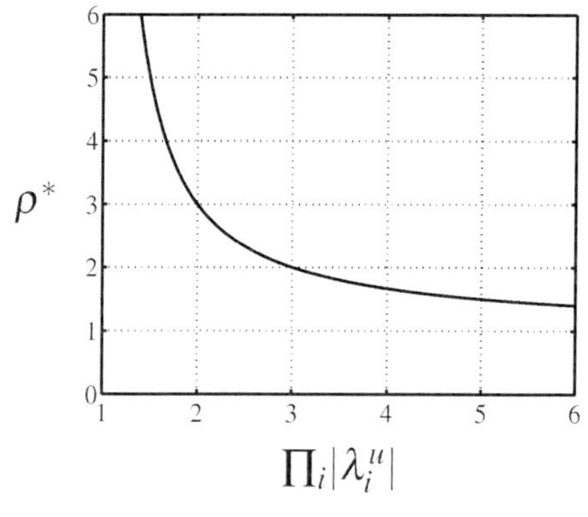

Fig. 8.9 The parameter ρ^* for the coarsest quantizer versus the product $\prod_i |\lambda_i^u|$ of plant unstable eigenvalues

The second step in the problem is to maximize the parameter ρ over all quadratic Lyapunov functions. Formally, this problem is stated as

$$\rho^* = \sup_{P>0} \rho. \tag{8.24}$$

This problem will definitely give us the coarsest quantizer. A closed form solution was given in [15]. The following theorem shows that the maximum ρ is determined only by the product of plant unstable poles.

Theorem 8.4. Consider the plant (8.1) under the quantized control (8.19). The quantizer Q that minimizes the density d_Q in (8.21) subject to quadratic stabilization is given by a logarithmic quantizer in (8.23) whose expansion ratio ρ^* is given by

$$\rho^* = \frac{\prod_i |\lambda_i^u| + 1}{\prod_i |\lambda_i^u| - 1}, \tag{8.25}$$

where λ_i^u are the unstable eigenvalues of the system matrix A of the plant. In the corresponding Lyapunov function $V(x) = x^T Px$, the matrix $P > 0$ is the solution to the following Riccati equation

$$A^T PA - P - \frac{A^T PBB^T PA}{B^T PB + 1} = 0.$$

We remark that the coarsest quantizer specified by ρ^* in the theorem means that ρ^* is the supremum of ρ for quadratic stabilization. That is, to obtain quadratic stability for the closed-loop system, the parameter ρ should be chosen smaller than ρ^*.

This result is illustrated in Figure 8.9, where the relation between the maximum expansion ratio ρ^* and the product of the unstable eigenvalues of the plant matrix A is shown. As the plant becomes more unstable, we need finer quantization to guarantee quadratic stabilization. The implication is similar to the minimum data

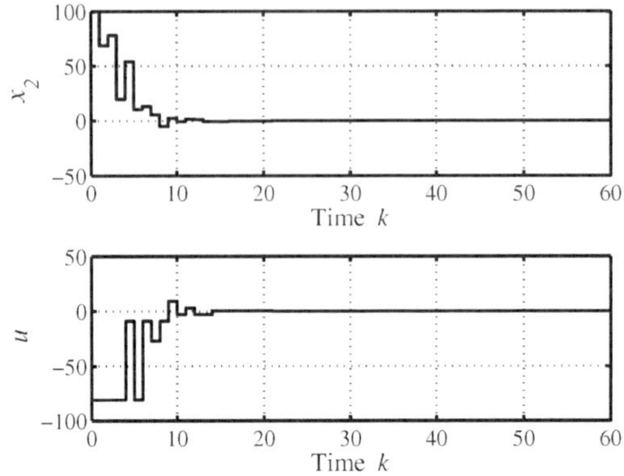

Fig. 8.10 Time responses: The state x_2 (top) and the control input u (bottom)

rate result that we have presented earlier, providing another form of limitation in the context of networked control.

Example 8.3. We present a numerical example to demonstrate the utility of the control scheme based on the logarithmic quantizer. As the plant, we considered the second-order system given by

$$x(k+1) = \begin{bmatrix} 0 & 1 \\ 1.8 & -0.3 \end{bmatrix} x(k) + \begin{bmatrix} 0 \\ 1 \end{bmatrix} u(k). \tag{8.26}$$

The system is unstable and has two poles 1.2 and -1.5. To achieve quadratic stabilization, Theorem 8.4 implies that the maximum coarseness for the logarithmic quantizer is given by $\rho^* = 3.5$. Here, we chose $\rho = 3$ and, as a result, we obtained the state feedback gain given by $K = \begin{bmatrix} -1.80 & 0.520 \end{bmatrix}$.

In Figure 8.10, the time responses of the state x_2 and the quantized control input u with the initial state $x(0) = \begin{bmatrix} 100 & 100 \end{bmatrix}^T$ are shown. We confirm the closed-loop stability. The responses decay exponentially, resulting from the quadratic stability attained by the control scheme. ▽

8.4.2 The Coarsest Quantizer for Stabilization over Lossy Channels

In this subsection, we extend the results on the coarsest quantizer and consider the case where the channel is unreliable and data loss may occur. We show that the limitation observed for coarse quantization can be generalized. The material is from [55].

When the reliability of the communication channel is insufficient, packets containing control-related signals may get lost during the transmissions. If this occurs often, the control performance degrades and, even worse, stability may not be guaranteed. We employ a packet loss model with the assumption that the probability of a loss is constant for all transmissions. That is, for example, 10 percent of the packets sent do not reach the receiver side. The loss probability will be utilized in the control scheme as a priori known information. This is clearly one of the most simple models for unreliable channels (see, *e.g.*, [9]).

In recent years, much attention has been devoted to the research on issues related to packet losses in networked control. The stochastic stabilization of scalar systems has been reported in [22], and then state estimation problems have been addressed in [50]. For controller design, LQ control schemes are given in [27] for remote control and in [21] for a general network topology setup. Approaches based on H^∞ control are studied in [48, 29].

It is interesting to note that in the presence of probabilistic packet losses, limitations in networked control can be derived. In particular, the following two points are known: (i) To guarantee stability in a stochastic sense, it is necessary and sufficient that the loss probability is smaller than a certain critical bound. (ii) This bound depends only on the level of instability of the plant. The objective of this subsection is to present an extension of the coarsely quantized control approach to the case when the channels are lossy.

Consider the networked control system in Figure 8.6. The problem setup is almost the same as in the previous case: The plant is as given in (8.1), where the state x is measured; on the sensor side is the state feedback K and the memoryless quantizer Q, and thus the input to the channel is $\hat{v}(k) = Q(Kx(k))$ as in (8.18).

However, in the current case, the channel is assumed to be unreliable, and the packet transmitted from the sensor side may not reach the actuator side; when such a loss occurs, we necessarily have $u(k) \neq \hat{v}(k)$. We assume that packet losses occur with probability $\alpha \in (0,1)$ at each time k, independently of other times. The random process that represents the losses is denoted by $\theta(k)$, $k \in \mathbb{Z}_+$, whose probability distribution is given by

$$\mathrm{Prob}\{\theta(k) = i\} = \begin{cases} \alpha & i = 0, \\ 1-\alpha & i = 1. \end{cases} \tag{8.27}$$

In the case of a loss, the convention is that the control input is set to zero, and otherwise the received data will be applied. Hence, using the process $\theta(k)$, we can write the control input as $u(k) = \theta(k)\hat{v}(k)$[2].

Consequently, the state-space equation of the closed-loop system becomes

$$x(k+1) = Ax(k) + B\theta(k)\hat{v}(k). \tag{8.28}$$

This system is a nonlinear system with stochastic switching. Hence, as the notion of stability, we need to employ a stochastic version of quadratic stability defined as

[2] We note that, in the case without quantization, other control methods have been studied in the literature such as using the previous input $u(k-1)$ if a loss occurs at time k (*e.g.*, [28, 38]); however, the results to be presented on stabilization will not change even if such methods are employed.

follows: For the system (8.28), the origin is said to be stochastically quadratically stable if there exists a positive-definite function $V(x) = x^T Px$ in a quadratic form and a positive-definite matrix R such that

$$\begin{aligned} \Delta V &:= E[V(x(k+1))|x(k)] - V(x(k)) \\ &\leq -x(k)^T Rx(k), \qquad \forall x(k) \in \mathbb{R}^n. \end{aligned} \tag{8.29}$$

It is known that the condition above is sufficient to guarantee the origin of the system (8.28) to be mean-square stable (see, *e.g.*, [8]): For every initial state $x(0)$, it holds that

$$\lim_{k \to \infty} E\left[\|x(k)\|^2\right] = 0. \tag{8.30}$$

The problem that we consider here is to find the coarsest quantizer such that the quantized control system with random packet losses in (8.28) is stochastically quadratically stable. As the measure of coarseness for the quantizer Q, we again employ the density d_Q given in (8.21). Note that this problem is a clear generalization of the one in the last subsection; the latter problem is the special case where the loss probability α is set to zero, and we have seen in Theorem 8.4 the limitation regarding quantization.

On the other hand, another special case of this problem is when quantization is not present. That is, in the system in Figure 8.6, $\hat{v}(k) \equiv v(k)$. Then, another form of limitation in networked control is known for the loss probability α. Note that the closed-loop system is now linear, which in turn makes the two stability notions of stochastically quadratic stability and mean-square stability equivalent. It has been shown in [14, 28] that there exists a state feedback gain K such that the closed-loop system in (8.28) is mean-square stable if and only if

$$\alpha < \frac{1}{\prod_i |\lambda_i^u|^2}. \tag{8.31}$$

This obviously implies that from the viewpoint of mean-square stability, there is an upper bound on the loss probability, having a tight relation with the unstable poles of the plant. Hence, for more unstable plants, more reliable channels are necessary, which is a natural and reasonable implication. Related results can be found in [30] for the case when the controller is remotely located; the bound above is shown to be critical for the two channels on the sensor and actuator sides.

Going back to the problem of this subsection, we consider the effects of both quantization and packet losses. Our goal is to clarify the relationship among the three key parameters: ρ^*, α, and $\prod_i |\lambda_i^u|$. A complete solution to this problem is provided in the theorem below due to [55].

Theorem 8.5. Consider the quantized control system with unreliable communication in (8.28). The quantizer Q that minimizes the density d_Q in (8.21) subject to stochastic quadratic stabilization is given by the logarithmic quantizer in (8.23). Its expansion ratio ρ^* is expressed as

$$\rho^* = \frac{1+\delta^*}{1-\delta^*}, \quad \delta^* = \sqrt{\frac{\frac{1}{\prod_i |\lambda_i^u|^2} - \alpha}{1-\alpha}}, \tag{8.32}$$

where λ_i^u are the unstable eigenvalues of the system matrix A of the plant and the loss probability α is chosen such that the inequality (8.31) holds.

Theorem 8.5 generalizes the two previous results in (8.25) from [15] and (8.31) from [14, 28]. In particular, just as in the case without packet losses, the coarsest quantizer is of logarithmic type represented by the maximum expansion ratio ρ^*.

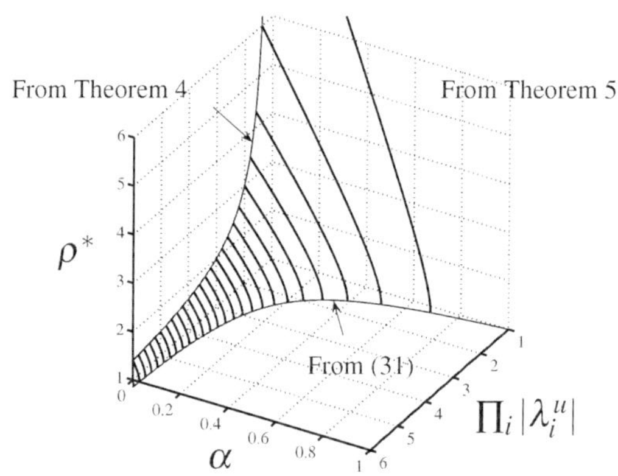

Fig. 8.11 The relationship between the product of unstable poles $\prod_i |\lambda_i^u|$, the loss probability α, and the coarseness ρ^* of the quantizer

Figure 8.11 summarizes the results mentioned above. The relations in (8.25) and (8.31) are found, respectively, on the $\prod_i |\lambda_i^u|$–ρ^* plane and on the $\prod_i |\lambda_i^u|$–α plane. The curved surface represents (8.32) in Theorem 8.5. It is easy to see that (8.32) unifies the two previous results. The result also shows a tradeoff between α and ρ^*, *i.e.*, to achieve closed-loop quadratic stability, high packet loss probabilities require quantizers with fine resolution and vice versa.

The coarse quantizer obtained above is memoryless and further logarithmic. As a consequence, the output of the quantizer cannot be transmitted over a finite data rate channel. Indeed, as seen in the expression in (8.23), the quantizer takes an infinite number of values around the origin. Moreover, in the discussion so far, we have not introduced any constraints on the size of the state and the control input. In [55], a dynamic quantizer is proposed for achieving stochastic stability based on the logarithmic quantizer. Though dynamic, this control law has a simple structure compared to those presented in Section 8.3.3.

Example 8.4. We continue with Example 8.3 in Section 8.4.1. As the plant, we employ the unstable one in (8.26). In Figure 8.11, this system corresponds to the cross section obtained by cutting the surface at $\prod_i |\lambda_i^u| = 1.8$. From the upper bound given in (8.31), we must select a communication channel with a packet loss rate $\alpha < 0.31$.

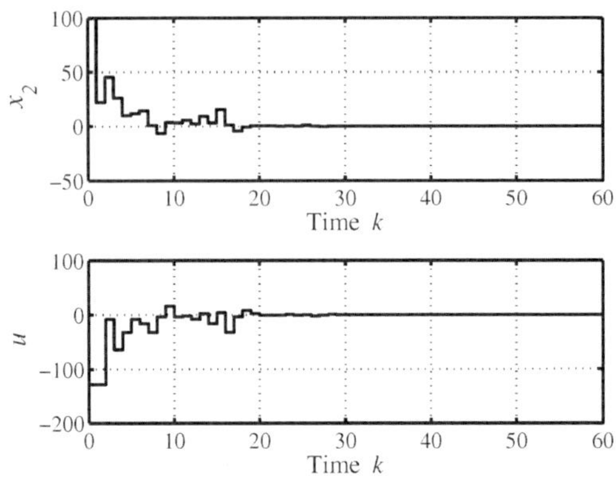

Fig. 8.12 Time responses of the system designed for the lossy channel: The state x_2 (top) and the control input u (bottom)

Here, we chose a channel with $\alpha = 0.2$, in which case, the maximum ρ is about $\rho^* = 2.17$. We took the quantizer parameter as $\rho = 2$. The proposed method then yields the feedback gain $K = \begin{bmatrix} -1.80 \; 0.651 \end{bmatrix}$. Notice that this gain is similar to the one that was found before.

In Figure 8.12, the sample paths of the state x_2 and the quantized control input u with the same initial state $x(0) = \begin{bmatrix} 100 \; 100 \end{bmatrix}^T$ are shown. We confirm the closed-loop stability. The responses are somewhat oscillatory, but decay almost exponentially. This good transient is due to the notion of mean-square stability, which takes account of the control performance in the average sense.

To exhibit the difference from the previous design in Example 8.3, we applied the control law from there to the current setting with the lossy channel. By simulating the system under the same loss process as above, the sample paths of x_2 and u shown in Figure 8.13 were obtained. The difference in the performance is clear. The trajectories for the current case, where the controller does not take account of the losses, start to oscillate with large magnitudes after about 10 time steps. $\triangledown$

8.4.3 Quantized Adaptive Control for Uncertain Systems

In this subsection, we consider the quantized stabilization of an uncertain system whose uncertainty bounds are unknown. In particular, we present an extension of [15] based on adaptive control; the material is from [24]. Under this approach, to ensure system performance, the controller adjusts its feedback gain as well as the coarseness in quantization in response to the plant outputs.

The system setup is depicted in Figure 8.14. The plant is LTI with uncertain parameters. The controller is on the sensor side, and the control input is quantized to be sent over the channel; we assume that the channel is noiseless, and hence the quantized signal can be fully recovered at actuator side. The quantizer is time varying, and at each time instant, its parameters are determined and adjusted corresponding

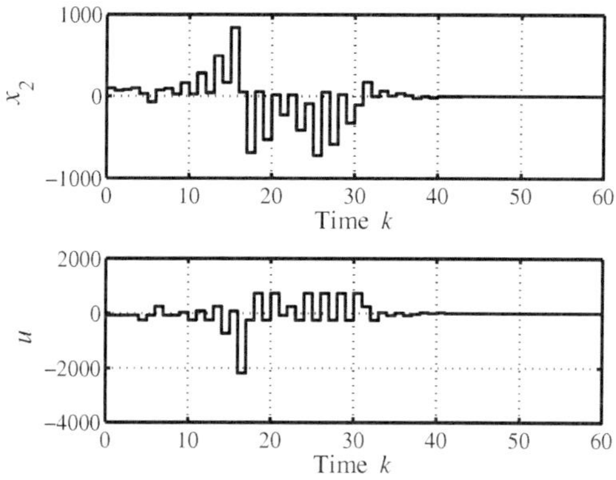

Fig. 8.13 Time responses of the system designed without accounting for the losses: The state x_2 (top) and the control input u (bottom)

to the updates in the feedback gain. Here, we employ a logarithmic quantizer and aim at keeping it as coarse as possible at each moment.

The proposed control scheme has three features as follows. First, our adaptive approach is a generalization of Theorem 8.4 from [15] in that if the system matrices are completely known, the controller and quantizer coincide with the optimal ones given Theorem 8.4. Second, the coarseness in quantization is time varying and, in particular, it must be fine while the controller gain is large, and vice versa. In general, this implies that systems that are more unstable would require more information for stabilization. This may appear contradictory to the fact that the unknown plant parameters, which determine the coarsest quantization levels according to Theorem 8.4, are constant. However, it is noted that the coarsest quantizer requires a specific gain, and in the adaptive case, this gain is unknown. Third, since the coarseness of the quantization varies with time, it is necessary to send over the channel the information specifying the quantizer being used. We also note that an adaptive quantized control method for uncertain nonlinear plants is developed in [25].

From the adaptive control viewpoint, we emphasize that the proposed approach is Lyapunov-based. In particular, it employs the method of [23], to which without quantization the scheme in this subsection reduces. This method guarantees Lyapunov stability of the plant state and the adaptive gain and attraction with respect to the plant state. We note that such an approach is standard in the continuous-time case, but for discrete-time systems, recursive least squares and least mean square algorithms are typically used (*e.g.*, [20]); the primary focus there is on state convergence rather than stability in the sense of Lyapunov.

Problem Settings

We introduce the quantized adaptive control problem for linear uncertain plants.

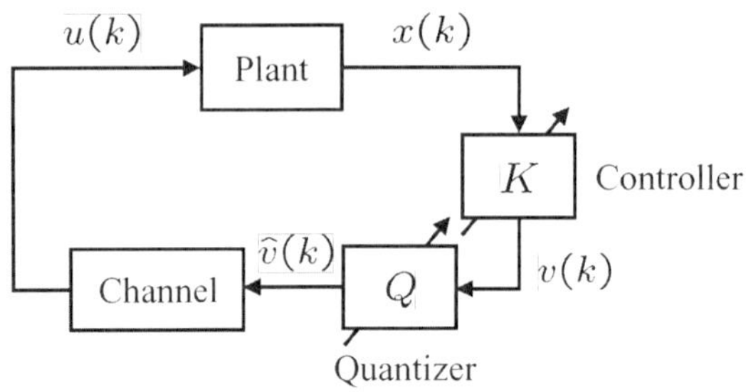

Fig. 8.14 Quantized adaptive control scheme

Consider the networked control system in Figure 8.14. The state-space equation of the plant is given by

$$x(k+1) = Ax(k) + Bu(k), \tag{8.33}$$

where $x(k) \in \mathbb{R}^n$ is the state, and $u(k) \in \mathbb{R}$ is the control input. We assume that the pair (A,B) is stabilizable, but the matrix A is unknown.

The state $x(k)$ is measured and the control input to be quantized is generated through

$$v(k) = K(k)x(k), \tag{8.34}$$

where $K(k)$ is the adaptive feedback gain. Then, the signal $v(k)$ is to be sent over the channel and quantized via

$$u(k) = Q_k(v(k)). \tag{8.35}$$

Here, Q_k represents the time-varying logarithmic quantizer of the form

$$Q_k(v) = \begin{cases} q_i(k) & \text{if } v \in \left(\frac{\rho(k)+1}{2}q_{i-1}(k), \frac{\rho(k)+1}{2}q_i(k)\right], \\ -q_i(k) & \text{if } v \in \left[-\frac{\rho(k)+1}{2}q_i(k), -\frac{\rho(k)+1}{2}q_{i-1}(k)\right), \\ 0 & \text{if } v = 0, \end{cases} \tag{8.36}$$

where the quantized values are $q_i(k) = \rho(k)^i q_0$ for $i \in \mathbb{Z}$ with $q_0 > 0$. Note that the expansion ratio $\rho(k)$ determines coarseness of the quantizer at time k.

The logarithmic quantizer (8.36) can be viewed as a time-varying sector-bounded memoryless nonlinearity. Letting

$$\delta(k) := \frac{\rho(k)-1}{\rho(k)+1},$$

we can write the sector condition for Q_k as

$$(1-\delta(k))v^2 \le Q_k(v)v \le (1+\delta(k))v^2, \quad v \in \mathbb{R}. \tag{8.37}$$

Thus, the parameter $\delta(k)$ specifies the bounds. A notable difference from the static quantizer case is that the information on the quantizer being used at each time k must be shared by both the sensor and the actuator. Here, we realize this by quantizing $\delta(k)$ and then transmitting it over the channel as well. We will later explain more on how to perform this quantization.

Our objective is to design a stabilizing adaptive controller in the form of (8.34) and a coarse logarithmic quantizer Q_k. We present a control law that ensures stability of the closed-loop system. Here, the results are limited to the case that the matrix B is known, but this assumption may be relaxed; for details, see [24].

The system matrix A is unknown under the following assumption: Let $\mathscr{A} := \{A + BK_{\mathrm{g}}^{(1)} : K_{\mathrm{g}}^{(1)} \in \mathbb{R}^{1\times n}\}$. Assume that there is a known matrix $\tilde{A} \in \mathscr{A}$ in this set, which may be unstable. This assumption holds for the class of systems in the controllable canonical form.

The Proposed Control Law

We now provide the design procedure of the adaptive controller and then present the proposed scheme.

Let $R \in \mathbb{R}^{n\times n}$ be a positive-definite matrix and let $\gamma > 0$. Find the positive-definite solution $P \in \mathbb{R}^{n\times n}$ of the Riccati equation

$$P = \tilde{A}^T P\tilde{A} + R - \tilde{A}^T PB(B^T PB)^{-1}B^T P\tilde{A} \tag{8.38}$$

with $P \geq I$. Such a P always exists. Then, let $A_{\mathrm{s}} := \tilde{A} + BK_{\mathrm{g}}^{(2)}$ with $K_{\mathrm{g}}^{(2)} := -B^T P\tilde{A}/(B^T PB)$. This matrix A_{s} is stable. Finally, take $\sigma \in (0,2)$.

The proposed adaptive control law is given as

$$v(k) = K(k)x(k).$$

Here, the gain $K(k) \in \mathbb{R}^{1\times n}$ is given by the update law

$$\begin{aligned} K(k+1) = K(k) - \frac{\sigma}{1 + x^T(k)Px(k)} B^{\dagger}\big\{x(k+1) - A_{\mathrm{s}}x(k) \\ - B\big[Q_k(v(k)) - v(k)\big]\big\}x^T(k), \end{aligned} \tag{8.39}$$

where $B^{\dagger}$ denotes the pseudo inverse of B. The quantizer Q_k in (8.35) is chosen such that the parameter $\delta(k)$ satisfies the inequality

$$R - 2\delta(k)^2 K^T(k)B^T PBK(k) \geq \gamma I. \tag{8.40}$$

We can establish closed-loop stability under this adaptive control law as follows.

Theorem 8.6. Consider the linear uncertain system in (8.33), where $A \in \mathbb{R}^{n\times n}$ is an unknown matrix, but $\tilde{A} \in \mathscr{A}$ is known. Then, the adaptive control law given above guarantees that the solution $(x(k), K(k)) \equiv (0, K_{\mathrm{g}})$, where $K_{\mathrm{g}} := -B^T PA/(B^T PB)$, of the closed-loop system is Lyapunov stable. Furthermore, $x(k) \to 0$ as $k \to \infty$ for each initial state $x(0) \in \mathbb{R}^n$.

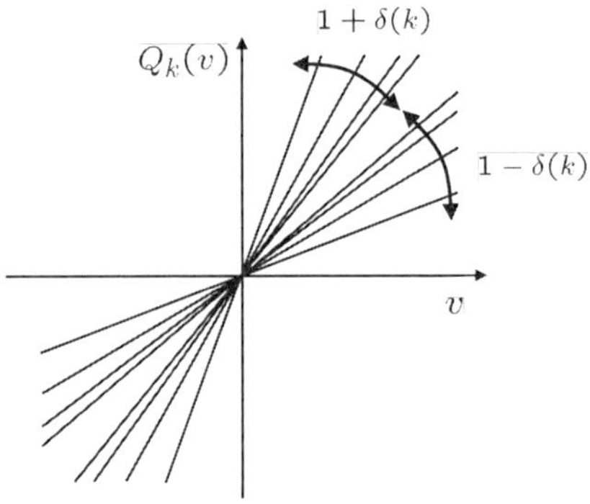

Fig. 8.15 Sector bounds for a time-varying logarithmic quantizer

This theorem shows that the solution $(x(k), K(k)) \equiv (0, K_g)$ of the overall closed-loop system is Lyapunov stable and that the state $x(k)$ converges to the origin. This stability notion is known as partial asymptotic stability. The convergence of the gain $K(k)$ follows from (8.39) though the gain may not go to K_g.

We have several remarks on the adaptive quantization scheme. As shown in (8.37), the coarseness in quantization is determined by the sector bounds and in particular the parameter $\delta(k)$. This must be chosen so that the bound in (8.40) holds at all times. In (8.40), we observe that quantization must be fine while the controller gain is large, and vice versa. This is in contrast to the nonadaptive case in the previous subsections, where specific gains are used for a known system to maximize the coarseness in quantization.

To meet the requirement in (8.40), the information regarding $\delta(k)$ must be communicated to the actuator side and hence must also be quantized. A simple scheme for this quantization is to employ a logarithmic one as follows (see Figure 8.15): Let $\delta(k)$ take values of the form $\mu^{-\ell}$ with $\ell \in \mathbb{Z}_+$ and $\mu > 1$. Then, (8.40) can always be satisfied since $\delta(k)$ can be arbitrarily small. In this scheme, it suffices to send the index ℓ over the channel.

The theorem has a close connection with the result in Section 8.4.1 in the special case when perfect knowledge of the plant is available. This can be seen as follows. Since the plant is known, we may take $\tilde{A} = A$ and thus

$$K(k) \equiv K_g = -B^T PA/(B^T PB), \tag{8.41}$$

where P is the solution of the Riccati equation (8.38). Then, with $\sigma = 0$, the update law (8.39) of the gain becomes unnecessary. Further, in this case, the sector condition (8.40) can be relaxed to $R - \delta^2 K_g^T B^T PBK_g > 0$. By (8.41), this condition can be rewritten to provide the maximum value of δ as

$$\delta = \sqrt{\frac{B^T PB}{B^T PAR^{-1}A^T PB}}. \tag{8.42}$$

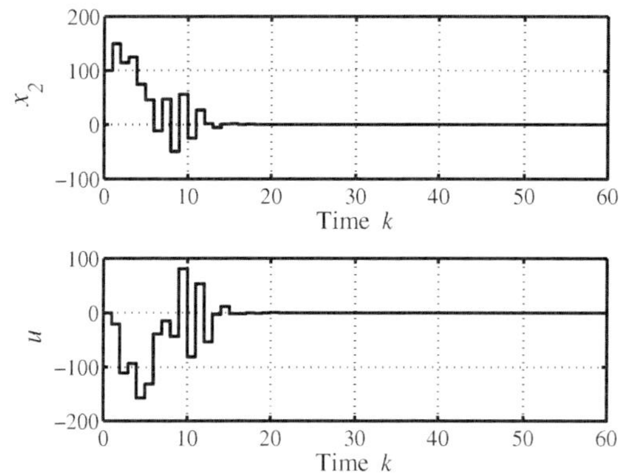

Fig. 8.16 Time responses: The state x_2 (top) and the control input u (bottom)

This is precisely the result given in (8.22), characterizing the coarsest possible quantizer for the given matrices A, B, and R. Moreover, it is shown there that properly choosing R in (8.42) further leads to the coarsest possible quantizer, which is determined solely by the unstable eigenvalues of A.

Example 8.5. We present a numerical example to demonstrate the utility of the proposed adaptive control scheme. We consider the same plant (8.26) in Example 8.3:

$$x(k+1) = \begin{bmatrix} 0 & 1 \\ -a_0 & -a_1 \end{bmatrix} x(k) + \begin{bmatrix} 0 \\ 1 \end{bmatrix} u(k),$$

where the parameters a_0 and a_1 are assumed to be unknown. The true values are $a_0 = -1.8$ and $a_1 = 0.3$. Since the plant is in the controllable canonical form, we may take $\tilde{A} \in \mathscr{A}$ as

$$\tilde{A} = A + BK_g^{(1)} = \begin{bmatrix} 0 & 1 \\ \theta_0 & \theta_1 \end{bmatrix} \quad \text{with } K_g^{(1)} = \begin{bmatrix} \theta_0 + a_0 & \theta_1 + a_1 \end{bmatrix},$$

where θ_0, $\theta_1 \in \mathbb{R}$ can be arbitrary. Here, we chose them as $\theta_0 = -0.167$ and $\theta_1 = 0.833$. In the design of the quantized adaptive controller, we chose $R = I$, $\sigma = 0.6$, and $\gamma = 0.25$. The coarseness of the quantizer is determined via the parameter $\delta(k)$, which is quantized and takes discrete values as

$$\delta(k) = 1.1^{-\ell(k)} \text{ with } \ell(k) \in \mathbb{Z}_+. \tag{8.43}$$

Figure 8.16 shows the time responses of $x_2(k)$ and the control input $u(k)$ with the initial conditions $x(0) = [100\ 100]^T$ and $K(0) = [0\ 0]$. Furthermore, Figure 8.17 displays the responses of the gains, where $K(k) = [K_1(k)\ K_2(k)]$, as well as the expansion ratio $\rho(k)$. It is interesting to note that the gain $K(k)$ converges as $\lim_{k\to\infty} K(k) = [-1.80\ 0.138]$; this final vector takes a similar form as the gains used in earlier examples in Sections 8.4.1 and 8.4.2. Further, it can be seen that $\rho(k)$

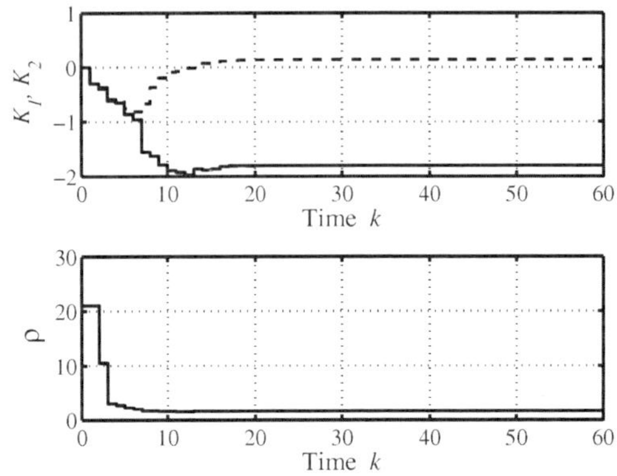

Fig. 8.17 Time responses: The gains K_1 (top: solid line), K_2 (top: dashed line) and ρ (bottom)

remains large for several time steps in the beginning, but then it gradually decreases. This shows that the communication rate for control is low while the adaptive gains are small. The final value of $\rho(k)$ is 1.63. In view of the earlier examples, this value is not small. We also note that for the quantized values of $\delta(k)$, the parameter $\ell(k)$ in (8.43) ranged between 1 and 15. ▽

8.5 Information Theoretic Approach to Bode's Integral Formula

This section presents the second part of the chapter and is mainly based on [43]. It outlines an approach to the so-called Bode's integral formula using information theory.

For a long time, the close relation between control theory and information theory has been pointed out. Indeed, both theories are concerned with signals and dynamical systems in general. However, the difference lies in the target applications as well as the approaches. While information theory focuses on signals involved in communication systems, control theory deals more with feedback control from the viewpoint of systems, representing the relation between their inputs and outputs.

The recent attention devoted to networked control has motivated studies on the interactions of the two theories. On the one hand, as we have seen in earlier sections, notions from information theory play essential roles in the context of networked control. In particular, the notions of channel capacities and communication rates have been employed in order to quantify the amount of information required for the stabilization of feedback control systems over channels. On the other hand, some problems in information theory can be considered as control problems. The work of [13] shows that the design of an optimal encoder and decoder pair for a communication scheme based on feedback can be reduced to the design of a feedback controller.

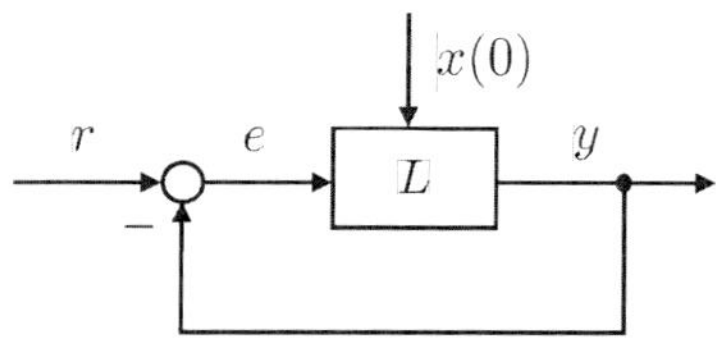

Fig. 8.18 Discrete-time feedback control system

The information theoretic approach has the characteristic of focusing on signals rather than the input-output relations of systems. As a consequence, in certain cases, we may impose fewer assumptions on systems, which then enables us to generalize conventional control results limited to, *e.g.*, LTI systems. One such result can be found in [39], where a sensitivity property of feedback systems is analyzed using information theory; a fundamental limitation of sensitivity functions is presented in relation to the poles of the plants.

In this section, we present a characterization a complementary sensitivity property in a feedback system by measuring the entropy of the signals. In particular, following the approach of [39], we derive a limitation of the complementary sensitivity function with respect to the unstable zeros of the open-loop system. This result corresponds to Bode's integral formula for the complementary sensitivity [51].

We emphasize that the advantage of this information theory based approach, as exhibited in [39], is that it is capable to consider general nonlinear systems in the closed loop. This means that networked systems consisting of nonlinear elements such as encoders and decoders can be dealt with. Hence, the approach will lead us to a networked control version of Bode's integral formula where channel capacity is critical. The discussion in this section is, however, limited to the linear case for the ease of presentation; general nonlinear extensions can be found in [44].

8.5.1 Bode's Integral Formula for Complementary Sensitivity Functions

We briefly introduce Bode's integral formula for the discrete-time system case. This famous formula first derived by Bode is for the case of the sensitivity function [5]. Consider the feedback system depicted in Figure 8.18. Suppose that the open-loop system L is discrete time, single-input single output, LTI, strictly proper, and stable. Let its transfer function be $L(z)$. The sensitivity function $S(z)$ is then given by $S(z) := 1/(1+L(z))$.

Bode's integral formula shows that if the closed-loop system is stable, then it holds that

$$\frac{1}{2\pi}\int_{-\pi}^{\pi} \log_2 |S(e^{j\omega})| d\omega = 0.$$

This formula clarifies that the integration of the logarithm of the sensitivity gain over all frequencies is a constant, and in fact it is always zero. This equality exhibits a form of fundamental limitations in feedback systems. Due to its significance, the formula has been generalized by various researchers (*e.g.*, [18, 59, 49]).

Complementary sensitivity functions are equally important, and the constraint corresponding to Bode's formula for this case is given in [51]. We describe this case next. For the open-loop system L in Figure 8.18, let its state-space representation be given by

$$\begin{aligned} x(k+1) &= Ax(k) + Be(k) \\ y(k) &= Cx(k), \end{aligned} \tag{8.44}$$

where $x(k) \in \mathbb{R}^n$ is the state, $y(k) \in \mathbb{R}$ is the output, $r(k) \in \mathbb{R}$ is the reference input, and $e(k) \in \mathbb{R}$ is the error signal. Denote by ν the relative degree of the transfer function $L(z)$; this satisfies $\nu \geq 1$. Then, letting $D_0 := CA^{\nu-1}B$, we have that D_0 is nonzero and is equal to the ratio of the leading coefficients of $L(z)$. Also, define the set of unstable zeros of $L(z)$ as $\mathscr{U}\mathscr{Z}_L := \{z \,|\, L(z) = 0, |z| \geq 1\}$.

Let $T(z)$ be the complementary sensitivity function: $T(z) := L(z)/(1 + L(z))$. Bode's formula corresponding to this function is as follows [51].

Proposition 8.1. If the closed-loop system in Figure 8.18 is stable, then the complementary sensitivity function $T(z)$ satisfies

$$\frac{1}{2\pi} \int_{-\pi}^{\pi} \log_2 |T(e^{j\omega})| d\omega = \sum_{\beta \in \mathscr{U}\mathscr{Z}_L} \log_2 |\beta| + \log_2 |D_0|. \tag{8.45}$$

This result shows that in the complementary sensitivity case, the integral of the gain over the whole frequency range is constrained by the unstable zeros. This relation can be established by applying Jensen's formula from complex analysis. In what follows, we derive a constraint similar to (8.45), independently from this method, by evaluating the entropies and mutual information of signals in the feedback system.

8.5.2 *Entropy and Mutual Information*

We introduce some notation and basic results from information theory that will be used later [9].

We adopt the following notation. For a stochastic process $\{x(k)\}_{k=0}^{\infty}$, the sequence of random variables from $k = \ell$ to $k = m$ $(m \geq \ell)$ is represented by $x_\ell^m := \{x(k)\}_{k=\ell}^{m}$. In particular, when $\ell = 0$, we write x_ℓ^m simply as x^m. When clear from the context, x is used instead of $\{x(k)\}_{k=0}^{\infty}$.

Entropy represents a measure of uncertainty of a random variable. For a continuous random variable $x \in \mathbb{R}$ with the probability density function p_x, its (differential) entropy $h(x)$ is defined by

$$h(x) := -\int_{\mathbb{R}} p_x(\xi) \log_2 p_x(\xi) d\xi.$$

Given two random variables, mutual information provides a measure of the amount of information contained in one random variable about the other. The mutual information $I(x;y)$ between $x \in \mathbb{R}$ and $y \in \mathbb{R}$ is defined as

$$I(x;y) := h(x) - h(x|y),$$

where $h(\cdot|\cdot)$ denotes the conditional entropy.

For stochastic processes, entropy rates play a key role in our analysis. The entropy rate of the process x is the time average of the entropies of its truncated processes x^{k-1} given by

$$h_\infty(x) := \limsup_{k\to\infty} \frac{h(x^{k-1})}{k}.$$

The random processes appearing in this section belong to the following class; see [39] and also [45]. Consider a zero-mean stochastic process x ($x(k) \in \mathbb{R}$). This process x is said to be asymptotically stationary if the following limit exists for every $\gamma \in \mathbb{Z}$:

$$\overline{R}_x(\gamma) := \lim_{k\to\infty} E\left[x(k)x(k+\gamma)\right].$$

Further, for an asymptotically stationary process x, define the asymptotic power spectral density $\overline{S}_x$ as

$$\overline{S}_x(\omega) := \sum_{\gamma=-\infty}^{\infty} \overline{R}_x(\gamma)\mathrm{e}^{-j\gamma\omega}.$$

The following inequality provides a relation between the entropy rate and the asymptotic power spectral density. For an asymptotically stationary process x, it holds that

$$h_\infty(x) \le \frac{1}{4\pi}\int_{-\pi}^{\pi} \log_2(2\pi\mathrm{e}\overline{S}_x(\omega))d\omega. \tag{8.46}$$

The equality holds if, in addition, x is a Gaussian process.

8.5.3 Characterization of Complementary Sensitivity Properties

Consider the system depicted in Figure 8.18. Suppose that the input r is a stochastic process. We characterize the complementary sensitivity property from r to y by evaluating the entropy of signals.

Here, it is assumed that the feedback system is stable in the mean-square sense, that is, $\sup_k E[x(k)^T x(k)] < \infty$. We further assume that r^k and $x(0)$ are independent for every $k \in \mathbb{Z}_+$. Moreover, regarding the initial state $x(0)$, we assume $|h(x(0))| < \infty$. This implies that $x(0)$ is neither completely known nor completely unknown.

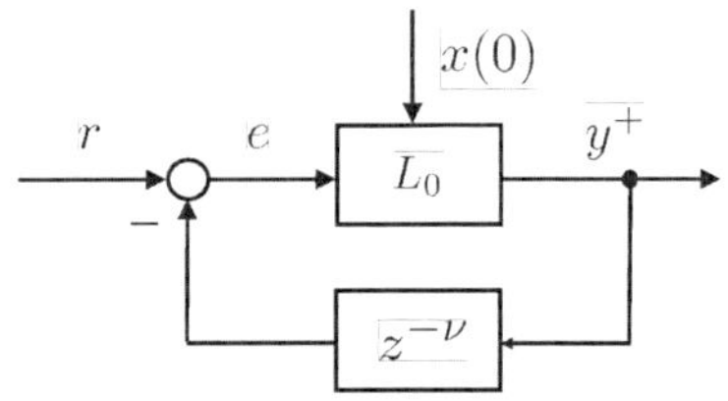

Fig. 8.19 Equivalent system with the biproper system L_0

As we will see later, this assumption is characteristic to the information theoretic approach.

Since we deal with asymptotically stationary processes, we introduce a complementary sensitivity-like property using the asymptotic power spectral densities of the input and output of T. Specifically, if r and y are asymptotically stationary, then the complementary sensitivity-like function $\overline{T}$ is given by

$$\overline{T}(\omega) := \sqrt{\frac{\overline{S}_y(\omega)}{\overline{S}_r(\omega)}}.$$

This complementary sensitivity-like function $\overline{T}(\omega)$ can be related to the transfer function $T(z)$, if the initial state is $x(0) = 0$, as $\overline{T}(\omega) = \left|T(\mathrm{e}^{j\omega})\right|$. This can be shown by the well-known relation between an LTI system with a stable transfer function and the power spectral densities of its input and output signals [45]. We however note that, in this case, $x(0)$ is deterministic and thus its entropy is $h(x(0)) = -\infty$. This means that the assumption $|h(x(0))| < \infty$ is not satisfied. In this respect, the problem setup of this paper is different from that based on transfer functions.

We consider the property of $\overline{T}$ instead of T and will obtain a constraint similar to (8.45). The derivation will be carried out in three steps.

Step 1: Conservation Law of Entropies

Because of the relation given in (8.46), power spectral densities can be expressed in terms of entropy rates. Hence, we first analyze the relation between the entropy rates $h_\infty(r)$ and $h_\infty(y)$ of the input and output of $\overline{T}$.

First, we have to consider how at each time k, the entropy $h(r(k))$ of the input $r(k)$ affects $h(y(k))$. However, since the open-loop transfer function $L(z)$ is strictly proper, there is time delay of ν steps due to the relative degree of L. Hence, $r(k)$ has an influence on the output y only after time $k+\nu$. To deal with this problem, we introduce the auxiliary system $L_0(z) := z^\nu L(z)$ by adding a time forward element of ν steps. Also, let $y^+(k) := y(k+\nu)$, *i.e.*, it is the output $y(k)$ forwarded by ν steps. By using L_0 and y^+, the system in Figure 8.18 can be expressed as that in Figure 8.19.

Now, we obtain the following proposition.

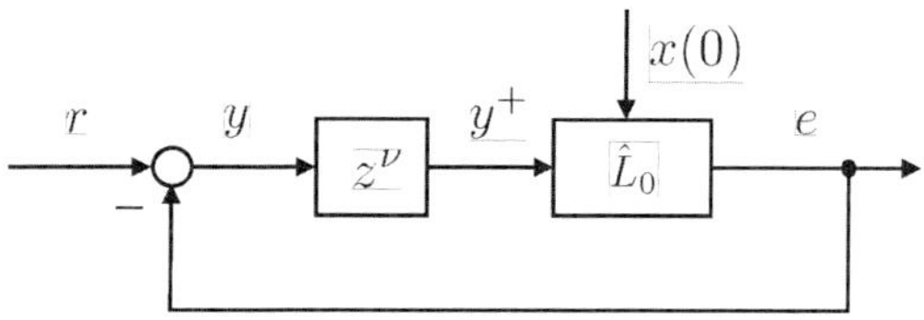

Fig. 8.20 Equivalent system with the inverse system $\hat{L}_0$ of L_0

Proposition 8.2

$$h(y_\nu^{k+\nu}) = h(r^k) + I(y_\nu^{k+\nu}; x(0)) + (k+1)\log_2|D_0|. \tag{8.47}$$

This result shows a conservation law of entropy between r and y. Intuitively, one can understand that $\log_2|D_0|$ reflects the scaling caused by the input r going through the system L. Moreover, $I(y_\nu^{k+\nu}; x(0))$ shows the effect of the initial state $x(0)$, which can be viewed as an external input entering y.

Next, we arrive at an inequality relating the entropy rates of r and y. Divide (8.47) by $k+1$ and take the limsup as $k \to \infty$ on both sides to obtain

$$h_\infty(y) \geq h_\infty(r) + \log_2|D_0| + \liminf_{k\to\infty} \frac{I(y_\nu^{k+\nu}; x(0))}{k}. \tag{8.48}$$

Step 2: Mutual Information and Unstable Zeros

In the next step, we relate the mutual information term in (8.48) and the unstable zeros of the open-loop transfer function $L(z)$.

Mutual information is a quantity in the time domain, and thus it is difficult to show its connection with zeros of a transfer function. Hence, we take an approach to view the zeros of L as the poles of the inverse system of L. Poles are easier to analyze because they can be expressed as the eigenvalues of the state matrix of the system. However, we must note that the inverse system of L is improper. For this reason, we consider the inverse system of the biproper system L_0 that was introduced earlier. Let $\hat{L}_0$ denote the inverse of L_0. The system in Figure 8.19 can then be converted to that in Figure 8.20.

Using this system, we obtain the following result.

Proposition 8.3. For the system depicted in Figure 8.20, the following inequality holds:

$$\liminf_{k\to\infty} \frac{I(y_\nu^{k+\nu}; x(0))}{k} \geq \sum_{\beta \in \mathscr{U}\mathscr{Z}_L} \log_2|\beta|. \tag{8.49}$$

This proposition exhibits that the output y and the initial state $x(0)$ have the mutual information which is a function of the unstable zeros of L. In general, from the viewpoint of the open-loop system L, when the system is unstable, the system amplifies the initial state at a level depending on the size of the unstable poles. Hence, the proposition can be interpreted as saying that in systems having more unstable dynamics, the output y contains more information about the initial state.

Furthermore, we note that if the initial condition $x(0)$ is exactly known, then from the definition of mutual information, it follows that $I(y_{\nu}^{k+\nu};x(0)) = 0$ for all k. In this case, the lower bound in (8.49) would be replaced with zero, which is obviously more conservative. Hence, we see that in the proposition above, the assumption $|h(x(0))| < \infty$ is critical in order to keep the dependence of the bound on the unstable poles.

Step 3: Complementary Sensitivity Property

Based on the results in Steps 1 and 2, we are ready to present the main result regarding the complementary sensitivity-like function $\overline{T}$.

Theorem 8.7. Consider the system depicted in Figure 8.18. Assume that the system is stable in the mean-square sense and that the input r is an asymptotically stationary Gaussian process. Then, $\overline{T}$ satisfies the following inequality:

$$\frac{1}{2\pi}\int_{-\pi}^{\pi} \log_2 \left|\overline{T}(j\omega)\right| d\omega \geq \sum_{\beta\in\mathscr{U}\mathscr{Z}_L} \log_2 |\beta| + \log_2 |D_0|. \tag{8.50}$$

Comparing the formulae (8.45) and (8.50), we observe that similar results are derived via two independent methods. Theorem 8.7 is a weaker result than Proposition 8.1 in the sense that the constraint is an inequality. We however note that the entropy rate of a signal is a notion in the time domain and thus can be defined for systems which do not have transfer function forms. This generalization is an important consequence of the information theoretic approach here.

As discussed earlier, the approach based on information theory is useful for further extensions to NCSs with capacity constraints. In fact, in [39], an integral-type constraint for the case of sensitivity property with an LTI plant and an arbitrary causal nonlinear controller is given. Such a controller may be composed of encoders, decoders, and channels under noise, thus leading us to a networked control version of Bode's formula.

For the complementary sensitivity case, such systems including nonlinear controllers have been studied in [44]. We however remark that the nonlinear extension is not straightforward. The reason can be explained as follows. In the complementary sensitivity case, for the input $r(k)$ at time k to reach the output $y(k)$, it must go through the system L. Hence, it is necessarily subject to the scaling caused by L. For the sensitivity case (from $r(k)$ to $e(k)$), this scaling is simply -1, which makes the analysis easier.

8.6 Conclusion

In this chapter, motivated by the new applications in NCSs, we have presented several approaches involving both control and communication and in particular control methods taking account of capacity constraints. We have shown recent theoretical results exhibiting fundamental limitations arising in networked control.

The first and the main part has been devoted to control problems with quantization, where the focus is on reducing the amount of communication as much as possible. Fundamental results on the minimum data rate for stabilization have been first highlighted with some historical backgrounds. Then, the quantized control approach based on the notion of coarsest quantizers has been discussed along with several variants considering uncertainties in the channel as well as in the plant.

The second part has dealt with a control analysis method by employing tools from information theory. Here, we have provided an extension of Bode's integral formula, by measuring the entropy of the signals within control systems. Though the results as presented do not involve communication channels explicitly, it is stressed that they have been obtained with a clear motivation from NCSs.

There are several interesting directions for future research. One is towards more decentralized and distributed control strategies for networked control employing multiple controllers communicating to each other; this direction is also related to the area of multi-agent systems and cooperative control; see, *e.g.*, [2, 7]. Another direction is to further explore the information theoretic approach for networked control problems.

References

1. Special Section on Complex Networked Control Systems. IEEE Control Systems Magazine, 27(4) (2007)
2. Antsaklis, P.J., Baillieul, J. (Guest eds.): Special Issue on the Technology of Networked Control Systems. Proc. IEEE 95(1) (2007)
3. Baillieul, J.: Feedback designs in information-based control. In: Pasik-Duncan, B. (ed.) Stochastic Theory and Control. LNCIS, vol. 280, pp. 35–57. Springer, Berlin (2002)
4. Bertram, J.E.: The effect of quantization in sampled-feedback systems. Trans. Amer. Inst. Elec. Engineers 77(2), 177–182 (1958)
5. Bode, H.W.: Network Analysis and Feedback Amplifier Design. D. Van Nostrand, New York (1945)
6. Brockett, R.W., Liberzon, D.: Quantized feedback stabilization of linear systems. IEEE Trans. Autom. Control 45, 1279–1289 (2000)
7. Bullo, F., Cortés, J., Piccoli, B., (Guest eds.): Special Issue on Control and Optimization in Cooperative Networks. SIAM J. Contr. Optim. 48(1) (2009)
8. Costa, O.L.V., Fragoso, M.D., Marques, R.P.: Discrete Time Markov Jump Linear Systems. Springer, London (2005)
9. Cover, T.M., Thomas, J.A.: Elements of Information Theory. Wiley, New York (1991)
10. Curry, R.E.: Estimation and Control with Quantized Measurements. MIT Press, Cambridge (1970)
11. De Persis, C.: n-bit stabilization of n-dimensional nonlinear systems in feedforward form. IEEE Trans. Autom. Control 50, 285–297 (2005)
12. Delchamps, D.F.: Stabilizing a linear system with quantized state feedback. IEEE Trans. Autom. Control 35, 916–924 (1990)
13. Elia, N.: When Bode meets Shannon: Control-oriented feedback communication schemes. IEEE Trans. Autom. Control 49, 1477–1488 (2004)

14. Elia, N.: Remote stabilization over fading channels. Systems & Control Letters 54, 238–249 (2005)
15. Elia, N., Mitter, S.K.: Stabilization of linear systems with limited information. IEEE Trans. Autom. Control 46, 1384–1400 (2001)
16. Fagnani, F., Zampieri, S.: Quantized stabilization of linear systems: Complexity versus performance. IEEE Trans. Autom. Control 49, 1534–1548 (2004)
17. Franklin, G.F., Powell, J.D., Emami-Naeini, A.: Feedback Control of Dynamic Systems, 5th edn. Prentice Hall, Uppersaddle River (2006)
18. Freudenberg, J.S., Looze, D.P.: Frequency Domain Properties of Scalar and Multivariable Feedback Systems. Springer, Berlin (1988)
19. Fu, M., Xie, L.: The sector bound approach to quantized feedback control. IEEE Trans. Autom. Control 50, 1698–1711 (2005)
20. Goodwin, G.C., Sin, K.S.: Adaptive Filtering Prediction and Control. Prentice Hall, Englewood Cliffs (1984)
21. Gupta, V., Dana, A.F., Hespanha, J.P., Murray, R.M.: Data transmission over networks for estimation and control. IEEE Trans. Autom. Control 54, 1807–1819 (2009)
22. Hadjicostis, C.N., Touri, R.: Feedback control utilizing packet dropping network links. In: Proc. 41st IEEE Conf. on Decision and Control, pp. 1205–1210 (2002)
23. Hayakawa, T., Haddad, W.M., Leonessa, A.: A Lyapunov-based adaptive control framework for discrete-time nonlinear systems with exogenous disturbances. Int. J. Control 77, 250–263 (2004)
24. Hayakawa, T., Ishii, H., Tsumura, K.: Adaptive quantized control for linear uncertain discrete-time systems. Automatica 45, 692–700 (2009)
25. Hayakawa, T., Ishii, H., Tsumura, K.: Adaptive quantized control for nonlinear uncertain systems. Systems & Control Letters 58, 625–632 (2009)
26. Hristu-Varsakelis, D., Levine, W.S. (eds.): Handbook of Networked and Embedded Control Systems. Birkhäuser, Boston (2005)
27. Imer, O.Ç., Yüksel, S., Başar, T.: Optimal control of LTI systems over unreliable communication links. Automatica 42, 1429–1439 (2006)
28. Ishii, H.: Stabilization under shared communication with message losses and its limitations. In: Proc. 45th IEEE Conf. on Decision and Control, pp. 4974–4979 (2006) See also arXiv:0806.3609
29. Ishii, H.: H^∞ control with limited communication and message losses. Systems & Control Letters 57, 322–331 (2008)
30. Ishii, H.: Limitations in remote stabilization over unreliable channels without acknowledgements. Automatica 45, 2278–2285 (2009)
31. Ishii, H., Başar, T.: Remote control of LTI systems over networks with state quantization. Systems & Control Letters 54, 15–31 (2005)
32. Ishii, H., Başar, T., Tempo, R.: Randomized algorithms for quadratic stability of quantized sampled-data systems. Automatica 40, 839–846 (2004)
33. Ishii, H., Francis, B.A.: Limited Data Rate in Control Systems with Networks. LNCIS, vol. 275. Springer, Berlin (2002)
34. Ishii, H., Ohyama, C., Tsumura, K.: Performance analysis of control systems under limited data rates. Trans. SICE 44, 396–404 (2008)
35. Khalil, H.K.: Nonlinear Systems, 2nd edn. Prentice-Hall, Uppersaddle River (1996)
36. Liberzon, D.: On stabilization of linear systems with limited information. IEEE Trans. Autom. Control 48, 304–307 (2003)
37. Liberzon, D., Nesic, D.: Input-to-state stabilization of linear systems with quantized state measurements. IEEE Trans. Autom. Control 52, 767–781 (2007)

38. Ling, Q., Lemmon, M.D.: Power spectral analysis of networked control systems with data dropouts. IEEE Trans. Autom. Control 49, 955–960 (2004)
39. Martins, N.C., Dahleh, M.A., Doyle, J.C.: Fundamental limitations of disturbance attenuation in the presence of side information. IEEE Trans. Autom. Control 52, 56–66 (2007)
40. Matveev, A.S., Savkin, A.V.: Estimation and Control over Communication Networks. Birkhäuser, Boston (2008)
41. Nair, G., Fagnani, F., Zampieri, S., Evans, R.J.: Feedback control under data constraints: An overview. Proc. IEEE 95(1), 108–137 (2007)
42. Nair, G.N., Evans, R.J.: Stabilizability of stochastic linear systems with finite feedback date rates. SIAM J. Contr. Optim. 43, 413–436 (2004)
43. Okano, K., Hara, S., Ishii, H.: Characterization of a complementary sensitivity property in feedback control: An information theoretic approach. Automatica 45, 504–509 (2009)
44. Okano, K., Ishii, H., Hara, S.: Sensitivity analysis of networked control systems via an information theoretic approach. In: Proc. 47th IEEE Conf. on Decision and Control, pp. 3360–3365 (2008)
45. Papoulis, A., Pillai, S.U.: Probability, Random Variables and Stochastic Processes, 4th edn. McGraw Hill, New York (2002)
46. Petersen, I.R., Savkin, A.V.: Multi-rate stabilization of multivariable discrete-time linear systems via a limited capacity communication channel. In: Proc. 40th IEEE Conf. on Decision and Control, pp. 304–309 (2001)
47. Renesas Technology Corp., `http://www.renesas.com/`
48. Seiler, P., Sengupta, R.: An H^∞ approach to networked control. IEEE Trans. Autom. Control 50, 356–364 (2005)
49. Seron, M.M., Braslavsky, J.H., Goodwin, G.C.: Fundamental Limitations in Filtering and Control. Springer, London (1997)
50. Sinopoli, B., Schenato, L., Franceschetti, M., Poolla, K., Jordan, M.I., Sastry, S.S.: Kalman filtering with intermittent observations. IEEE Trans. Autom. Control 49, 1453–1464 (2004)
51. Sung, H., Hara, S.: Properties of complementary sensitivity function in SISO digital control systems. Int. J. Control 50(4), 1283–1295 (1989)
52. Tatikonda, S.: Some scaling properties of large distributed control systems. In: Proc. 42nd IEEE Conf. on Decision and Control, pp. 3142–3147 (2003)
53. Tatikonda, S., Mitter, S.K.: Control under communication constraints. IEEE Trans. Autom. Control 49, 1056–1068 (2004)
54. Tatikonda, S., Sahai, A., Mitter, S.K.: Stochastic linear control over a communication channel. IEEE Trans. Autom. Control 49, 1549–1561 (2004)
55. Tsumura, K., Ishii, H., Hoshina, H.: Tradeoffs between quantization and packet loss in networked control of linear systems. Automatica 45, 2963–2970 (2009)
56. Wong, W.S., Brockett, R.W.: Systems with finite communication bandwidth constraints II: Stabilization with limited information feedback. IEEE Trans. Autom. Control 44, 1049–1053 (1999)
57. Yüksel, S., Başar, T.: Minimum rate coding for LTI systems over noiseless channels. IEEE Trans. Autom. Control 51, 1878–1887 (2006)
58. Yüksel, S., Başar, T.: Optimal signaling policies for decentralized multi-controller stabilizability over communication channels. IEEE Trans. Autom. Control 52, 1969–1974 (2007)
59. Zhang, G., Iglesias, P.A.: Nonlinear extension of Bode's integral based on an information-theoretic interpretation. Systems & Control Letters 50, 11–19 (2003)

Chapter 9
Event-Triggered Feedback in Control, Estimation, and Optimization

Michael Lemmon

Abstract. Networked control systems often send information across the communication network in a periodic manner. The selected period, however, must assure adequate system performance over a wide range of operating conditions and this conservative choice may result in significant over-provisioning of the communication network. This observation has motivated the use of *sporadic* transmission across the network's feedback channels. *Event-triggering* represents one way of generating such sporadic transmissions. In event-triggered feedback, a sensor transmits when some internal measure of the novelty in the sensor information exceeds a specified threshold. In particular, this means that when the gap between the current and the more recently transmitted sensor measurements exceeds a state-dependent threshold, then the information is transmitted across the channel. The state-dependent thresholds are chosen in a way that preserves commonly used stability concepts such as input-to-state stability or $\mathscr{L}_2$ stability. This approach for threshold selection therefore provides a systematic way of triggering transmissions that provides some guarantees on overall control system performance. While early work in event-triggering focused on control applications, this technique can also be used in distributed estimation and distributed optimization. This chapter reviews recent progress in the use of state-dependent event-triggering in embedded control, networked control systems, distributed estimation, and distributed optimization.

9.1 Introduction

Embedded and networked control systems often rely on the periodic sampling and transmission of data. This periodic data abstraction is advantageous from the design standpoint. It permits real-time system engineers and control system engineers to pursue their design objectives in relative isolation from each other. While this

Michael Lemmon
University of Notre Dame, Notre Dame, Indiana, USA
e-mail: lemmon@nd.edu

A. Bemporad, M. Heemels, and M. Johansson: Networked Control Systems, LNCIS 406, pp. 293–358.
springerlink.com

so-called *separation-of-concerns* has proven advantageous from a designer's perspective, it does not necessarily lead to cost effective implementations of the control system. By separating the concerns of the control engineer from the real-time system engineer, one forces each designer to adopt a conservative viewpoint that may lead to unnecessary over-provisioning in the system implementation and hence to higher system costs. When one applies these traditional design principles to extremely large-scale systems, then the cost of enforcing the periodic data abstraction may become prohibitive.

As a result of these scaling issues, there has been recent interest in developing *co-design* frameworks where the concerns of real-time systems and control systems engineers are treated in a unified manner. One of the first statements of the co-design problem was given by Seto et al. [66]. This work presented co-design as an optimization problem that sought to minimize a traditional quadratic integral measure of control cost subject to task schedulability constraints. Seto's problem was an off-line design approach since the optimization problem was solved prior to system deployment. Since that time, a number of other co-design approaches have been suggested. A number of promising methods were listed in a paper by Arzen et al. [3]. Since that time a number of research scientists have investigated the methods on this list. These methods include feedback modification of task attributes [8, 45, 9, 12], anytime controllers [7, 21], and event-triggered sampling [2, 70, 83].

One approach to co-design involves adjusting task attributes through feedback. An example of this is found in the elastic scheduling method [8] of Buttazzo et al. This method uses measured task execution times to adaptively adjust task periods. Lu et al. [45] presented a feedback control approach to real-time scheduling. This idea was later applied to the scheduling of control tasks by Caccamo et al. [9] and Cervin et al. [12]. This work clearly demonstrated that feedback control principles could be used to reduce the sensitivity of real-time systems to uncertainties in control task period, jitter, and execution time. The reduction in real-time system sensitivity also leads to improved control system performance, since one no longer needs to design the real-time system for the worst-case variation in jitter and execution time.

While these early schemes used feedback about the real-time system's performance to adjust task attributes, this feedback was not directly based on the control system's measured performance. A more direct link between real-time system and control system performance will be found in recent work examining *anytime* controllers and *event-triggered* sampling. Anytime controllers are control systems that adjust their structure based on the performance of the real-time system [7, 21]. In other words, if the real-time system becomes overloaded, then the application will select a less complex (though stabilizing) controller to execute. In this way, the controller's performance is directly tied in an intelligent way to the real-time system's performance.

Event-triggered controllers, on the other hand, adapt the real-time system's task period directly in response to the application's performance [2] . Under event-triggering the control task is only executed when the application's error signal exceeds a specified threshold. Ostensibly, this error provides a measure of how valuable the current state is to the overall system's closed-loop behavior. In this way the real-time system is only used when it is essential for maintaining the system's performance. Since the system state is always changing, this approach generates a sporadic sequence of controller invocations. In general, the hope is that the average rate of this sporadic task set will be much lower than the rate of a comparable periodic task set.

There is experimental evidence to support the assertion that event-triggered feedback improves overall control system performance while reducing the real-time system's use of computational resources. Two examples are shown in figure 9.1 which shows results from [4] and [63, 65].

The left-hand plot in figure 9.1 shows a plot from [4]. This paper considers a controlled scalar diffusion process of the form,

$$dx = axdt + udt + dw,$$

where a is a real constant and w is a standard Brownian motion. The signal u is the control signal generated by a full-state controller. This control is computed in either a periodic or event-triggered manner. Under event-triggering, the control is updated whenever the state magnitude, $|x|$, exceeds a specified threshold. The performance of the system is characterized by the steady-state variance of the system state. The variance of the periodically triggered system is denoted as V_R whereas the variance of the event-triggered system is denoted as V_L. The left-hand plot in figure 9.1 plots the ratio V_R/V_L as a function of the mean sampling period, T. Note that for all choices of the system constant, a, this performance ratio is greater than one, thereby showing that the event-triggered system has better performance than periodically triggered systems operating at the same mean sampling period.

The right-hand side of figure 9.1 shows another example in which an event-triggered system demonstrates lower usage of computational resources. This result is taken from [63, 24] which considers the control of a linear plant under a PID

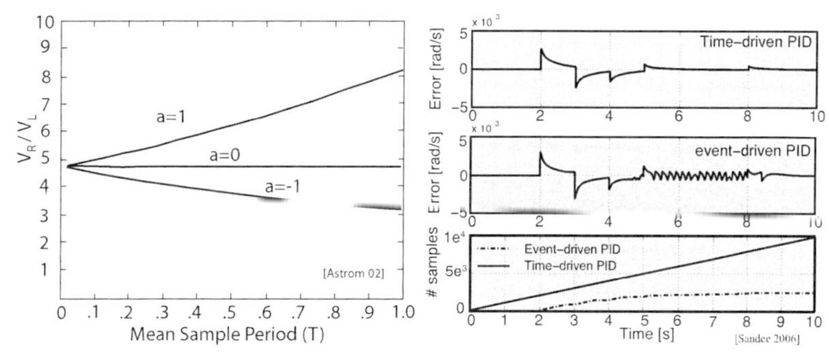

Fig. 9.1 Experimental results demonstrating that event-triggered feedback reduces a real-time system's use of computational resources while providing good overall control system performance

controller. This controller is discretized at a specified sampling rate and the resulting tracking error is plotted as a function of time in the top plot on the right-hand side of figure 9.1. The middle plot shows the tracking error for a comparable event-triggered implementation of the system. In this case the control is recomputed when the *gap* between the current system state and the last sampled system state exceeds a specified threshold, e_T,

$$\text{gap} = |x(t) - x(r_j)| \leq e_T = \text{threshold},$$

where r_j denotes the jth consecutive time when the state was sampled. For the simulation shown in the middle plot, the threshold e_T was chosen to match the peak error of the periodically triggered system. This means that these first two plots are comparing the behavior of an event-triggered and periodically triggered system having similar performance levels. The bottom plot in the figure shows the number of samples that were generated by the periodically triggered (time-driven) and event-triggered PID control. As can be seen from this plot, the number of event-triggered samples is smaller than the time-driven control. Moreover, as the system approaches its equilibrium point, the number of samples begins to level off, thereby suggesting that as the information content within the error signal decreases, the controller needs to be invoked less often.

The left-hand example shown in figure 9.1 suggests an event-triggered system will perform better than a periodically-triggered systems with similar computational usage. The right-hand example suggests an event-triggered system will use fewer computational resources than a periodically triggered system with similar performance levels. These results, unfortunately, are only empirical in nature. The objective of this chapter is to review prior work that provides a more complete analysis of the relationship between performance and computation in event-triggered feedback systems.

Event-triggering samples the system state when some measure of the *novelty* in the state exceeds a given threshold. This approach to sampling has been around for quite awhile. Early examples of event-triggered systems may be found in relay [73] and pulse-width modulated feeback [54]. Event-triggered feedback has been used in reaction jet control of spacecraft. More recent examples have examined event-triggering in systems using motors [23, 65, 64, 24]. A comparison of the performance of event-triggered systems against periodically triggered systems may be found in [26]. Event-triggering has also appeared under a variety of other names such as interrupt-based feedback [29], Lebesgue sampling [4], asynchronous sampling [76], self-triggered feedback [75], state-triggered feedback [71], and level crossing sampling [39].

While event-triggering has been around for quite awhile it has only been in recent years [2] that researchers have made significant advances in understanding the event-triggering process. Sampled stochastic differential equations have been used to study event-triggered sampling [4]. This model has also been used to study event-triggered control [5, 27]. Optimal control and estimation in these event-triggered stochastic systems was studied in [87] for infinite horizons. The results from [87] determine event triggers that maximize control/estimator performance subject to a soft constraint on communication usage. The resulting optimal event-triggers take the form of static thresholds. These results were extended to finite horizon event-triggered systems for control [31, 62] and estimation [30, 58, 57, 60]. These finite-horizon results optimize control/estimator performance subject to hard communication constraints. For estimation problems, the resulting optimal event-triggers are time-varying and for control problems, the event-triggers are time-varying functions of the initial system state.

Much of the aforecited work, however, focused only on scalar systems due to the computational complexity associated with solving the associated dynamic programming equations. Research scientists have recently been using state-based methods that can be more easily applied to vector systems. Making use of the emulation method [50] in sampled-data systems, recent work has identified sufficient sampling conditions that preserve closed-loop stability concepts such as input-to-state stability [70] or $\mathscr{L}_2$ stability [83]. A similar state-based approach has also been proposed in [40]. This recent work again derives state-dependent thresholds for the event-triggers. While the recent state-based methods do not explicitly constrain communication usage, experimental studies suggest that state-dependent event-triggers can be very effective in reducing an embedded system's usage of computational and communication resources.

This chapter discusses how event-triggering can be used in a wide range of networked control applications; ranging from control to estimation to optimization. In all of these application areas, event-triggering appears to greatly reduce the communication and/or computational effort required of the supporting real-time system. The remainder of this chapter is organized as follows. The chapter first reviews some mathematical preliminaries in section 9.2. Section 9.3 examines state-based event-triggering in embedded single processor control systems. The results from this section are then extended to networked control systems in section 9.4. The controllers in sections 9.3-9.4 all use full state feedback. As a first step towards developing output-feedback controls, section 9.5 examines a recent approach to event-triggered state estimation. Finally section 9.6 presents a novel application of event-triggering in distributed optimization of networked systems. Event-triggered control is still an active research area and a number of promising future research directions are discussed in section 9.7.

9.2 Mathematical Preliminaries

The event-triggers in this chapter are designed to enforce a variety of *stability concepts* found in the system science literature. This section reviews those stability concepts primarily to establish notational conventions that are followed throughout the rest of this chapter. In particular, this section reviews stability concepts such as asymptotic stability, input-to-state stability, and $\mathscr{L}_2$ stability. Much of this material may be found in textbooks [32, 74, 37].

This chapter adopts the following notational conventions. The function x mapping elements on the real line, $\mathbb{R}$, onto elements of Euclidean n-space, $\mathbb{R}^n$, is denoted as $x : \mathbb{R} \to \mathbb{R}^n$. Let $x(t)$ denote the value that this function takes at time $t \in \mathbb{R}$ and let $\dot{x}(t) = dx(t)/dt$ denote the time derivative of x at time t. This function is said to solve an initial value problem of the form

$$\dot{x}(t) = f(x(t)), \quad x(0) = x_0 \tag{9.1}$$

if the above equations are satisfied for almost all $t \geq 0$. In equation (9.1), $f : \mathbb{R}^n \to \mathbb{R}^n$ is a function mapping Euclidean n-space back onto itself. To assure the above equation has unique solutions, one often requires that f be *Lipschitz continuous*. Namely, there exists a positive real constant L such that

$$\|f(x) - f(y)\| \leq L\|x - y\|$$

for all x and y in $\mathbb{R}^n$. The vector $x(t) \in \mathbb{R}^n$ is an element of a normed vector space where $\|x(t)\|$ denotes the usual Euclidean 2-norm. The function x is a member of a normed linear space. Important norms used for such functions are the supremum norm, $\|x\|_{\mathscr{L}_\infty} = \operatorname{ess\,sup}_t \|x(t)\|$ and the 2-norm $\|x\|_{\mathscr{L}_2} = \sqrt{\int_0^\infty \|x(\tau)\|^2 d\tau}$. Let $\mathscr{L}_2$ denote the linear space of all measurable functions with bounded 2-norms. $\mathscr{L}_\infty$ denotes the linear space of all measurable functions with bounded supremum norm. A function $\alpha : \mathbb{R} \to \mathbb{R}$ is said to be class $\mathscr{K}$ if it is continuous, strictly increasing, and $\alpha(0) = 0$. A function $\beta : \mathbb{R} \times \mathbb{R} \to \mathbb{R}$ is said to be of class $\mathscr{K}\mathscr{L}$ if it is a continuous function that is class $\mathscr{K}$ with respect to the first argument and decreasing asymptotically to zero with respect to the second argument.

With these notational conventions established one can now define a variety of stability concepts. One of the best known stability concepts is *Lyapunov stability*. This concept applies to homogeneous systems characterized in equation (9.1). Given such as system, one says that a point $\overline{x} \in \mathbb{R}^n$ is an *equilibrium point* if $0 = f(\overline{x})$; in other words $\overline{x}$ represents a fixed point of the system . Without loss of generality, one can presume the equilibrium point $\overline{x} = 0$ lies at the origin.

The concept of Lyapunov stability is a property of the system's equilibrium point. In particular one says that the equilibrium point, $\overline{x} = 0$, is stable in the sense of Lyapunov if for all $\varepsilon > 0$ there exists $\delta > 0$ such that for all $t \geq 0$

$$\|x(0)\| < \delta \Rightarrow \|x(t)\| < \varepsilon.$$

Essentially, this means that the equilibrium point is Lyapunov stable if there always exists an initial condition that permits us to confine the system state within an arbitrarily small neighborhood of the equilibrium point.

A somewhat stronger (and better known) notion of Lyapunov stability is *asymptotic stability*. An equilibrium point is said to be asymptotically stable if the point is Lyapunov stable and if the state $x(t)$ asymptotically approaches the equilibrium point as t goes to infinity.

The existence of a *Lyapunov function* provides a well known sufficient condition for Lyapunov (asymptotic) stability. Consider a homogeneous system $\dot{x}(t) = f(x(t))$

with equilibrium point $\bar{x} = 0$. One says a continuously differentiable function $V : \mathbb{R}^n \to \mathbb{R}$ is a Lyapunov function for the system if V is a positive definite function and its directional derivative, $\dot{V} = \frac{\partial V}{\partial x} f(x)$, is negative semi-definite. The existence of a Lyapunov function V is sufficient to show that the equilibrium point is stable in the sense of Lyapunov. Moreover, if one can strengthen the condition on $\dot{V}$ to be negative definite, then this suffices to establish that the equilibrium point is asymptotically stable.

While the Lyapunov stability concept has been widely used, it cannot be directly used to characterize the behavior of inhomogeneous systems whose state trajectories $x : \mathbb{R} \to \mathbb{R}^n$ satisfy the initial value problem,

$$\dot{x}(t) = f(x(t), w(t)), \quad x(0) = x_0. \tag{9.2}$$

where $w : \mathbb{R} \to \mathbb{R}^m$ is an external disturbance. In this case $f : \mathbb{R}^n \times \mathbb{R}^m \to \mathbb{R}^n$ maps the current system state, $x(t)$, and an external disturbance, $w(t)$, onto the state's time derivative. Because this system is driven by an external disturbance, one cannot usually identify a single equilibrium point for the system. Without this equilibrium point, one cannot use Lyapunov stability concepts to study the system's behavior. This observation motivates a variety of other stability concepts for such inhomogeneous systems. Two such stability concepts are input-to-state stability and $\mathscr{L}_2$ stability.

The system in equation (9.2) is input-to-state stable (ISS) if there exists a class $\mathscr{KL}$ function β and a class $\mathscr{K}$ function γ such that for any initial condition, $x(0) = x_0$, the response under any input $w \in \mathscr{L}_\infty$ satisfies

$$\|x(t)\| \leq \beta(\|x_0\|, t) + \gamma(\|w\|_{\mathscr{L}_\infty})$$

for all $t \geq 0$. An alternative and equivalent characterization of ISS is that the system response satisfies

$$\|x(t)\| \leq \max\{\beta(\|x_0\|, t), \gamma(\|w\|_{\mathscr{L}_\infty})\}$$

for all $t \geq 0$. Both definitions essentially require that the transient and steady state behaviors of the system are appropriately bounded. This view of the ISS-concept is illustrated in figure 9.2. The dashed line shows the bound due to the class $\mathscr{KL}$ function β acting on the initial transient portion of the system's response. The dotted line shows the bound due to the class $\mathscr{K}$ function γ acting on the steady-state portion of the system's response. To be ISS, the system's response must lie below the point-wise maximum of both of these comparison functions.

Input-to-state stability can also be characterized using Lyapunov-type functions. In particular, one says that a continuously differentiable function $V : \mathbb{R}^n \to \mathbb{R}$ is an *ISS-Lyapunov* function for the system in equation (9.2) if there exist class $\mathscr{K}$ functions $\underline{\alpha}$, $\overline{\alpha}$, γ, and β such that

$$\begin{gathered}\underline{\alpha}(\|x\|) \leq V(x) \leq \overline{\alpha}(\|x\|)\\ \dot{V}(x, w) \leq -\gamma(\|x\|) + \beta(\|w\|)\end{gathered}$$

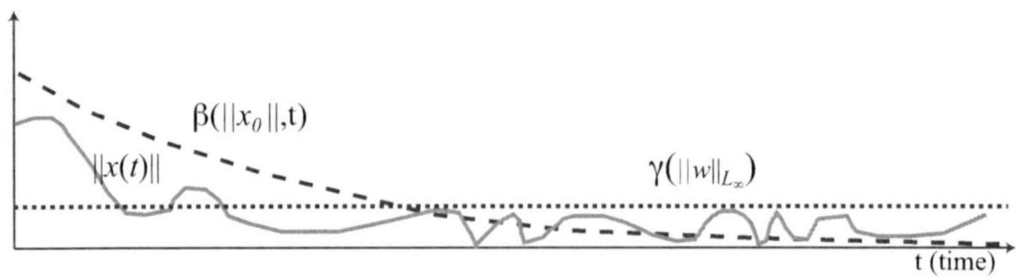

Fig. 9.2 Input-to-State Stability bounds the response's transient and steady-state behavior

hold for $x \in \mathbb{R}^n$ and all $w \in \mathbb{R}^m$. The existence of an ISS-Lyapunov function for the system in equation (9.2) is necessary and sufficient for that system to be ISS.

$\mathcal{L}_2$ stability is another useful stability concept for inhomogeneous systems. In this case one usually thinks of the system as a mapping, $G : \mathcal{L}_2 \to \mathcal{L}_2$ between two normed linear spaces, $\mathcal{L}_2$. This means that if one is given an input $w \in \mathcal{L}_2$ then the system's output function Gw will also be a function in $\mathcal{L}_2$. The system G is finite-gain $\mathcal{L}_2$ stable (or just $\mathcal{L}_2$ stable) if there exist finite positive real constants γ and β such that

$$\|Gw\|_{\mathcal{L}_2} \leq \gamma \|w\|_{\mathcal{L}_2} + \beta. \tag{9.3}$$

The right-hand side of the above inequality represents an affine function that overbounds the norm of the actual system's output. In particular, one can think of γ as a *gain* and β as an offset or *bias*. The so-called *induced gain* of G is then taken as the greatest lower bound on all of the possible γ's for which the above inequality holds. This *induced gain* is often denoted as $\|G\|$ and can be formally defined as

$$\|G\| = \inf\left\{\gamma \in \mathbb{R} \; : \; \|Gw\|_{\mathcal{L}_2} \leq \gamma \|w\|_{\mathcal{L}_2} + \beta\right\}$$

for all $w \in \mathcal{L}_2$.

The induced gain provides an important way of defining a control system's performance. Many control synthesis problems can be formulated as so-called *regulator* problems in which the objective is to minimize the *gain* from the closed-loop system's uncontrolled external input to some output function. By making the induced gain of the closed-loop system sufficiently small, one provides some guarantee on the control system's performance level. The induced gain therefore becomes a direct way of characterizing overall control system performance.

When the inhomogeneous system in equation (9.2) has a special affine form then there is a useful characterization of the $\mathcal{L}_2$ induced gain. This characterization will be used later to design event-triggered systems that enforce the $\mathcal{L}_2$ stability concept. In particular, let's consider a special form of the inhomogeneous control system in which the state trajectory satisfies

$$\begin{aligned} \dot{x}(t) &= A(x(t)) + B_1(x(t))w(t) + B_2(x(t))u(t) \\ z(t) &= \begin{bmatrix} x(t) \\ u(t) \end{bmatrix}^T \end{aligned} \tag{9.4}$$

where $x(0) = x_0$, $w : \mathbb{R} \to \mathbb{R}^p$ is an external $\mathscr{L}_2$ disturbance and $u : \mathbb{R} \to \mathbb{R}^m$ is a *control* signal that is generated by a controller $K : \mathbb{R}^n \to \mathbb{R}^m$. The functions $A : \mathbb{R}^n \to \mathbb{R}^n$, $B_1 : \mathbb{R}^n \to \mathbb{R}^{n\times p}$ and $B_2 : \mathbb{R}^n \to \mathbb{R}^{n\times m}$ define how the system state, external input, and control map into the state's time derivative. The other signal $z : \mathbb{R} \to \mathbb{R}^{n+m}$ represents the system's *output* signal. The objective is to find a controller K such that the induced $\mathscr{L}_2$ gain from the external input w to the output z is less than a specified amount, γ.

The main result characterizing such a controller makes use of the so-called *Hamilton-Jacobi Inequality* (HJI). In particular, assume there exist a real constant $\gamma \geq 0$ and a positive definite continuously differentiable function $V : \mathbb{R}^n \to \mathbb{R}$ that satisfy the HJI,

$$\frac{\partial V}{\partial x}A(x) + \frac{1}{2}\frac{\partial V}{\partial x}\left[\frac{1}{\gamma^2}B_1(x)B_1^T(x) - B_2(x)B_2^T(x)\right]\frac{\partial V^T}{\partial x} + \frac{1}{2}x^T x \leq 0 \tag{9.5}$$

for all $x \in \mathbb{R}^n$. If one then selects the control output, u, so that

$$u = K(x) = -B_2^T(x)\frac{\partial V(x)^T}{\partial x} \tag{9.6}$$

then one can show that the closed-loop system's $\mathscr{L}_2$ gain is less than or equal to γ.

The bound on $\|G\|$ can be obtained as follows. The directional derivative of V is

$$\dot{V} = \frac{\partial V}{\partial x}A(x(t)) + \frac{\partial V}{\partial x}B_1(x(t))w(t) + \frac{\partial V}{\partial x}B_2(x(t))u(t).$$

Completing the square on the cross-term $\frac{\partial V}{\partial x}B_1(x)w$ and using the fact that $u = -B_2^T\frac{\partial V}{\partial x}^T$, yields

$$\dot{V} = \frac{\partial V}{\partial x}A - \frac{1}{2}\left\|\gamma w - \frac{1}{\gamma}B_1^T\frac{\partial V}{\partial x}^T\right\|^2 + \frac{1}{2\gamma^2}\frac{\partial V}{\partial x}B_1B_1^T\frac{\partial V}{\partial x} + \frac{1}{2}\gamma^2\|w\|^2 - \|u\|^2.$$

Making use of the Hamilton-Jacobi inequality, one can bound $\dot{V}$ as

$$\begin{aligned}\dot{V} &\leq -\frac{1}{2}\left\|\gamma w - \frac{1}{\gamma}B_1^T\frac{\partial V}{\partial x}^T\right\|^2 - \frac{1}{2}\|u\|^2 + \frac{1}{2}\gamma^2\|w\|^2 - \frac{1}{2}\|x\|^2 \\ &\leq -\frac{1}{2}(\|u\|^2 + \|x\|^2 - \gamma^2\|w\|^2).\end{aligned}$$

Since $z = \begin{bmatrix} x \\ u \end{bmatrix}$, this implies that

$$\dot{V} \leq -\frac{1}{2}\|z\|^2 + \frac{1}{2}\gamma^2\|w\|^2.$$

If one then integrates the above inequality from 0 to infinity, one can readily use the definition of the $\mathscr{L}_2$-norm to see that

$$\|z\|_{\mathscr{L}_2} \leq \gamma \|w\|_{\mathscr{L}_2} + \sqrt{2V(x(0))}.$$

This inequality is precisely what was seen in equation (9.3) which is sufficient to imply the closed-loop system is $\mathscr{L}_2$ stable with an induced gain less than or equal to γ. In other words, if one choses the control as stated in equation (9.6), then one can guarantee that the $\mathscr{L}_2$ performance level achieved by the closed-loop system is less than or equal to γ.

As mentioned at the opening of this section, this chapter derives event-triggers that preserve input-to-state stability or $\mathscr{L}_2$ stability concepts. The preceding definitions and derivations will be used later in deriving these event-triggers. Let's now turn to see precisely how such event-triggers would be derived for both embedded control and networked control systems.

9.3 Event-Triggered Feedback in Embedded Control Systems

This section discusses the design of event-triggering schemes for embedded control systems. The main idea is to first design a continuous-time controller that guarantees a stability concept such as input-to-state stability or $\mathscr{L}_2$ stability. The section then develops an *event-triggering threshold* such that the associated sporadically triggered control system preserves this underlying stability concept. This approach is sometimes called the *emulation-based* method [50].

Sampled-Data System Model: Let's first consider how a sampled-data system might be configured. Figure (9.3) shows a block diagram for the system under study. The *plant* (G) has two types of inputs. There is an external *uncontrolled* disturbance, $w : \mathbb{R} \to \mathbb{R}^q$ and a control input, $u : \mathbb{R} \to \mathbb{R}^m$. The plant's state, $x : \mathbb{R} \to \mathbb{R}^n$, satisfies the inhomogeneous differential equation

$$\dot{x}(t) = f(x(t), u(t), w(t))$$

for $t \geq 0$ with initial condition $x(0) = x_0 \in \mathbb{R}^n$. The output of the plant is the system state, x.

Rather than working directly with the continuous-time state, x, the controller works with a sampled version of the state trajectory. In particular, let's introduce a *sampler* (S) system that is characterized by a monotone increasing sequence of *sampling instants*. This sequence of sampling instants is denoted as $\{r_j\}_{j=0}^{\infty}$ where $r_j > r_{j-1}$ for $j = 1, 2, \ldots, \infty$. The time $r_j \in \mathbb{R}$ denotes the jth consecutive sampling instant. The output of the sampler is therefore a sequence of sampled states, $\{\hat{x}_j\}$, in which $\hat{x}_j = x(r_j)$. A state-feedback controller, $K : \mathbb{R}^n \to \mathbb{R}^m$, maps the sampled state onto a control vector $\hat{u}_j \in \mathbb{R}^m$. The resulting sequence $\{\hat{u}_j\}_{j=0}^{\infty}$ of controls is then transformed into a continuous-time signal through a zero-order hold (H) without any delay. The control signal, $u \in \mathbb{R} \to \mathbb{R}^m$, used by the plant is a piecewise

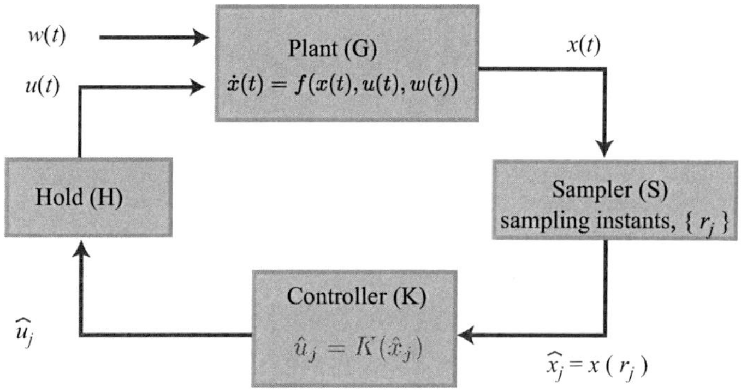

Fig. 9.3 Sampled Data Control System

constant function. In particular, let's introduce a sequence of functions $\tilde{u}_j : \mathbb{R} \to \mathbb{R}^m$ that have support over the time interval $[r_j, r_{j+1})$. The value of $\tilde{u}_j$ at time $t \in \mathbb{R}$ is

$$\tilde{u}_j(t) = \begin{cases} \hat{u}_j & \text{for } t \in [r_j, r_{j+1}) \\ 0 & \text{otherwise} \end{cases}$$

for $j = 0, 1, \dots, \infty$. With regard to this sequence the controlled input feeding the plant simply becomes $u(t) = \sum_{j=0}^{\infty} \tilde{u}_j(t)$.

ISS Event-Triggers: Under the *emulation-based* approach for developing sampled-data systems, one assumes that the controller, K, enforces a specified stability concept. In particular, let's confine our attention to input-to-state stability and let's consider the *continuously* sampled closed-loop system,

$$\dot{x}(t) = f(x(t), K(x(t) + e(t)), w(t))$$

where $e : \mathbb{R} \to \mathbb{R}^n$ and $w : \mathbb{R} \to \mathbb{R}^m$ are $\mathscr{L}_\infty$ input disturbances. Let's assume that the controller K leaves this closed-loop system ISS with respect to the two inputs w and e.

From our earlier discussion in section (9.2), the ISS assumption implies the existence of an ISS-Lyapunov function $V : \mathbb{R}^n \to \mathbb{R}$ with class $\mathscr{K}$ functions $\underline{\alpha}$, $\overline{\alpha}$, γ, β_1 and β_2 such that

$$\underline{\alpha}(\|x\|) \leq V(x) \leq \overline{\alpha}(\|x\|) \tag{9.7}$$

$$\frac{\partial V}{\partial x} f(x, K(x+e), w)) \leq -\gamma(\|x\|) + \beta_1(\|e\|) + \beta_2(\|w\|). \tag{9.8}$$

The inequalities in equation (9.7) essentially require that V is positive definite. The inequality in equation (9.8) is a dissipative relation on the ISS-Lyapunov function's directional derivative.

Now let's consider the sampled-data version of this system. The sampler generates a sequence of sampling instants, $\{r_j\}_{j=0}^{\infty}$. The time r_j is referred to as the jth

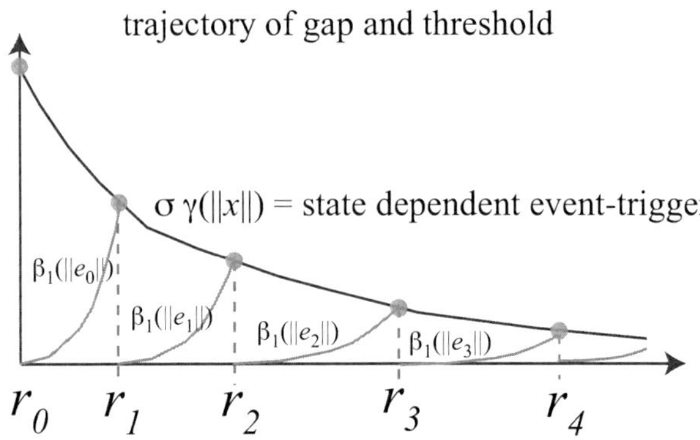

Fig. 9.4 Time history of gap and event threshold

consecutive *release time* of the system. (This term refers to the fact that in a real-time computer system, state sampling is implemented through a task that is released for execution at time r_j). The sampled states $\{\hat{x}_j\}_{j=0}^{\infty}$ form a sequence in which $\hat{x}_j = x(r_j)$. Let's define the *gap* function associated with the jth sampling time as a function $e_j : [r_j, r_{j+1}) \to \mathbb{R}^n$ in which

$$e_j(t) = \hat{x}_j - x(t)$$

for $t \in [r_j, r_{j+1})$ where $j = 0, 1, \ldots, \infty$.

The sampled data system's controller uses $\hat{x}_j = e_j(t) + x(t)$, rather than $x(t)$, so the sampled-data system's state must satisfy

$$\dot{x}(t) = f(x(t), K(x(t) + e_j(t)), w(t))$$

for all $t \in [r_j, r_{j+1})$ and all $j = 0, 1, \ldots, \infty$. Under the ISS assumption, one knows that

$$\dot{V} \leq -\gamma(\|x(t)\|) + \beta_1(\|e_j(t)\|) + \beta_2(\|w(t)\|). \tag{9.9}$$

Let's assume the gap can be restricted so that for some $\sigma \in (0, 1)$

$$\beta_1(\|e_j(t)\|) \leq \sigma\gamma(\|x(t)\|) \tag{9.10}$$

for all $t \geq 0$ and all $j = 0, 1, \ldots, \infty$. Inserting equation (9.10) into equation (9.9) implies

$$\dot{V} \leq -(1 - \sigma)\gamma(\|x(t)\|) + \beta_2(\|w(t)\|).$$

In light of the characterization of input-to-state stability (Sec. 9.2) and since $0 < \sigma < 1$, it should be apparent that enforcing the constraint on the gap (equation (9.10)) leaves the sampled-data system input-to-state stable with respect to the input w.

The constraint in equation (9.10) can be viewed as a state-dependent threshold condition. In particular, one knows that at the beginning of the interval $[r_j, r_{j+1})$, that the gap $e_j(r_j) = 0$. After that, one expects the norm of the gap to increase. When the gap satisfies the inequality $\beta_1(\|e_j(t)\|) > \sigma\gamma(\|x(t)\|)$, then the system state is again *sampled* by setting $\hat{x}_j = x(t)$, thereby forcing the gap to zero again. In this way the condition in equation (9.10) can be viewed as an *event-trigger*. This event-trigger would be realized by the sampler, S. In particular, one would require the sampler to continuously monitor the inequality in equation (9.10). Upon detecting a violation of the inequality, the sampler would trigger the sampling of the system state. The resulting time history of the threshold $\gamma(\|x(t)\|)$ and the gap is shown below in figure 9.4. As the state asymptotically approaches the origin, the threshold gets smaller. This *state-dependent* threshold idea and the above analysis underlying the ISS event-trigger was first discussed by Tabuada [70].

Let's look at a simple example to see how well the ISS event-trigger works. Consider a process model (without the external disturbance w) of the form

$$\begin{aligned} \dot{x}(t) &= f(x(t)) + u(t) \\ u(t) &= -2f(\hat{x}_j) \end{aligned}$$

for $t \in [r_j, r_{j+1})$. The release times r_j are selected as the times when the *event-trigger* is violated. The ISS event-trigger is chosen to have the following form,

$$\beta_1(\|e_j(t)\|) = e_j^2(t) \leq x^2(t) = \gamma(\|x(t)\|).$$

The system function $f : \mathbb{R} \to \mathbb{R}$ is chosen so the proposed control leaves the closed-loop system input-to-state stable. In particular, let's consider three different types of system dynamics. The chosen system dynamics have sublinear, linear, and superlinear f functions of the forms,

sublinear	linear	superlinear
$f(x) = \mathrm{sgn}(x)\sqrt{\lvert x\rvert}$	$f(x) = x$	$f(x) = x^3$

The plots below in figure 9.5 show the system response for the linear and superlinear choices for f. The top graphs plot the gap, $\beta_1(\|e(t)\|)$, and the threshold $\gamma(\|x(t)\|)$ as a function of time for both cases. This response is plotted on a logarithmic axis. For both linear and superlinear cases one sees that the gap satisfies the basic form seen earlier in figure 9.4. The bottom plots show the intersample time, $T_j = r_j - r_{j-1}$, for both cases. For the linear f, the choice of event trigger and f yields a periodic sampling of the system state. The case with the superlinear case f shows that the intersample time gets longer as the system state approaches the equilibrium point of the unforced system.

An interesting behavior is seen if f is the sublinear function $f(x) = \mathrm{sgn}(x)\sqrt{|x|}$. The gap and intersample time histories for this case are shown in figure 9.6. In this case, the intersample times get shorter and shorter as the system approaches its equilibrium point. Asymptotically these time intervals go to zero at a finite time around 3.5 seconds into the simulation. This type of behavior is sometimes called a

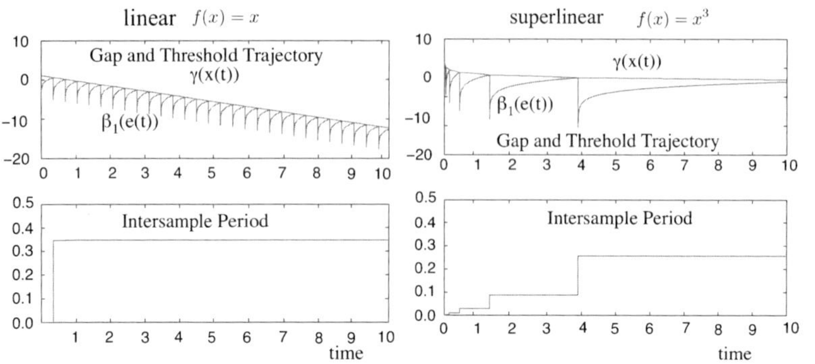

Fig. 9.5 Two examples showing gap, threshold, and intersampling time history. (left) linear function $f(x) = x$. (right) superlinear function $f(x) = x^3$

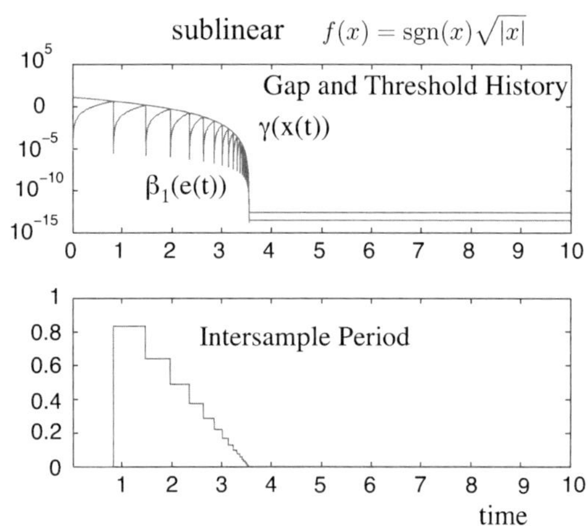

Fig. 9.6 Gap, threshold, and intersampling time history for a sublinear dynamical system where $f(x) = \mathrm{sgn}(x)\sqrt{|x|}$

Zeno behavior. Zeno-sampling is highly undesirable in real-time control for it would require the computer to eventually sample infinitely fast.

Avoiding Zeno-sampling: To better understand when Zeno-sampling might occur, let's try to derive a lower bound on the event-triggered system's intersample time. Let's first assume that the closed-loop system is Lipschitz with respect to the state x and the gap e. In other words, there exists a positive constant L such that for all x and e in $\mathbb{R}^n$,

$$\|f(x, K(x+e))\| \leq L\|x\| + L\|e\|.$$

If the jth gap function, e_j, violates the event-trigger in equation (9.10) at time r_{j+1}, then

$$\beta_1(\|e_j(r_{j+1})\|) > \sigma\gamma(\|x(r_{j+1})\|).$$

Let's assume there exists a positive constant P such that

$$P\|e_j(r_{j+1})\| \geq \gamma^{-1}\left(\frac{1}{\sigma}\beta_1(\|e_j(r_{j+1})\|)\right) \geq \|x(r_{j+1})\|.$$

It can therefore be concluded that with these constraints, the ratio of the gap and the system state must be greater than a positive constant $1/P$. In other words, the next sample occurs when

$$\frac{1}{P} \leq \frac{\|e_j(t)\|}{\|x(t)\|}. \tag{9.11}$$

This condition is a more conservative than the original event-trigger in equation (9.10). It is useful, however, because it provides an analytically tractable method of bounding the earliest time when the event-trigger can occur. As long as this earliest sampling time can be shown to be bounded away from zero, then one can assure that Zeno-sampling doesn't occur.

For any $j = 0, 1, \ldots, \infty$, the trajectory for $\|e_j(t)\|/\|x(t)\|$ can be bounded through the use of differential inequalities. A direct computation of the ratio's time derivative shows that

$$\frac{d}{dt}\frac{\|e_j(t)\|}{\|x(t)\|} \leq \left(1+\frac{\|e_j(t)\|}{\|x(t)\|}\right)\frac{L\|x(t)\|+L\|e_j(t)\|}{\|x(t)\|} = L\left(1+\frac{\|e_j(t)\|}{\|x(t)\|}\right)^2.$$

This differential inequality is used in the Comparison principle [37] to obtain an upper bound on the time history of the event quotient $\|e_j(t)\|/\|x(t)\|$. This bound takes the form

$$\frac{\|e_j(t)\|}{\|x(t)\|} \leq \frac{tL}{1-tL}$$

for t beween 0 and $r_{j+1} - r_j$ where $j = 0, 1, \ldots, \infty$.

So one merely needs to see when the right-hand side of the above inequality triggers the event quotient condition in equation (9.11). This occurs if the next release time r_{j+1} satisfies

$$\frac{1}{P} \leq \frac{\|e_j(r_{j+1})\|}{\|x(r_{j+1})\|} \leq \frac{T_jL}{1-T_jL}$$

where $T_j = r_{j+1} - r_j$ is the intersample time interval. Solving the right-hand inequality for T_j yields a lower bound of the form

$$r_{j+1} - r_j = T_j \geq \frac{1}{L + LP}.$$

Note that this is a lower bound on the intersample time. So as long as the bound is non-zero one can guarantee that the event-triggered system won't exhibit Zeno-sampling. Clearly this bound goes to zero when L is unbounded. In other words, this occurs when the system function f fails to be Lipschitz. In reviewing the sublinear example where Zeno sampling occurs, it is apparent that the sublinear function $\text{sgn}(x)\sqrt{|x|}$ is not Lipschitz. These results show that ISS event-triggering can guarantee non-Zeno sampling of the system state whenever f is Lipschitz.

The sampling generated under these conditions is sporadic rather than aperiodic. Aperiodic sampling simply means that the intersample interval T_j is not a constant. Being aperiodic, however, doesn't require that the minimum intersample interval is positive. Following notational conventions in real-time computing, the term *sporadic* is reserved for systems whose intersample intervals need not be constant and whose minimum intersample intervals are positive.

$\mathscr{L}_2$ **Event-Triggers:** The prior subsection derived sporadic event-triggers that preserve the input-to-state stability of the original non-sampled control system. This framework [70] places relatively few assumptions on the nature of the controller. It only requires that the controller has an ISS-Lyapunov function to ensure input-to-state stability with regard to both the external disturbance, w, and the state gap, e_j. If one makes some assumptions about the structure of the controller, it is possible to say a bit more about the robustness of the closed-loop system's stability concept with regard to non-zero delays, or what is sometimes referred to as *jitter* in the real-time systems community.

This subsection derives event-triggers that preserve the $\mathscr{L}_2$ stability of the closed-loop system. In particular, these so-called $\mathscr{L}_2$ event triggers guarantee that the closed-loop system's induced $\mathscr{L}_2$ gain is preserved (up to a user-defined scaling factor). The so-called $\mathscr{L}_2$ event-trigger was first proposed by Lemmon et al. [41] and then formally analyzed by Wang et al. [83]. Since these $\mathscr{L}_2$ event-triggers preserve the original non-sampled system's closed-loop gain, one can say that these event-triggers are *performance preserving* since the $\mathscr{L}_2$ gain is a commonly used measure of a regulator's performance.

By focusing on the $\mathscr{L}_2$ stability concept, one can use the aforecited results relating the closed-loop system $\mathscr{L}_2$ gain to a Hamilton-Jacobi inequality. To use this relationship, let's narrow our attention to systems that are affine in the external disturbance, w, and the control u. In particular, let's assume that the system state, $x : \mathbb{R} \to \mathbb{R}^n$ satisfies the following differential equation

$$\dot{x}(t) = A(x(t)) + B_1(x(t))w(t) + B_2(x(t))u(t) \tag{9.12}$$

for $t \geq 0$ and initial condition $x(0) = x_0 \in \mathbb{R}^n$. As before $w : \mathbb{R} \to \mathbb{R}^q$ is an external disturbance that is assumed to lie in $\mathscr{L}_2$. The control signal, $u : \mathbb{R} \to \mathbb{R}^m$ is assumed to be a special control of the form

$$u(t) = -B_2^T(x(t))\left[\frac{\partial V(x(t))}{\partial x}\right]^T = K(x(t))$$

where $V : \mathbb{R}^n \to \mathbb{R}$ is a continuously-differentiable positive definite function (sometimes called the *storage function* [37]) that satisfies the Hamilton-Jacobi inequality (9.5) for some $\gamma > 0$. For this particular control, one can show that the $\mathscr{L}_2$ gain of the system from the input w to the output $z = \begin{bmatrix} x \\ u \end{bmatrix}$ is less than or equal to γ. These results were summarized in the earlier section on mathematical preliminaries.

The event-triggered version of the above system starts by introducing a sequence of *release* or *sampling* times $\{r_j\}_{j=0}^{\infty}$ where $r_j \in \mathbb{R}$ denotes the jth consecutive time when the system state has been sampled. In this case the control law uses the *sampled* state instead of the true state so that $u(t)$ becomes

$$u(t) = -B_2^T(\hat{x}_j)\left[\frac{\partial V(\hat{x}_j)}{\partial x}\right]^T = K(\hat{x}_j)$$

for $t \in [r_j, r_{j+1})$ and $j = 0, 1, \ldots, \infty$. In the above equation $\hat{x}_j \in \mathbb{R}^n$ denotes the jth consecutive sampled state

$$\hat{x}_j = x(r_j).$$

As before let's introduce the *gap* between the current state $x(t)$ and the previously sampled state. The jth gap function therefore is

$$e_j(t) = \hat{x}_j - x(t)$$

for $t \in [r_j, r_{j+1})$ and all $j = 0, 1, \ldots, \infty$. Assume that the controller $K : \mathbb{R}^n \to \mathbb{R}^m$ is Lipschitz with respect to the gap. In other words, there exists a non-negative real constant L such that

$$\|K(x) - K(\hat{x}_j)\| = \|K(x) - K(x + e_j)\| \le L\|e_j\|. \tag{9.13}$$

This assumption is satisfied in many applications. In particular, the assumption is valid when the controller is affine with respect to the gap signal.

Let's now examine the time rate of change of the storage function V under the sampled control law. The directional derivative of V is

$$\dot{V} = \frac{\partial V(x)}{\partial x}A(x) + \frac{\partial V(x)}{\partial x}B_1(x)w + \frac{\partial V(x)}{\partial x}B_2(x)K(\hat{x}_j).$$

Since $K(x) = -B_2^T \frac{\partial V}{\partial x}^T$, one can rewrite $\dot{V}$ as

$$\dot{V} = \frac{\partial V}{\partial x}A(x) + \frac{\partial V}{\partial x}B_1(x)w - K^T(x)K(\hat{x}_j).$$

Completing the square for the cross-term $\frac{\partial V}{\partial x} B_2(x) w$ yields,

$$\dot{V} = \frac{\partial V}{\partial x} A(x) - \frac{1}{2} \left\| \gamma w - \frac{1}{\gamma} \frac{\partial V}{\partial x}^T \right\|^2 + \frac{\gamma^2}{2} \|w\|^2 + \frac{1}{2\gamma^2} \left\| B_1^T(x) \frac{\partial V}{\partial x}^T \right\|^2 - K^T(x) K(\hat{x}_j).$$

Applying the Hamilton-Jacobi inequality bounds $\dot{V}$ as was done in section 9.2 yields

$$\begin{aligned} \dot{V} &\leq -\frac{1}{2} \left\| \gamma w - \frac{1}{\gamma} \frac{\partial V}{\partial x}^T \right\|^2 + \frac{\gamma^2}{2} \|w\|^2 - \frac{1}{2} \|x\|^2 - K^T(x) K(\hat{x}_j) \\ &\leq -\frac{1}{2} \|x\|^2 + \frac{\gamma^2}{2} \|w\|^2 - K^T(x) K(\hat{x}_j). \end{aligned} \tag{9.14}$$

The above cross-term, $K^T(x)K(\hat{x}_j)$, in equation (9.14) must still be dealt with. From the Lipschitz assumption on K in equation (9.13), one rewrites this cross-term as

$$\begin{aligned} K^T(x) K(\hat{x}_j) &= \frac{1}{2} \|K(x) - K(\hat{x}_j)\|^2 - \frac{1}{2} \|K(x)\|^2 - \frac{1}{2} \|K(\hat{x}_j)\|^2 \\ &\leq \frac{1}{2} L^2 \|e_j\|^2 - \frac{1}{2} \|K(x)\|^2 - \frac{1}{2} \|K(\hat{x}_j)\|^2 \\ &\leq \frac{1}{2} L^2 \|e_j\|^2 - \frac{1}{2} \|K(\hat{x}_j)\|^2 \end{aligned}$$

where $e_j = \hat{x}_j - x$ is the gap. Substituting this bound for $K^T(x)K(\hat{x}_j)$ into equation (9.14) yields the following bound on the directional derivative of the storage function,

$$\begin{aligned} \dot{V} &\leq -\frac{1}{2} \|x\|^2 + \frac{\gamma^2}{2} \|w\|^2 - \frac{1}{2} \|K(\hat{x}_j)\|^2 + \frac{1}{2} L^2 \|e_j\|^2 \\ &= -\frac{\beta^2}{2} \|x\|^2 + \frac{\gamma^2}{2} \|w\|^2 + \left(-\frac{1-\beta^2}{2} \|x\|^2 - \frac{1}{2} \|K(\hat{x}_j)\|^2 + \frac{1}{2} L^2 \|e_j\|^2 \right) \end{aligned} \tag{9.15}$$

for some user-defined parameter $\beta \in [0,1]$. Note that the above inequality will be a dissipative inequality for $\dot{V}$ provided one can guarantee that the last three terms within the parentheses are collectively negative definite. If this is the case, then

$$\dot{V} \leq -\frac{\beta^2}{2} \|x\|^2 + \frac{\gamma^2}{2} \|w\|^2$$

for all x and w. As noted in the mathematical preliminaries section, satisfaction of this inequality is sufficient to establish that the sampled-data system's induced $\mathscr{L}_2$ gain from the input w to the output x is less than or equal to γ/β. Note that the user-defined parameter β becomes a parameter that controls how close the gain of the sampled-data system will be to the original gain of the continuously-sampled system.

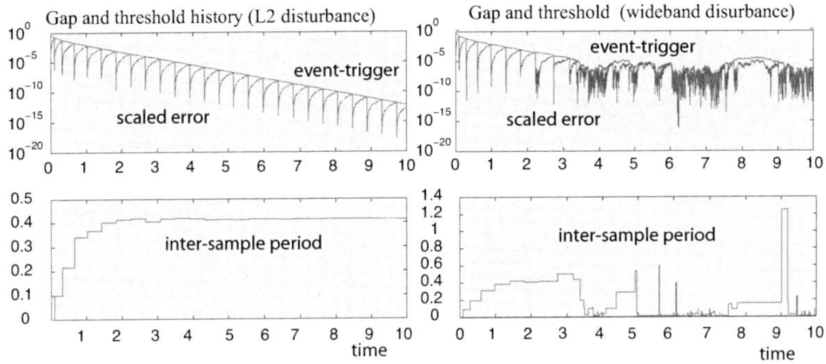

Fig. 9.7 Gap, threshold, and intersample time history using an $\mathscr{L}_2$ event-trigger. (left) $\mathscr{L}_2$ noise case. (right) wideband noise case

In summary, if the last three terms on the right-hand side of equation (9.15) are negative, then the sampled-data system is $\mathscr{L}_2$ stable with a gain less than γ/β. This inequality is satisfied for all times, t, if one can guarantee

$$L^2\|e_j(t)\|^2 \leq (1-\beta^2)\|x(t)\|^2 + \|K(\hat{x}_j)\|^2. \tag{9.16}$$

The left-hand side of this inequality is simply the size of the system gap, e_j. The right-hand side of the inequality is a state-dependent threshold that is very similar to the ISS event threshold derived earlier. In other words equation (9.16) is an $\mathscr{L}_2$ preserving event-trigger. The event-trigger is used in the same way the ISS event-trigger was used. Namely, the sampler, S, monitors the gap against the threshold on the right-hand side. When the inequality is violated (or about to be violated), then the state is sampled and the next release time r_{j+1} is generated.

Note that the event-trigger depends on the user-defined parameter β. This parameter controls how close the event-triggered system's gain will be to γ, the gain of the original continuous-time system. If β is close to one then the sampled system achieves the original gain of γ. As β gets smaller, the gain of the system increases, thereby reducing the event-triggered system's $\mathscr{L}_2$ performance. In other words, the smallest thresholds and hence the most frequent sampling occurs when β is close to one. As β gets smaller, the intersampling periods will get longer at the cost of a higher closed-loop system gain. This is a tradeoff between the system gain and how frequently the state must be sampled.

The following example illustrates the use of the $\mathscr{L}_2$ event trigger on a variation of the superlinear system examined in figure (9.5). In this case let's consider the controlled system characterized by the following equations,

$$\begin{aligned}\dot{x}(t) &= x^3(t) + u(t) + w(t)\\ u(t) &= -\alpha\hat{x}_j^3 - \hat{x}_j\end{aligned}$$

for $t \in [r_j, r_{j+1})$ and all $j = 0, 1, \ldots, \infty$. The sequence of release times $\{r_j\}_{j=0}^{\infty}$ is generated by the violation of the $\mathscr{L}_2$ event-trigger in equation (9.16). Choose a special type of $\mathscr{L}_2$ disturbance of the form,

$$w(t) = e^{-2t} v(t)$$

for $t \geq 0$ and where v is white noise process. The left-hand side of figure 9.7 plots the gap and threshold time histories for this system (top plot) and the intersample time (bottom plot), T_j, that was generated. As can be seen the sampling period is initially very small (about 0.1 sec) at the beginning of the simulation when the disturbance is largest. As the disturbance decays, the sampling period stabilizes to a relatively long period (0.4 sec) that is four times longer than the initial sampling period.

The right-hand side of figure 9.7 shows the gap, threshold, and intersample time histories for the same $\mathscr{L}_2$ event-triggered system in which the disturbance is no longer guaranteed to go to zero as our system approaches the equilibrium point. In this case let the disturbance be

$$w(t) = (e^{-2t} + 0.1) v(t)$$

where $v(t)$ is again a white noise process. In this case, the disturbance amplitude does not go to zero as time goes to infinity. In particular, this $w(t)$ is referred to as a wide band disturbance. The top plot shows the gap and threshold time histories. The bottom plot shows the resulting intersample times. When the system state is far from the equilibrium point, the system's response is similar to the earlier $\mathscr{L}_2$ disturbance case. As the system state approaches the equilibrium point, however, the periodic nature of the intersampling time disappears with sampling times that can become arbitrarily short. In this case, therefore, event-triggering only yields aperiodic, rather than sporadic, sampling of the system state.

This type of behavior is common in both the $\mathscr{L}_2$ event-triggered and ISS event-triggered systems. It essentially results because the state-dependent threshold gets very small as the system state approaches the origin. With such a small threshold, the introduction of noise into the disturbance makes the system's sampling-events trigger much more often. This example therefore shows that state-dependent event-triggered system may be sensitive to wide-band disturbances. One way to address this sensitivity is to place a lower bound on the event-triggering threshold of the form (in the $\mathscr{L}_2$ event-trigger case)

$$L^2 \|e_j(t)\|^2 \leq \max\left\{\overline{T}, (1-\beta^2)\|x(t)\|^2 + \|K(\hat{x}_j)\|^2\right\}.$$

Assuming that $\|e_j(t)\|$ and $\|x(t)\|$ have bounded rates of growth, this modified event-trigger prevents the sampling period from being arbitrarily close to zero and can therefore assure the sporadic nature of event-triggered sampling.

Impact of Delays: The preceding analysis for the $\mathscr{L}_2$ and ISS event-triggers assumed that both the system state and control update are generated at the same time. In other words, the control signal, based on the system state sampled at time r_j, is applied to the plant at the same time. This means that our prior analysis ignored delays. Real-life implementations of such systems will always exhibit some delay due to the amount of time it takes to compute the control signal. It would be highly desirable to show that the performance of the event-triggered system (as measured by the closed-loop system's $\mathscr{L}_2$ gain) is preserved under such delays. The following analysis from [83] shows how robust $\mathscr{L}_2$ performance will be under event-triggering with delays.

Before starting the analysis, let's discuss the modeling of event-triggered sampling with delays. In particular, one now needs to consider two sequences of times. The sequence of release times $\{r_j\}_{j=0}^{\infty}$ is defined as before. Release time r_j represents the time when 1) the state was sampled and 2) the control task was released for execution by the central processing unit (CPU). The other sequence of interest is the sequence of *finishing times*, $\{f_j\}_{j=0}^{\infty}$. The time $f_j \in \mathbb{R}$ denotes the time when the control signal computed by the control task is actually used by the plant. This time also marks the finish of the control job that had been released at time r_j. In general one make the *small delay* assumption which states that $r_j \leq f_j < r_{j+1}$ for all $j = 0, 1, \ldots, \infty$. In other words, the sample taken at time r_j is used at a time, f_j, which always occurs before the next invocation of the control task.

Figure 9.8 shows the timing relationships assumed in this analysis. The figure is a timeline in which the black rectangles indicate when the control task job is being executed. With regard to this diagram, let's define two measures of real-time system performance. The first measure is the task period $T_j = r_{j+1} - r_j$. This is the interval of time between any two consecutive invocations of the control task. As in the case of the ISS event-trigger, there is great interest in obtaining lower bounds on T_j, thereby identifying the smallest sampling period required by the real-time computer. The other measure of interest is the *delay*, D_j, of the jth job. This is the time between the finishing time and release time, *i.e.* $D_j = f_j - r_j$. The control task *deadline* $\overline{D}$ is a real-time constraint that one might place on these delays. In particular, a real-time system that is functioning properly will have all delays less than the deadline ($D_j \leq \overline{D}$ for all $j = 0, 1, \ldots, \infty$). The choice of the deadline is an important constraint. In our case, the deadline is chosen to ensure the $\mathscr{L}_2$ performance of the control application. In particular, this means that our analysis would like to derive upper bounds on the maximally allowable delay (MAD) that any task can tolerate before losing our guarantee on $\mathscr{L}_2$ performance. This upper bound then becomes the deadline quality-of-service (QoS) constraint on the real time system.

To obtain tight bounds on the maximally allowable delays (MAD) and intersample intervals, let's confine our attention to linear time-invariant control systems of the form

$$\begin{aligned} \dot{x}(t) &= Ax(t) + w(t) + Bu(t) \\ u(t) &= -B^T P \hat{x}_j = K(\hat{x}_j) \end{aligned}$$

for all $t \in [f_j, f_{j+1})$ and $j = 0, 1, \ldots, \infty$. A and B are are suitably dimensioned real matrices. In terms of the earlier system model considered in equation 9.12, we let $B_1 = I$ and $B_2 = B$. This is simply done for notational convenience. The sampled

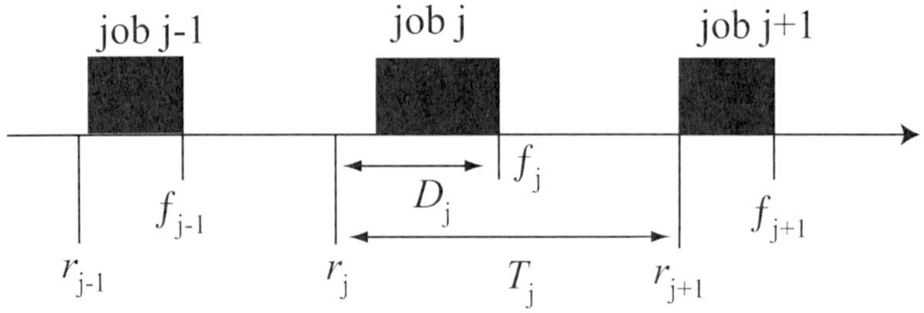

Fig. 9.8 Definition of Timing Relationships used in Studying the Real-time Implementations of $\mathscr{L}_2$ event-triggered control

state, $\hat{x}_j = x(r_j)$, is the state that occurs at the jth consecutive release time. Note that the above equation holds between *finishing* times, rather than between release times. This is in accordance with the fact that control signals can only be changed after the control task's job has finished executing.

By confining our attention to linear systems one can use a storage function of the form $V(x) = x^T Px$ where P is a real valued n by n matrix. With this choice of V the Hamilton-Jacobi inequality reduces to an algebraic Riccati inequality where P is a symmetric positive definite matrix that for some real $\gamma > 0$ satisfies the Riccati inequality

$$A^T P + PA - P(BB^T - \gamma^{-2} I)P + I \leq 0.$$

With this choice of control, the induced $\mathscr{L}_2$ gain of the continuously sampled closed-loop system is guaranteed to be less than or equal to γ.

The $\mathscr{L}_2$ event-trigger is derived in the same way it was for the nonlinear system. The difference is that now in establishing the dissipative inequality, one makes use of the algebraic Riccati inequality rather than the Hamilton-Jacobi inequality. The resulting $\mathscr{L}_2$ event-trigger (derived in [83]) is

$$e_j^T(t) M e_j(t) < \delta x^T(r_j) N x(r_j)$$

where

$$\begin{aligned} M &= (1-\beta^2)I + PBB^T P \\ N &= \frac{1}{2}(1-\beta^2)I + PBB^T P \end{aligned}$$

and β and δ are user supplied constants between 0 and 1. Note that the earlier $\mathscr{L}_2$ triggering threshold was a function of $x(t)$ and $x(r_j)$. The preceding threshold for the LTI case is only a function of $x(r_j)$. This means that the above threshold is *weaker* than our earlier result (in other words it would cause the system to sample sooner). This particular form of the $\mathscr{L}_2$ threshold was used in [83] because it rendered the analysis of delays more tractable. With this weaker threshold it was possible to obtain specific bounds on the acceptable delays and minimum periods that could be

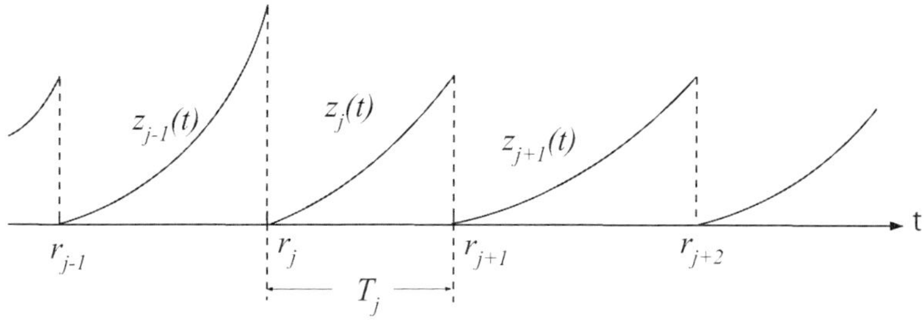

Fig. 9.9 Time history of normalized gap $\|z_j(t)\|$ when there are no delays ($r_j = f_j$)

tolerated by the event-triggered system. As mentioned above, the bounds on delays are useful because they serve as deadlines for the real-time computer implementing the event-triggered control system. The bounds on period are useful in verifying whether the system can exhibit Zeno-sampling.

Let's first examine the problem of obtaining a lower bound on the sampling period under the assumption of no delays. For notational convenience, rewrite the earlier $\mathscr{L}_2$ event-trigger in terms of a *normalized* gap function,

$$z_j(t) = \sqrt{M}e_j(t)$$

so that the triggering inequality takes the form

$$\|z_j(t)\| < \sqrt{x^T(r_j)Nx(r_j)} = \rho(x(r_j))$$

where the function $\rho : \mathbb{R}^n \to \mathbb{R}$ is defined in the above equation. Figure 9.9 shows the time history of the gap functions when there are no delays; in other words the controller job's finishing time equals the release time. As was done in the earlier analysis regarding sampling periods for ISS event-triggers, let's examine the normalized gap function's rate of growth over the interval $[r_j, r_{j+1})$.

The analysis starts by bounding the time derivative of $\|z_j(t)\|$,

$$\begin{aligned}\frac{d}{dt}\|z_j(t)\| &\leq \left\|\sqrt{M}\dot{e}_j(t)\right\| = \left\|\sqrt{M}\left(Ax(t) - BB^TPx(r_j) + w(t)\right)\right\| \\ &< \left\|\sqrt{M}Ae_j(t)\right\| + \left\|\sqrt{M}(A - BB^TP)x(r_j)\right\| + \left\|\sqrt{M}\right\| \|w(t)\|\end{aligned}$$

for $t \in [r_j, r_{j+1})$. Let's assume the disturbance is bounded by the norm of the system state. In other words, let's assume there exists a positive real constant W such that $\|w(t)\| \leq W\|x(t)\|$. In this case the preceding upper bound on $\frac{d}{dt}\|z_j(t)\|$ may be simplified to the form,

$$\frac{d}{dt}\|z_j(t)\| \leq \alpha\|z_j(t)\| + \mu_0(x(r_j)) \tag{9.17}$$

where α is a real constant such that

$$\alpha = \left\| \sqrt{M} A \sqrt{M^{-1}} \right\| + W \left\| \sqrt{M} \right\| \left\| \sqrt{M^{-1}} \right\|$$

and $\mu_0 : \mathbb{R}^n \to \mathbb{R}$ is a function such that

$$\mu_0(x(r_j)) = \left\| \sqrt{M}(A - BB^T P)x(r_j) \right\| + W \left\| \sqrt{M} \right\| \|x(r_j)\|.$$

The differential inequality in equation (9.17) bounds the rate of growth of the normalized gap function over the interval between two consecutive release times, r_j and r_{j+1}. Since the state is sampled at time r_j, the gap is zero at that time. So the initial condition for the differential inequality is $\|z_j(r_j)\| = 0$. One can therefore use the Comparison principle to show that for all $t \in [r_j, r_{j+1})$ that

$$\|z_j(t)\| \leq \frac{\mu_0(x(r_j))}{\alpha} \left(e^{\alpha(t - r_j)} - 1 \right).$$

This is an upper bound on the normalized gap between two consecutive release times. Clearly, the next release r_{j+1} must occur before the right-hand side of the above inequality violates the $\mathcal{L}_2$ event threshold. In other words, the following inequality must hold

$$\frac{\mu_0(x(r_j))}{\alpha} (e^{\alpha T_j} - 1) \geq \rho(x(r_j)).$$

This inequality can be solved for the sampling period $T_j = r_{j+1} - r_j$ to obtain

$$T_j \geq \frac{1}{\alpha} \ln \left(1 + \alpha \frac{\rho(x(r_j))}{\mu_0(x(r_j))} \right)$$

when the next release occurs. The above inequality represents a lower bound on the intersample time intervals generated by $\mathcal{L}_2$ event-triggers. It can be shown [83] that this bound is always bounded away from zero, so as in the ISS event-trigger case Zeno-sampling does not occur. A major reason for this lies in the requirement that the external disturbance has a norm that goes to zero as the system state approaches its equilibrium. This is a particularly strong assumption that can be justified if the source of the disturbance arises from modeling uncertainty. In general, however, if this assumption does not hold, then $\mathcal{L}_2$ event-triggers can lead to Zeno-sampling as was seen in figure 9.7. One can avoid these undesirable behaviors by imposing some additional constraints on the event-triggers. This particular approach was discussed in more detail in [84].

The usefulness of the prior analysis is limited by the no task delay assumption. Let's now examine how this assumption might be relaxed. In the case of delays, the normalized gap's evolution changes as shown in figure 9.10. With non-zero delays, the individual gap functions overlap as shown in the figure. This means that one should partition the time interval $[r_j, f_{j+1})$ into two subintervals $[r_j, f_j)$ and $[f_j, f_{j+1})$. Over the first subinterval, the system state evolves according to the differential equation

$$\dot{x}(t) = Ax(t) - BB^T Px(r_{j-1}) + w(t)$$

in which the state used in the controller is the state at sample time r_{j-1}. After the jth control job finishes, the control is updated with the most recent sampled state. This means that over the time interval $[f_j, f_{j+1})$, the system state evolves according to the differential equation

$$\dot{x}(t) = Ax(t) - BB^T Px(r_j) + w(t).$$

In a manner similar to what was done in the no-delay case, differential inequalities can be used to bound $\|z_j(t)\|$ for all $t \in [r_j, f_{j+1})$.

The analysis of the non-zero delay case is done by viewing the event threshold $\rho(x(r_j))$ as a *budget* that is allocated to the normalized gap. In particular we partition this budget between the two subintervals $[r_j, f_j)$ and $[f_j, f_{j+1})$. Let's require that over the first subinterval $[r_j, f_j)$ is constrained so the normalized gap doesn't get bigger than $\varepsilon\rho(x(r_j))$ where ε is a user-defined constant between 0 and 1. One can again use differential inequalities to show that the normalized gap is bounded as

$$\|z_j(t)\| \leq \frac{\mu_1(x(r_j), x(r_{j-1}))}{\alpha}\left(e^{\alpha(t-r_j)} - 1\right) = \Phi(x(r_j), x(r_{j-1}); t - r_j) \quad (9.18)$$

for all $t \in [r_j, f_j)$ and where the function $\mu_1 : \mathbb{R}^n \times \mathbb{R}^n \to \mathbb{R}$ is defined as

$$\mu_1(x(r_j), x(r_{j-1})) = \left\|\sqrt{M}(Ax(r_j) - BB^T Px(r_{j-1}))\right\| + W\|\sqrt{M}\|\,\|x(r_j)\|.$$

The right-hand side of the above inequality (9.18) represents the solution to the differential equation

$$\frac{d}{dt}\|z_j\| = \alpha\|z_j(t)\| + \mu_1(x(r_j), x(r_{j-1}))$$

where α is a real constant. The solution to this differential equation is characterized by the function $\Phi : \mathbb{R}^n \times \mathbb{R}^n \times \mathbb{R} \to \mathbb{R}$. This function returns the normalized gap $z_j(t)$ at time t as a function of the system states $x(r_j)$ and $x(r_{j-1})$. The dependence of Φ on the system state at times r_j and r_{j-1} is a consequence of the fact that the

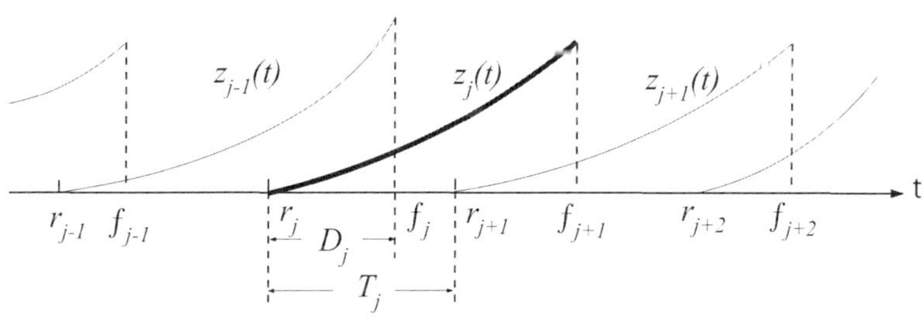

Fig. 9.10 Time history of normalized gap $\|z_j(t)\|$ when there are task delays ($r_j < f_j$)

differential equations governing the evolution of the system state are different over time intervals $[r_j, f_j)$ and $[f_j, f_{j+1})$.

Note that the duration of the first subinterval, $[r_j, f_j)$ is the delay, D_j. To ensure the normalized gap gets no larger than $\varepsilon\rho(x(r_j))$ over this first subinterval, equation (9.18) implies that

$$\frac{\mu_1(x(r_j),x(r_{j-1}))}{\alpha}\left(e^{\alpha D_j}-1\right)\leq \varepsilon\rho(x(r_j)).$$

Solving for D_j yields

$$0\leq D_j\leq \frac{1}{\alpha}\ln\left(1+\varepsilon\alpha\frac{\rho(x(r_j))}{\mu_1(x(r_j),x(r_{j-1}))}\right).$$

The above equation represents an upper bound on the *delay* that ensures the gap at the end of the first subinterval is less than the allocated budget of $\varepsilon\rho(x(r_j))$.

The analysis is completed by examining the behavior of the gap over the second subinterval $[f_j, f_{j+1})$. At the beginning of this interval,

$$\|z_j(f_j)\|\leq \Phi(x(r_j),x(r_{j-1});D_j)\leq \varepsilon\rho(x(r_j))\leq \rho(x(r_j)). \tag{9.19}$$

The system state over the interval $[f_j, f_{j+1})$ satisfies the differential equation

$$\dot{x}(t)=Ax(t)-BB^TPx(r_j)+w(t).$$

Using an argument similar to that employed in analyzing the gap over the first subinterval, one can show that

$$\frac{d}{dt}\|z_j(t)\|\leq \alpha\|z_j(t)\|+\mu_0(x(r_j)).$$

Solutions to this differential inequality over the interval $t\in[f_j, f_{j+1}]$ are bounded above by solutions to the associated differential equality

$$\frac{d}{dt}\|\tilde{z}_j(t)\|=\alpha\|\tilde{z}_j(t)\|+\mu_0(x(r_j))$$

with the initial condition $\tilde{z}_j(f_j)=z(f_j)$. This bounding solution, $\tilde{z}_j(t)$, is used to predict when the release time should be selected to ensure the $\mathscr{L}_2$ stability of the event-triggered system.

This result is proven in [83]. In particular, this paper states that the closed-loop system is $\mathscr{L}_2$ stable with a gain less than or equal to γ/β if the task's $(j+1)^{\text{st}}$ release time is generated by

$$r_{j+1}=f_j+\frac{1}{\alpha}\ln\left(1+\alpha\frac{\delta\rho(x(r_j))-\Phi(x(r_j),x(r_{j-1});D_j)}{\mu_0(x(r_j))+\alpha\Phi(x(r_j),x(r_{j-1});D_j)}\right) \tag{9.20}$$

and the delay D_{j+1} satisfies

$$D_{j+1} \leq \frac{1}{\alpha}\left(1+\varepsilon\alpha\frac{(1-\delta)\rho(x(r_j))}{\alpha\delta\rho(x(r_j))+\mu_0(x(r_j))}\right). \tag{9.21}$$

In the above function α is the real constant defined earlier, ρ and μ_0 are the class $\mathscr{K}$ functions defined above and Φ bounds a function of the system state as it evolves over the delay time D_j.

Self-Triggered Feedback Control: The results in the prior section do more than suggest that an event-triggered system's $\mathscr{L}_2$ stability will be robust with respect to task delays. Equation (9.20) is interesting in that it computes the *future* release time given the current release and finishing times. This equation therefore provides a prediction of the next release time and it suggests that it should be possible to develop a *software* implementation of event-triggered systems. This software version of event-triggering has sometimes been referred to as *self-triggered* feedback control. Such software versions of event-triggering may be preferred in applications where the cost of adding event-detection hardware is deemed unacceptable.

The concept of self-triggered task models was original presented in [75]. Simulation results [41] suggested that self-triggering systems exhibit a robustness to delays that is consistent with what one might exhibit from event-triggered systems. In these earlier works the computation of the next release time was usually done in a heavy handed fashion that was not computationally efficient. This has changed recently with results in [83] which allow a more computationally efficient way of selecting the next release time. More recently, it has been noted that for *homogenous* systems [1], release times enforcing input-to-state stability satisfy certain scaling relationships. These relationships can be used to reduce the computation of the next release time to a table look-up. Another important aspect of these analytical bounds on acceptable delay and release times is that they can be used as quality-of-service constraints for real-time schedulers. Since it now becomes possible to predict when the next control job should be released, one can use these estimates as the period and deadline that govern how real-time scheduling services adjust task priorities. In other words, the aforecited analytical bounds provide a formal way of connecting real-time scheduling constraints to the application's (*i.e.* control system's) actual performance.

Event-triggering was proposed [2, 3] as a means to co-design controllers and schedulers in embedded systems. The analysis of *state-dependent* event-triggers [70] formally characterized the stability properties of event-triggering and the later analyses of intersample intervals and maximally allowable deadlines [83] provided bounds that could be used in adapting the embedded system's controller tasks. These methods raise the possibility of moving away from the traditional hard real-time task models that have dominated embedded control. While the use of event-triggering has focused on conserving the embedded system's computational resources, it is anticipated that event-triggering may be used to conserve other types of shared resources. Of particular interest are embedded sensor-actuator control systems where communication resources are highly constrained. The next section examines the application of state-dependent event-triggering to such networked control systems.

9.4 Event-Triggered Feedback in Networked Control Systems

Many of the results for event-triggered control of embedded systems can be extended to networked control systems. A networked control system or NCS is a set of controllers that coordinate their actions over a communication network. For NCS, event-triggering is used to decide when to *transmit* or *broadcast* the system state to a local controller's neighbors. Using events to trigger communication actually provides a much stronger motivation for event-triggered control. The reason for this is that in many cases, the energy or cost associated with the transmission of a bit of information is much more than the energy associated with using that bit to compute the control law. Event-triggering, therefore, provides a realistic way of reducing traffic congestion in communication networks used by NCS. The objective of this section is to show how the earlier results from event-triggered control of embedded systems can be extended to networked control systems. The section first discusses the NCS architecture under study and then it derives event-triggers assuring the NCS is ISS. As in the case of embedded systems, the NCS implementation introduces a number of so-called *network artifacts* that complicate the analysis of the idealized model. These network artifacts include delays in the transmission of information as well as dropped information packets. This section studies the impact of such network artifacts and demonstrates that event-triggered NCS stability is robust to such network artifacts in a quantifiable manner.

While there is a great deal of literature [11, 25, 50, 51, 90] examining networked control systems, there is relatively little work pertaining to event-triggered NCS. Most of the results in this section are drawn from [81] and [80]. Related work will be found in [48].

Model of Networked Control System: Let's first describe a model of a networked control system or NCS. Consider a distributed NCS consisting of N agents. Figure 9.11 provides a graphic illustration of an NCS with three agents. Each agent consists of a *physical* component and a *cyber*-component. The physical components are interconnected as shown by the solid lines in the figure. The cyber-components are also interconnected through a communication network whose links are shown by the dashed lines in the figure.

This system may be more formally characterized using graph theoretic notation. In particular, let $\mathcal{N} = \{1, 2, \ldots, N\}$ denote the set of agents. A graph $\mathcal{G}_\mathrm{p} = (\mathcal{N}, \mathcal{E}_\mathrm{p})$ represents the physical coupling between the agents. $\mathcal{N}$ denotes the vertices of the graph and $\mathcal{E}_\mathrm{p} \subset \mathcal{N} \times \mathcal{N}$ denotes the set of edges in the graph. The edge (i, j) connects node $i \in \mathcal{N}$ to node $j \in \mathcal{N}$. The edge, therefore, is an ordered pair (i, j) of nodes. The ordered pair (i, j) is in $\mathcal{E}_\mathrm{p}$ if the dynamics of agent j's physical component are directly driven by agent i's local state. The graph $\mathcal{G}_\mathrm{c} = (\mathcal{N}, \mathcal{E}_\mathrm{c})$ models the interconnections between the cyber-components of the agents. As before $\mathcal{N}$ denotes the vertices (nodes) of the graph and $\mathcal{E}_\mathrm{c} \subset \mathcal{N} \times \mathcal{N}$ represents the edges of the graph.

In this section, the graphs for the physical and cyber-interconnections need not be the same. This requires us to define a number of special neighborhoods in the graph. In particular,

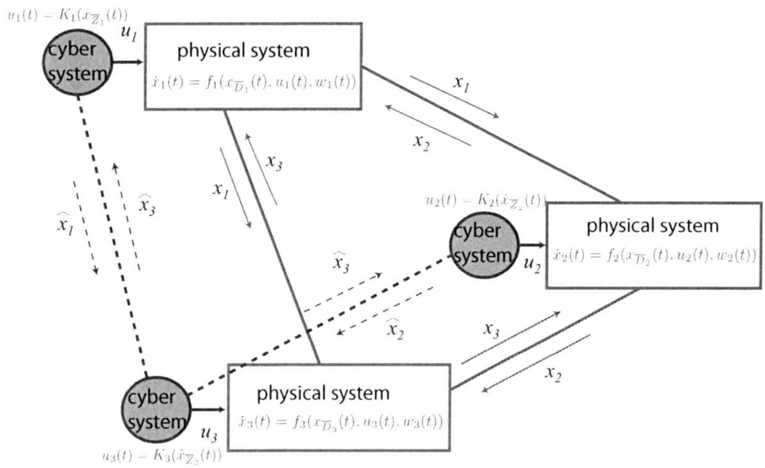

Fig. 9.11 Model of Event-Triggered Networked Control Systems

- $Z_i = \{j \in \mathcal{N} \mid (j,i) \in \mathcal{E}_c\}$ represents those agents whose cyber-components can send information to agent i's cyber-component.
- $U_i = \{j \in \mathcal{N} \mid (i,j) \in \mathcal{E}_c\}$ denotes those agents whose cyber-components can receive information from agent i's cyber-component.
- $D_i = \{j \in \mathcal{N} \mid (j,i) \in \mathcal{E}_p\}$ represents those agents whose physical components directly drive the dynamics of agent i's physical component.
- $S_i = \{j \in \mathcal{N} \mid (i,j) \in \mathcal{E}_p\}$ denotes those agents whose physical components are directly driven by the physical component of agent i.

For any set $\Sigma \subset \mathcal{N}$, let $|\Sigma|$ denote the number of elements in that set and let $\overline{\Sigma} = \Sigma \cup \{i\}$.

The physical component of agent i is characterized by a *local state* $x_i : \mathbb{R} \to \mathbb{R}^n$ where x_i satisfies the differential equation

$$\begin{aligned} \dot{x}_i(t) &= f_i(x_{\overline{D}_i}(t), u_i(t), w_i(t)) \\ x_i(t_0) &= x_{i0} \end{aligned}$$

where $x_{\overline{D}_i} = \{x_j\}_{j \in \overline{D}_i}$ are the local states of agent i's neighbors that are physically connected to it. The system dynamics are characterized by the function $f_i : \mathbb{R}^{n|\overline{D}_i|} \times \mathbb{R}^m \times \mathbb{R}^\ell \to \mathbb{R}^n$ which is locally Lipschitz and satisfies $f_i(0,0,0) = 0$. $u_i : \mathbb{R} \to \mathbb{R}^m$ is a control input generated by the cyber-component of the agent and $w_i : \mathbb{R} \to \mathbb{R}^\ell$ is an external disturbance. The above characterization assumes all subsystems have the same local state dimension, n. This is done for notational convenience. The model and subsequent analysis would also apply to subsystems with local states of different dimensionalities.

The control u_i is generated by agent i's cyber-component. Since these cyber-components exchange information over a digital communication network, local states are transmitted in a discrete manner. In particular, let $\{r_j^i\}_{j=1}^{\infty}$ denote the sequence of broadcast release times for the ith agent. So the *transmitted* state from agent i is denoted as

$$\hat{x}_i(t) = x_i(r_j^i)$$

for $t \in [r_j^i, r_{j+1}^i)$ and $j = 0, 1, \ldots, \infty$. Agent i's cyber-component uses the local state information received from all its neighbors in the set Z_i to compute the control u_i. So let $K_i : \mathbb{R}^{n|\overline{Z}_i|} \to \mathbb{R}^m$ denote the i'th agent's local controller so that

$$u_i(t) = K_i(\hat{x}_{\overline{Z}_i}(t)).$$

Following the same notational conventions as before, $\hat{x}_{\overline{Z}_i}$ denotes the broadcast states of all neighbors of agent i whose cyber-components send information directly to agent i.

ISS Event-Triggered Networked Control: Let's now derive ISS event-triggers for the NCS described above. In particular, let $e_i(t) = \hat{x}_i(t) - x_i(t)$ denote the local *gap* between agent i's current state and its last broadcast state. Assume there exist positive definite function $V : \mathbb{R}^{nN} \to \mathbb{R}$, controllers $K_i : \mathbb{R}^{|n\overline{Z}_i|} \to \mathbb{R}^m$, and class $\mathscr{K}$ functions γ_i, ψ_i, and β_i (for $i = 1, 2, \ldots, N$) such that

$$\begin{aligned}\dot{V} &= \sum_{i=1}^{N} \frac{\partial V}{\partial x_i} f_i(x_{\overline{D}_i}, K_i(x_{\overline{Z}_i} + e_{\overline{Z}_i}), w_i) \\ &\leq \sum_{i=1}^{N} (-\gamma_i(\|x_i\|) + \psi_i(\|e_i\|) + \beta_i(\|w_i\|))\end{aligned} \tag{9.22}$$

where $e_{\overline{Z}_i}$ is the gap of all agent i's cyber-neighbors. This assumption means that V is an ISS-Lyapunov function with respect to w when the the gap $e_i(r_j^i) = 0$. In view of our earlier discussion, this is sufficient to imply that the local controllers K_i leave the original continuously sampled version of the networked control system input-to-state stable.

So again, one selects a user-defined parameter $\sigma_i \in (0,1)$ and notes that if the local state and gap trajectories satisfy the inequality

$$-\sigma_i\gamma_i(\|x_i(t)\|) + \psi_i(\|e_i(t)\|) \leq 0 \tag{9.23}$$

for all $t \in \mathbb{R}$ and all $i = 1, 2, \ldots, N$, then the bound on $\dot{V}$ becomes

$$\dot{V} \leq \sum_{i=1}^{N} (-(1-\sigma_i)\gamma_i(\|x_i\|) + \beta_i(\|w_i\|)).$$

This is a dissipative inequality that was seen earlier to be sufficient to show that the event-triggered NCS is ISS with respect to the external input w_i.

As before in our study of the embedded event-triggered controllers, the inequality in equation (9.23) can be used as the basis of a state-dependent threshold test. In particular, the ith agent would check the validity of the following threshold test on the gap,

$$\psi_i(\|e_i(t)\|) \le \sigma_i \gamma_i(\|x_i(t)\|). \tag{9.24}$$

At the broadcast time r^i_j, the local gap, $e_i = 0$. This gap then grows until $\psi_i(\|e_i(t)\|)$ exceeds the state dependent threshold $\gamma_i(\|x_i(t)\|)$. The violation of that threshold triggers agent i to broadcast its state again. Note that this is a *cooperative* broadcast mechanism in that the violation of the threshold results in an agent sharing its local state information with its neighbors. In other words, the success of such an event-triggered broadcast scheme relies on all agent's agreeing to work in the same manner.

Note that the ISS event trigger given above is only a *local* function of the agent's state. This is important, for it means each agent is able to trigger its broadcast without relying directly on its neighbors. A key part of the prior analysis is the assumption that there exists an ISS Lyapunov function that is separable in the sense specified by the bounds in equation (9.22). Such a Lyapunov function may be constructed by identifying a set of N positive definite functions $V_i : \mathbb{R}^n \to \mathbb{R}$ for $i = 1, 2, \ldots, N$ with class $\mathscr{K}$ functions γ_i, η_i, ψ_i, and β_i such that

$$\begin{aligned}\frac{\partial V_i}{\partial x_i} f_i(x_{\overline{D}_i}, K_i(x_{\overline{Z}_i} + e_{\overline{Z}_i}), w_i) \le -\gamma_i(\|x_i\|) + \sum_{j \in D_i \cup Z_i} \eta_j(\|x_j\|) \\ + \sum_{j \in \overline{Z}_i} \psi_j(\|e_j\|) + \beta_i(\|w_i\|).\end{aligned} \tag{9.25}$$

As a specific example, let's consider class $\mathscr{K}$ functions that are quadratic so $\gamma_i(\|x\|)$ can be expressed as $\gamma_i\|x\|^2$ and similarly for the other functions, η_i, ψ_i, and β_i. In this case, one sees that by choosing $V = \sum_{i=1}^N V_i$, the following inequality is obtained

$$\begin{aligned}\dot{V} &\le \sum_{i=1}^N \left(-\gamma_i\|x_i\|^2 + \sum_{j \in D_i \cup Z_i} \eta_j\|x_j\|^2 + \sum_{j \in \overline{Z}_i} \psi_j\|e_j\|^2 + \beta_i^2\|w_i\|^2 \right) \\ &= \sum_{i=1}^N \left(-(\gamma_i - |S_i \cup U_i|\eta_i)\|x_i\|^2 + \psi_i|\overline{U}_i|\|e_i\|^2 + \beta_i\|w_i\|^2 \right).\end{aligned}$$

Note that this matches the conditions in equation (9.22) provided the first term on the right-hand side is negative definite. This term will be negative definite if

$$\gamma_i - |S_i \cup U_i|\eta_i > 0.$$

This condition places a restriction on the amount of coupling between physically interconnected physical systems. In particular, it says that if one can appropriately bound this physical coupling and if there exist candidate ISS-Lyapunov functions satisfying the bounds in equation (9.25), then it is always possible to construct a global V that is an ISS-Lyapunov function for the entire networked system. In this case, the associated ISS event-trigger is shown to have the form

$$\|e_i(t)\| \leq \sigma_i \sqrt{\frac{\gamma_i - |S_i \cup D_i|\eta_i}{|\overline{U}_i|\psi_i}} \|x_i(t)\| = \frac{\sigma_i}{\alpha_i}\|x_i(t)\|$$

which would ensure the $\mathscr{L}_2$ stability of the entire system.

The ability to construct V from smaller *local* candidate ISS-Lyapunov functions is important, for it allows us to distribute the design of the ISS event-triggers. This is particularly important in large-scale networked systems where agent subsystems may be added and modified in an *ad hoc* manner. Linear networked systems provide a particularly good example of when one can exploit this distributed strategy for constructing ISS event-triggers. For linear NCS, the parameters in the triggering conditions can be computed using linear matrix inequalities [82].

Simulation results for this approach to event-triggered broadcasting are shown in figure 9.12. This example was taken from [81]. It consists of N carts that are interconnected through soft springs. The local state of the ith cart is $x_i = \begin{bmatrix} y_i & \dot{y}_i \end{bmatrix}^T$ where y_i is the position of the ith cart with respect to the system's equilibrium point. Assuming soft spring coupling between the carts, the state equation for the ith cart can be written as

$$\dot{x}_i(t) = \frac{d}{dt}\begin{bmatrix} y_i \\ \dot{y}_i \end{bmatrix} = \begin{bmatrix} \dot{y}_i(t) \\ u_i(t) + k_i^1 \tanh(y_{i+1}(t) - y_i(t)) + k_i^2 \tanh(y_{i-1}(t) - y_i(t)) + w_i(t) \end{bmatrix}$$

for all $t \in \mathbb{R}$ where $i = 1, 2, \ldots, N$. The parameters k_i^1 and k_i^2 denote the spring constants for the springs on the right-hand and left-hand side of the ith cart, respectively. From the cart geometry shown in figure 9.12, one can see that these spring constants satisfy $k_i^1 = k_{i+1}^2$ for $i = 1, 2, \ldots, N-1$. The left-end cart's spring constant is $k_1^2 = 0$ and the right-end cart's spring constant in $k_N^1 = 0$. The function $u_i : \mathbb{R} \to \mathbb{R}$ denotes the control applied to the cart by its local controller.

In this example the communication network's links mirror the physical interactions between the carts so that $Z_i = D_i$. The sampled state is denoted as $\hat{x}_i(t) = \begin{bmatrix} \hat{y}_i(t) & \frac{d}{dt}\hat{y}_i(t) \end{bmatrix}^T$ where $\hat{y}_i(t) = y_i(r_j^i)$ and $\frac{d}{dt}\hat{y}_i(t) = \dot{y}_i(r_j^i)$ for all $t \in [r_j^i, r_{j+1}^i)$ and $j = 0, 1, \ldots, \infty$. The local control is computed from these sampled measurements as

$$u_i(t) = K_i\hat{x}_i(t) - k_i^1(\tanh(\hat{y}_{i+1}(t) - \hat{y}_i(t)) - k_i^2 \tanh(\hat{y}_{i-1}(t) - \hat{y}_i(t)).$$

In this case, the agents controlling the *end* cars use the ISS event-trigger $5.9\|e_i(t)\| < 0.2\|x_i(r_j^i)\|$ and the interior agents use the event-trigger $10.3\|e_i(t)\| < 0.2\|x_i(r_j^i)\|$. The results from this simulation are shown in figure 9.12.

The top plot on the left-hand side of figure 9.12 plots the state trajectories for all three carts. As can be seen, this event-triggered system is asymptotically stable since all points asymptotically approach their equilibrium points at zero. The bottom plot on the left-hand side of figure 9.12 plots the intersample time intervals that were generated by the proposed event-triggers. As can be seen, these intersample time intervals vary over time in a regular manner.

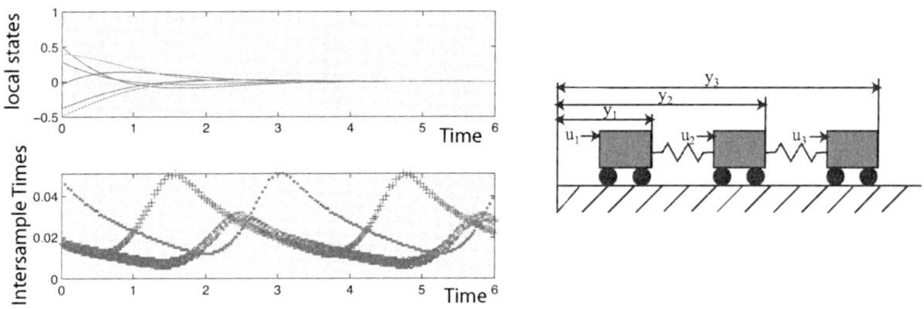

Fig. 9.12 Simulation Example of Event-Triggered Networked Control System consisting of three coupled carts

Impact of Network Artifacts: The prior analysis for the ISS event-triggers in networked control systems had two important assumptions that now need to be examined in more detail. The first assumption was that there was no delay between the transmission and reception of information over the communication network. The second important assumption was that all neighbors in the set U_i receive and use the broadcast data in a synchronized manner. Both assumptions are difficult to justify in real-life wireless sensor-actuator networks. This difficulty is a direct consequence of the unreliable and time-varying nature of wireless communication. The second assumption can be dealt with by making use of a *broadcast protocol* that essentially synchronizes the transmitted data across all neighbors in U_i. The use of such a protocol, however, introduces a number of *network artifacts* such as delays and dropped messages; both of which have a significant impact on the event-triggered system's performance. The objective of this subsection is to establish bounds on acceptable transmission delays and message dropout rates, thereby showing that the stability of the event-triggered NCS is robust to such network artifacts.

Let's first describe the network broadcast protocol used to ensure that the received broadcasts are synchronized between all neighbors in U_i. Such a broadcast protocol is illustrated graphically on the left-hand side of figure 9.13. In this case, the shaded agent represents the ith broadcasting agent at time instant r^i_j. This broadcast is made to the two neighboring agents. Since this is a broadcast, both neighboring agents receive the same sampled copy of the transmitting agent's local state. Upon receiving the ith agent's message, each agent acknowledges the receipt of that message through an ACK signal. When the ith agent receives ACKs from all of the neighbors in U_i, it then broadcasts a *permission* or PERM message to those neighbors. As soon as all neighbors receive the PERM message they use the previously received data in computing their controls. The delay between initial transmission and the final receipt of the PERM messages represents the delay between sampling and actuation. As long as this delay is sufficiently small, the overall networked system should still be stable.

ACKs and PERMs are control packets that are very short in length and can therefore be delivered with a high degree of reliability. The data packets, on the other hand are relatively long and will be more subject to unreliable transmission. Even if an ACK or PERM message were lost, the impact such lost information has on

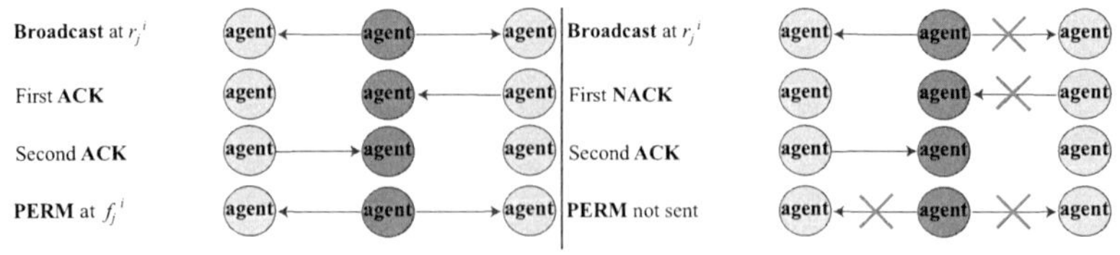

Fig. 9.13 Broadcast Protocol in Wireless NCS. (left) step-by-step description of broadcast protocol. (right) mechanism by which transmitted data is dropped

the overall system's performance can be detected and used to trigger additional data broadcasts. So one would expect the system's overall performance to be robust to such faults. Just how robust this system is to such faults, however, has yet to be fully studied.

It is relatively easy to see why the assumption that transmissions are received instantaneously is unreasonable. While the transmitted signal propagates at the speed of light, it takes time for a message to work its way though an agent's network stack. Moreover, it takes time to transmit, receive and acknowledge the ACKs of an agent's neighbors. As a result, the analysis cannot assume that messages are transmitted and received with zero delay.

The right-hand side of figure 9.13 shows another network artifact that can't be neglected. Wireless communication is inherently unreliable since there is a finite probability that a message will not be successfully transported across the channel. In this case, it is highly likely that a broadcast message may not be received by all neighbors in Z_i. When this occurs, ACK messages will only be sent by a subset of the agents in Z_i. Since the transmitting agent doesn't receive all of the ACKs it is expecting, it will not send the PERM message and so the neighboring agents will not use the information that was previously transmitted to them. In this situation, the data transmitted by the ith agent is actually *dropped*. An important question is whether or not event-triggered system stability is robust to such data dropouts.

Under the proposed broadcast protocol, one must therefore adopt a somewhat more complex view of the timing relations between message transmission and reception than was presented earlier. Figure 9.14 illustrates the underlying timing relationships assumed in the following analysis. As before, let's define a sequence $\{r_j^i\}_{j=0}^{\infty}$ which consists of the time instants when the ith agent *releases* a message for broadcast to its neighbors. If this agent receives ACKs from all of its neighbors then the broadcast is said to be successful. One can therefore introduce a subsequence of $\{r_j^i\}_{j=0}^{\infty}$ that consists of all the *successful* broadcast times. Let $\{b_k^i\}_{k=0}^{\infty}$ denote this sequence of successful broadcasts. If a broadcast is successful, there is a finite delay associated with informing all neighbors that the broadcast was successful (*i.e.* the time required to execute the broadcast protocol). One can therefore define a sequence of successful *finishing* times, $\{f_k^i\}_{k=0}^{\infty}$. The time instant f_k^i denotes the time when the broadcast that was released at time instant b_k^i was given permission for use by all agents in U_i. With regard to these timing relations, the *number of*

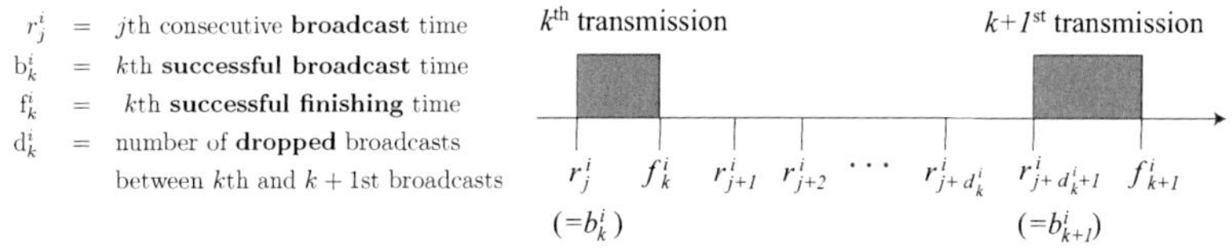

Fig. 9.14 Timing Relationships under NCS Broadcast Protocol

dropped broadcasts between the kth and $(k+1)^{\text{st}}$ successful broadcasts is denoted as d^i_k.

Analyzing the effect that such network artifacts have on the event-triggered system's performance can be done in a manner that is analogous to our earlier analysis of delays in event-triggered embedded systems. As before, one first considers a somewhat weaker version of the event-trigger in which the threshold is only a function of the last sampled state $\hat{x}_i(t)$, rather than the current local state $x_i(t)$. The original event-trigger has the form

$$\|e_i(t)\| \leq \sigma_i \sqrt{\frac{\gamma_i - |S_i \cup B_i|\eta_i}{|\overline{U}_i|\psi_i}} \|x_i(t)\| = \frac{\sigma_i}{\alpha_i}\|x_i(t)\|.$$

A sufficient condition that ensures the above inequality is satisfied would be

$$\|e_i(t)\| \leq \frac{\sigma_i}{\sigma_i + \alpha_i}\|\hat{x}_i(t)\| = c_i\|\hat{x}_i(t)\|.$$

In this weaker condition, the threshold is constant over the interval $[f^i_j, f^i_{j+1})$. As mentioned above, this event-trigger makes it easier to analyze the impact that delays and dropouts have on the overall event-triggered system's performance.

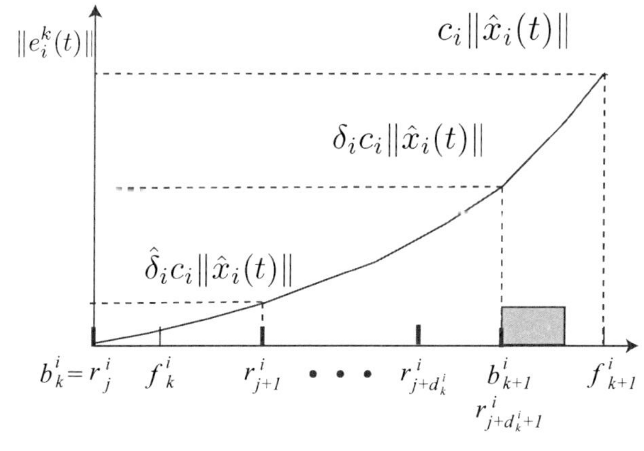

Fig. 9.15 Gap time history in the presence of network artifacts such as dropouts and delays

With this simplified event-triggering condition, one analyzes the impact of dropouts and delays by allocating some portion of the threshold condition to each network artifact. Figure 9.15 shows the gap $\|e_i(t)\|$ as a function of time between a successful broadcast at time b^i_k and the finishing time of the next successful broadcast f^i_{k+1}. The gap grows over this interval of time and in order to assure the performance of the event-triggered system, one requires that this gap always remains less than $c_i\|\hat{x}_i(t)\|$. The effect of the dropouts on the gap is confined to the first part of this interval between times b^i_k and b^i_{k+1}. The effect of the delay on the gap is confined to the last part of the interval from b^i_{k+1} to f^i_{k+1}. This means that one can separate the impact of dropouts and delays between these two subintervals. One exploits this separation by requiring that the next successful broadcast at time b^i_{k+1} occur before the gap gets larger than $\delta_i c_i\|\hat{x}_i(t)\|$ where $\delta_i \in (0,1)$ is a user specified constant. Once δ_i is selected, this determines how many dropouts the system can tolerate before violating the condition.

Is it possible to ensure the gap due to dropouts is no larger than than the allocated gap budget of $\delta_i c_i\|\hat{x}_i(t)\|$? This is done by simply triggering the event early. In particular, let's use an actual event-trigger of the form

$$\|e^k_i(t)\| \leq \hat{\delta}_i c_i\|\hat{x}_i(t)\| = \hat{\delta}_i c_i\|x_i(b^i_k)\|$$

where $e^k_i(t) = x_i(t) - x_i(b^i_k)$ for $t \in [b^i_k, f^i_{k+1})$, $\hat{\delta}_i \in (0,\delta_i)$ and b^i_k is the kth successful broadcast. As shown in figure 9.15, the use of such a smaller threshold will cause the system to trigger early, thereby providing some margin for dropouts or delays. With this threshold the next release of a transmission occurs when $\|x_i(t) - x_i(b^i_k)\| = \hat{\delta}_i c_i\|x_i(b^i_k)\|$. The transmitting agent, however, does not know if this transmission was successful, so when it tests for the next released transmission, it uses the threshold condition

$$\|x_i(t) - x_i(r^i_{j+1})\| < \hat{\delta}_i c_i\|x_i(r^i_{j+1})\|.$$

Note that in this inequality, the sampled system state that is used is taken at time r^i_{j+1} rather than $r^i_j = b^i_k$. So if one now looks at the total gap $\|e^k_i(t)\|$ that occurs for times after r^i_{j+1}, then one can write this as

$$\begin{aligned}\|e^k_i(t)\| &= \|x_i(t) - x_i(b^i_k)\| \\ &\leq \|x_i(t) - x_i(r^i_{j+1})\| + \|x_i(r^i_{j+1}) - x_i(b^i_k)\|.\end{aligned}$$

Each of the two terms can be bounded using the event-triggering conditions to obtain

$$\begin{aligned}\|e^k_i(t)\| &\leq \hat{\delta}_i c_i\|x_i(r^i_{j+1})\| + \hat{\delta}_i c_i\|x_i(b^i_k)\| \\ &\leq \hat{\delta}_i c_i\left(\|x_i(b^i_k)\| + \hat{\delta}_i c_i\|x_i(b^i_k)\|\right) + \hat{\delta}_i c_i\|x_i(b^i_k)\| \\ &= \left((1+\hat{\delta}_i c_i)^2 - 1\right)\|x_i(b^i_k)\|.\end{aligned}$$

The last relationship shows that under event-trigger $\|x_i(t) - x_i(r^i_j)\| < \hat{\delta}_i c_i \|x_i(r^i_j)\|$, that the first release r^i_{j+1} occurs when the gap reaches $\hat{\delta}_i c_i \|x_i(b^i_k)\|$. If that first release is unsuccessful, then the second release occurs at r^i_{j+2} when the gap equals $\left((1+\hat{\delta}_i c_i)^2 - 1\right)\|x_i(b^i_k)\|$. In a similar way one can show that if additional releases are unsuccessful then

$$\|e^k_i(t)\| \leq \left((1+\hat{\delta}_i c_i)^{d^i_k+1} - 1\right)\|x_i(b^i_k)\|$$

for all t where d^i_k is the number of consecutive unsuccessful releases (*i.e.* dropped transmissions) between times b^i_k and b^i_{k+1}. In order to assure the stability of this system let's require that the right-hand side of the above inequality be less than $\delta c_i \|x_i(b^i_k)\|$ or rather that

$$\left((1+\hat{\delta}_i c_i)^{d^i_k+1} - 1\right)\|x_i(b^i_k)\| \leq \delta_i c_i \|x_i(b^i_k)\|.$$

Solving this inequality for d^i_k determines an upper bound on the maximum number of successive dropouts that can be tolerated to assure overall system stability. This bound is called the *maximally allowed number of successive dropouts* (MANSD) and the bound is

$$d^i_k \leq \text{MANSD} = \left\lfloor \log_{1+\hat{\delta}_i c_i}(1+\delta_i c_i) \right\rfloor - 1.$$

This bound represents the maximum number of dropouts that our system can accept.

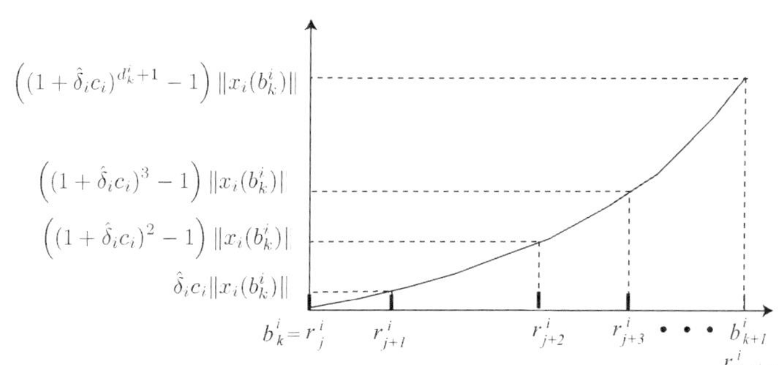

Fig. 9.16 Gap history in the presence of multiple dropouts

Delays only impact the gap in the subinterval between b^i_{k+1} and the finishing time f^i_{k+1}. This case must ensure that the gap does not get larger than the event threshold $c_i \|x_i(b^i_k)\|$. Bounding the allowable length of the interval $[b^i_{k+1}, f^i_{k+1})$ is done by bounding the gap's rate of growth of the gap by a constant

$$\frac{d}{dt}\|e^k_i(t)\| \leq p_i.$$

This assumption is reasonable if the gap can be shown to evolve over compact sets. Given this rate of growth, p_i, the bound on the admissible *delay* between broadcast and reception will be

$$f_{k+1}^i - b_{k+1}^i \leq \frac{(1-\delta_i)c_i}{p_i}\|x_i(b_k^i)\| = \text{upper bound on } \textit{deadline}.$$

This expression represents the admissible *deadline* by which a network transmission must be received to assure overall system stability.

This section has shown that state-dependent event-triggering can greatly reduce the usage of communication resources in networked control systems. A potential weakness in the existing results is their reliance on state-feedback controllers. How one might extend these formalisms to output feedback controllers is still an open question. One way to begin addressing this question is to first examine the use of event-triggering in state estimation. The following section reviews recent results in this direction.

9.5 Event-Triggered Estimation

This section examines a simple problem involving the use of event-triggering in state estimation. In this case, let's assume that a sensor is observing a discrete-time process over a finite horizon and computing a *local* estimate of the process state. The problem involves determining when this local estimate should be transmitted to a *remote* observer so that the remote observer's mean square estimation error is minimized subject to a constraint on the transmission rate between the local sensor and remote observer. These event-triggers are therefore referred to as MMSE (minimum mean square error) event-triggers. Problems of this sort are relevant to estimation over wireless sensor networks [91].

Early work on this problem focused on characterizing the impact that intermittently received observations had on the performance of the estimator [68, 47]. A solution to the MMSE event-triggering problem was presented by Imer et al. [30] and by Rabi et al. [59, 61, 57]. Rabi viewed the transmission problem as an optimal stopping problem [35], whereas Imer made use of dynamic programming concepts. This approach can also be applied to control systems [31]. An alternative approach to event-triggered estimation will be found in [67]. This section uses dynamic programming to rederive the results from Rabi's earlier work [59].

It should be noted that MMSE triggers differ in a significant manner from the earlier stability-based triggers derived in sections 9.3 and 9.4. The prior stability-based triggers preserved some desired stability concept such as input-to-state or $\mathscr{L}_2$ stability. The MMSE event-triggers, however, actually *optimize* the estimator's performance subject to a *constraint on transmission frequency*. Recall that one motivation for considering event-triggered systems is that experimental evidence suggests that event-triggering can greatly reduce communication and computational effort while maintaining overall system performance. None of the prior stability-based event-triggers, however, actually show why this should be the case. The MMSE event-triggers suggested in Imer's and Rabi's work, however, explicitly optimize overall estimator system performance subject to a constraint on communication effort. In this way, MMSE event-triggers may shed more light on why event-triggered systems appear to be more efficient in their use of limited computational and communication resources.

Remote Estimation Problem: The event-triggering problem considered in [61] assumes that a sensor is observing a scalar linear discrete-time process over a finite horizon of length $M+1$. The process state $x : [0,1,2,\ldots,M] \to \mathbb{R}$ satisfies the difference equation

$$x_{k+1} = ax_k + w_k$$

for $k \in [0,1,\ldots,M-1]$ where a is a real constant, $w : [0,1,\ldots,M-1] \to \mathbb{R}$ is a sample path for a zero mean white Gaussian noise process with variance Q. The initial state, $x_0 \in \mathbb{R}$, is chosen from a Gaussian distribution with mean μ_0 and variance Π_0. The sensor generates a measurement $y : [0,1,\ldots,M] \to \mathbb{R}$ that is a corrupted version of the process state. The sensor measurement at time k is

$$y_k = x_k + v_k$$

for any integer k between 0 and M where $v : [0,1,\ldots,M] \to \mathbb{R}$ is a sample path of a zero mean white Gaussian noise process with variance R that is uncorrelated with the process noise, w. The process and sensor blocks are shown on the left-hand side of figure 9.17. In this figure the output of the sensor feeds into a transmission subsystem that decides when to transmit information to a remote observer.

This transmission subsystem consists of three components; an *event detector*, a *filter*, and a *local observer*. The event detector decides when to transmit information to the remote observer. It is assumed that the detector is only allowed to transmit at B distinct time instants where B is an integer between 0 to $M+1$, inclusive. The particular transmission times form a sequence $\{\tau_\ell\}_{\ell=1}^{B}$ where $\tau_\ell \in [0,1,\ldots,M]$

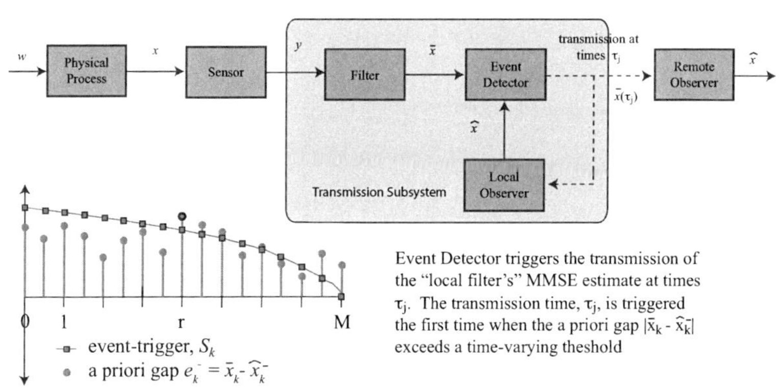

Fig. 9.17 Remote Estimation Problem

denotes the time when the ℓth consecutive transmission was made. The decision to transmit is based on estimates that are generated by the *filter* and the *local observer* shown in figure 9.17.

The *filter* and *local observer* shown in figure 9.17 generate state estimates that the *event-detector* uses to make its transmission decision. Let $\mathscr{Y}_k = \{y_0, y_1, \ldots, y_k\}$ denote the measurement information available at time k. The *filter* generates a state estimate $\overline{x} : [0, 1, \ldots, M] \to \mathbb{R}$ that minimizes the mean square estimation error (MSEE), $E\left[(x_k - \overline{x}_k)^2 \,|\, \mathscr{Y}_k\right]$, at each time step conditioned on all of the sensor information received up to and including time k. These estimates can be computed using a Kalman filter. For the scalar process under study these filter equations are

$$\begin{aligned}
\overline{x}_k &= E\left[x_k \,|\, \mathscr{Y}_k\right] = a\overline{x}_{k-1} + L_k\left(y_k - a\overline{x}_{k-1}\right) \\
\overline{P}_k &= E\left[(x_k - \overline{x}_k)^2 \,|\, \mathscr{Y}_k\right] \\
&= a^2\overline{P}_{k-1} + Q - L_k^2(a^2\overline{P}_{k-1} + Q + R)
\end{aligned}$$

where $\overline{x}_0 = \frac{\Pi_0}{\Pi_0+R}y_0 + \frac{R}{\Pi_0+R}\mu_0$, $\overline{P}_0 = \frac{\Pi_0 R}{\Pi_0+R}$ and $L_k = \frac{a^2\overline{P}_{k-1}+Q}{a^2\overline{P}_{k-1}+Q+R}$.

The event detector uses the *filter's* state estimate, $\overline{x}_k$ at time k, and another estimate generated by the *local observer* to decide when to transmit the filtered state $\overline{x}_k$ to the *remote observer*. Given a set of transmission times $\{\tau_\ell\}_{\ell=1}^{B}$, let $\overline{\mathscr{X}}_k$ denote the information received at the remote observer by time k. In particular, this information set is $\overline{\mathscr{X}}_k = \left\{\overline{x}_{\tau_1}, \overline{x}_{\tau_2}, \ldots, \overline{x}_{\tau_{\ell(k)}}\right\}$ where $\ell(k) = \max\{\ell : \tau_\ell \le k\}$. The remote observer generates an a posteriori estimate $\hat{x} : [0, 1, \ldots, M] \to \mathbb{R}$ of the process state that minimizes the MSEE, $E\left[(x_k - \hat{x}_k)^2 \,|\, \overline{\mathscr{X}}_k\right]$, at time k conditioned on the information received at the remote observer up to and including time k. The a priori estimate at the remote observer, $\hat{x}^- : [0, 1, \ldots, M] \to \mathbb{R}$, minimizes the $E\left[(x_k - \hat{x}_k)^2 \,|\, \overline{\mathscr{X}}_{k-1}\right]$, the MSEE at time k conditioned on the information received up to and including time $k-1$. Due to the scalar nature of the process, these estimates take the form,

$$\begin{aligned}
\hat{x}_k^- &= E\left[x_k \,|\, \overline{\mathscr{X}}_{k-1}\right] = a\hat{x}_{k-1} \\
\hat{x}_k &= E\left[x_k \,|\, \overline{\mathscr{X}}_k\right] = \begin{cases} \hat{x}_k^- & \text{don't transmit at time step } k \\ \overline{x}_k & \text{transmit at time step } k \end{cases}
\end{aligned}$$

where $\hat{x}_0^- = \mu_0$.

The event-detection strategy that is used to select the transmission times, τ_ℓ, is based on observing the *gap*, $e_k^- = \overline{x}_k - \hat{x}_k^-$ between the filter's estimate $\overline{x}$ and the remote observer's a priori estimate, $\hat{x}_k^-$. In the following it will be convenient to adopt the following notational conventions,

$$\begin{aligned}
\hat{e}_k &= x_k - \hat{x}_k && \text{estimation error at step } k, \\
\overline{e}_k &= x_k - \overline{x}_k && \text{filtered state error at step } k, \\
e_k^- &= \overline{x}_k - \hat{x}_k^- && \text{a priori gap at step } k, \\
e_k &= \overline{x}_k - \hat{x}_k && \text{a posteriori gap at step } k.
\end{aligned}$$

Note that even through the gap is a function of the remote observer's estimate, this signal will be available to the sensor. The sensor has access to this information because the sensor has access to all of the information, $\overline{\mathscr{X}}_k$, that it sent to the remote observer. As a result, the sensor can use another *local estimator* to construct a copy of $\hat{x}$ that can be used locally by the event-detector to compute the gap, e_k^-. This local estimator is shown as part of the transmission subsystem shown in figure 9.17.

The event detector's decision to transmit is triggered when the estimate's gap, e_k^-, leaves a time-varying trigger set, S_k^b, where $k \in [0, 1, \ldots, M]$ is the current time and b is the number of transmissions remaining at step k. In general, the trigger sets can be cast as threshold conditions on the estimate's gap. This is shown graphically in the lower left-hand side of figure 9.17. The event-triggers are marked with the squares. The actual gap e_k^- is shown by the solid bullets. Note that the event-triggers are time-varying and equal zero at the end of the time horizon M. Sampling is triggered the first time the gap violates the threshold as shown in figure 9.17. For a given time k, there can be at most $\min\{B, M+1-k\}$ transmissions remaining. The state of the event detector at given time r will be a function of the current a priori gap, e_r^-, and the number of remaining transmissions, T_r. This a priori information at the detector is denoted as the ordered pair, $I_r^- = (e_r^-, T_r)$. In a similar way, the a posteriori information at the detection is denoted as $I_r = (e_r, T_{r+1})$.

Backward Recursion for Value Function: One can use a backward recursion similar to that found in stochastic dynamic programming [20] to determine the triggering sets. Towards this end let's introduce two collections of triggering sets that will be used later. These collections are,

$$\begin{aligned} \mathscr{S}_r^b(k) &= \left\{ S_k^{\max\{0, b-k+r\}}, \ldots, S_k^{\min\{b, M+1-k\}} \right\} \\ \mathscr{S}_r^b &= \left\{ \mathscr{S}_r^b(r), \ldots, \mathscr{S}_r^b(M) \right\}. \end{aligned}$$

These two sets are shown in figure 9.18 for a problem with horizon $M = 4$ and total number of transmissions, $B = 3$. This figure shows the indices for the time steps r and the number of remaining transmissions, b. The collection $\mathscr{S}_1^2$ consists of those indices enclosed within the dotted line. That set is composed of four other collections; $\mathscr{S}_1^2(1)$, $\mathscr{S}_1^2(2)$, $\mathscr{S}_1^2(3)$, and $\mathscr{S}_1^2(4)$. Each of these subcollections is shown as a column of indices enclosed within the rectangles in the figure.

Denote the estimation error at the remote observer as $\hat{e}_k = x_k - \hat{x}_k$. The problem to be solved involves picking the event-triggers in collection $\mathscr{S}_0^B$ to minimize the total MSEE at the remote observer. Formally, the problem is stated as follows

$$\text{minimize: } J(\mathscr{S}_0^B) = E\left[\sum_{k=0}^M \hat{e}_k^2\right] \tag{9.26}$$

where the expectation is taken over $\hat{e}_0, \ldots, \hat{e}_M$. The optimal collection is the set $\mathscr{S}_0^{B*}$ such that $J(\mathscr{S}_0^{B*}) \leq J(\mathscr{S}_0^B)$ over all possible $\mathscr{S}_0^B$ and the resulting optimal cost is

$$J^* = \min_{\mathscr{S}_0^B} J(\mathscr{S}_0^B).$$

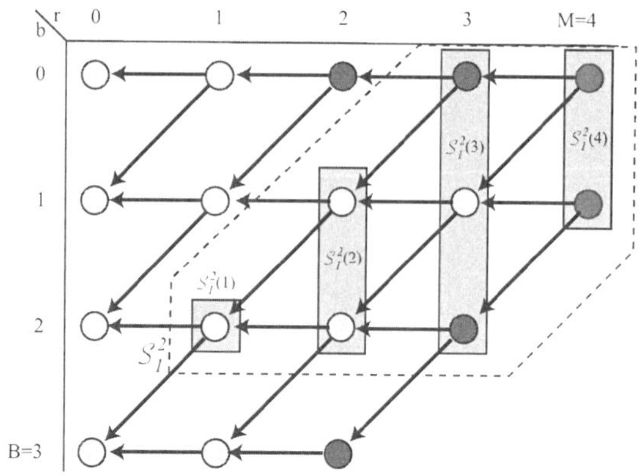

Fig. 9.18 Trigger set collection $\mathscr{S}_1^2 = \bigcup_{k=1}^{M} \mathscr{S}_1^2(k)$ for $B = 3$ and $M = 4$

The problem in equation (9.26) can be treated using results from optimal stochastic control. In our case, the control variables are the collection of triggering sets in $\mathscr{S}_0^B$, rather than some control signal. Since this is a dynamic optimization problem, it can be treated using stochastic dynamic programming. This requires a *value function* that represents the remote observer's total MSEE assuming one uses the optimal triggering sets and assuming the initial information set is $I_r^- = (\zeta, b)$ where ζ is the current value of random variable e_r^- and b is the number of remaining transmissions. In other words, the value function is

$$V(\zeta, b, r) = \min_{\mathscr{S}_r^b} E\left[\sum_{k=r}^{M} \hat{e}_k^2 \,|\, I_r^- = (\zeta, b)\right]. \tag{9.27}$$

This is the minimal expected cost conditioned on the information I_r^- available to the event detector at time r. The optimal cost achieved is $J^* = E\left[V(e_0^-, B, 0)\right]$. As will be shown below, this value function can be computed using a backward recursion often found in dynamic programming.

To develop the backward recursion, let's first consider that the event-detector starts at some time after the initial time step 0. In particular, consider an initial state, (ζ, b, r), at time step r in which the a priori gap, e_r^- equals ζ assuming there are b transmissions remaining to be made. Note that e_r^- is a random variable whereas ζ is a specific value for this random variable. From this initial condition, the collection of admissible trigger sets can be described as

$$\mathscr{S}_r^b = \{S_r^b\} \cup \mathscr{S}_r^b(r+1) \cup \cdots \cup \mathscr{S}_r^b(M).$$

This is seen from figure 9.18. The minimization in equation (9.27) may therefore be broken apart as

$$V(\zeta,b,r) = \min_{S_r^b}\left\{ \min_{\mathscr{S}_r^b(r+1),\dots,\mathscr{S}_r^b(M)} E\left[\sum_{k=r}^{M} \hat{e}_k^2 \,|\, I_r^- = (\zeta,b)\right]\right\}.$$

For notational convenience we'll denote the inner minimization shown above as $G(\zeta,b,r)$. In other words,

$$G(\zeta,b,r) = \min_{\mathscr{S}_r^b(r+1),\dots,\mathscr{S}_r^b(M)} E\left[\sum_{k=r}^{M} \hat{e}_k^2 \,|\, I_r^- = (\zeta,b)\right].$$

The computation of $G(\zeta,b,r)$ may be decomposed into two cases. The first case is when $\zeta \in S_r^b$ (*i.e.* the sensor decides *not* to transmit) and the other case occurs when $\zeta \notin S_r^b$ (*i.e.* the sensor decides to transmit). Let's outline the computation for the first case below. In particular, when $\zeta \in S_r^b$, one sees that

$$G(\zeta,b,r) = \min_{\mathscr{S}_r^b(r+1),\dots,\mathscr{S}_r^b(M)} E\left[\sum_{k=r}^{M} \hat{e}_k^2 \,|\, e_r^- = \zeta \in S_r^b, T_r = b\right].$$

When no transmission takes place, the information in the conditions in the above equation hold if and only if $I_r = (e_r, T_{r+1}) = I_r^- = (e_r^-, T_r) = (\zeta, b)$. $G(\zeta,b,r)$ may therefore be rewritten as

$$G(\zeta,b,r) = \min_{\mathscr{S}_r^b(r+1),\dots,\mathscr{S}_r^b(M)} E\left[\sum_{k=r}^{M} \hat{e}_k^2 \,|\, I_r = (\zeta,b)\right].$$

Since $T_{r+1} = b$ means that there are b transmissions remaining at step $r+1$, one can conclude that not all of the trigger sets in $\bigcup_{k=r+1}^{M} \mathscr{S}_r^b(k)$ will impact the minimization. In particular, one can disregard the sets S_k^p where p ranges from 0 to $b-1$ and $k = r+b-p$.

This means that the minimization is really done over the set $\mathscr{S}_{r+1}^b$. Let's now compute $G(\zeta,b,r)$ as a function of the value function at $V(e_{r+1}^-, b, r+1)$. In particular, $G(\zeta,b,r)$ may be written as

$$\begin{aligned} G(\zeta,b,r) &= \min_{\mathscr{S}_{r+1}^b} E\left[\sum_{k=r}^{M} \hat{e}_k^2 \,|\, I_r = (\zeta,b)\right] \\ &= E\left[\hat{e}_r^2 \,|\, I_r = (\zeta,b)\right] + \min_{\mathscr{S}_{r+1}^b} E\left[\sum_{k=r+1}^{M} \hat{e}_k^2 \,|\, I_r = (\zeta,b)\right] \\ &= \overline{P}_r + \zeta^2 + \min_{\mathscr{S}_{r+1}^b} E\left[\sum_{k=r+1}^{M} \hat{e}_k^2 \,|\, I_r = (\zeta,b)\right]. \end{aligned}$$

Since the information set sequence $\{I_k^-, I_k\}_{k=0}^M$ is Markov and e_{r+1}^- is independent from $\mathscr{S}_{r+1}^b$, the remaining minimization may be rewritten as

$$
\begin{aligned}
G(\zeta,b,r) &= \overline{P}_r+\zeta^2+\min_{\mathscr{S}^b_{r+1}} E\left[E\left[\sum_{k=r+1}^{M}\hat{e}_k^2 \,|\, I^-_{r+1}=(e^-_{r+1},b),I_r=(\zeta,b)\right] \,|\, I_r=(\zeta,b)\right]\\
&= \overline{P}_r+\zeta^2+\min_{\mathscr{S}^b_{r+1}} E\left[E\left[\sum_{r+1}^{M}\hat{e}_k^2 \,|\, I^-_{r+1}=(e^-_{r+1},b)\right] \,|\, I_r=(\zeta,b)\right]\\
&= \overline{P}_r+\zeta^2+E\left[\min_{\mathscr{S}^b_{r+1}} E\left[\sum_{k=r+1}^{M}\hat{e}_k^2 \,|\, I^-_{r+1}=(e^-_{r+1},b)\right] \,|\, I_r=(\zeta,b)\right]\\
&= \overline{P}_r+\zeta^2+E\left[V(e^-_{r+1},b,r+1) \,|\, I_r=(\zeta,b)\right].
\end{aligned}
$$

The preceding argument showed that if $\zeta\in S^b_r$, then the term

$$
\begin{aligned}
G(\zeta,b,r) &= \min_{\mathscr{S}^b_r(r+1),\ldots,\mathscr{S}^b_r(M)} E\left[\sum_{k=r}^{M}\hat{e}_k^2 \,|\, e^-_r=\zeta\in S^b_r, T_r=b\right]\\
&= \overline{P}_r+\zeta^2+E\left[V(e^-_{r+1},b,r+1) \,|\, I_r=(\zeta,b)\right].
\end{aligned}
$$

A similar argument can be used for the case when $\zeta\notin S^b_r$ (*i.e.* the sensor decides to transmit). In this case it can be shown that

$$
\begin{aligned}
G(\zeta,b,r) &= \min_{\mathscr{S}^b_r(r+1),\ldots,\mathscr{S}^b_r(M)} E\left[\sum_{k=r}^{M}\hat{e}_k^2 \,|\, e^-_r=\zeta\notin S^b_r, T_r=b\right]\\
&= \overline{P}_r+E\left[V(e^-_{r+1},b-1,r+1) \,|\, I_r=(0,b-1)\right].
\end{aligned}
$$

Combining both of these results determines the following backward recursion for the value function,

$$
V(\zeta,b,r)=\min_{S^b_r}\left\{V_{NT}(\zeta,b,r)\mathbf{1}_{\zeta\in S^b_r}+V_T(\zeta,b,r)\mathbf{1}_{\zeta\notin S^b_r}\right\} \tag{9.28}
$$

where $\mathbf{1}_{\zeta\in\Omega}$ is the indicator function that takes a value of 1 if ζ is in the set Ω and is zero otherwise. The choice implied in the above equation is between $V_{NT}(\zeta,b,r)$ and $V_T(\zeta,b,r)$ where

$$
\begin{aligned}
V_T(\zeta,b,r) &= \overline{P}_r+E\left[V(e^-_{r+1},b-1,r+1) \,|\, I_r=(0,b-1)\right]\\
&= \text{optimal cost if there is a transmission at time step } r\\
V_{NT}(\zeta,b,r) &= \overline{P}_r+\zeta^2+E\left[V(e^-_{r+1},b,r+1) \,|\, I_r=(\zeta,b)\right]\\
&= \text{optimal cost if there is no transmission at time step } r.
\end{aligned}
$$

Note that $V_T(\zeta,b,r)$ is independent of ζ. Because of the properties of the indicator function, the expression given in equation (9.28) is more naturally seen as a *choice* in which the sensor chooses between the smaller of two costs,

$$
V(\zeta,b,r)=\min_{S^b_r} G(\zeta,b,r)=\min\left\{V_{NT}(\zeta,b,r),V_T(\zeta,b,r)\right\}.
$$

In other words, we have a recursive expression for the optimal cost from the given state (ζ, b, r) and the sensor simply decides to transmit if the cost, V_T, is smaller than not transmitting, V_{NT}.

The computation shown in equation (9.28) is a backward recursion over two sets of indices: the time steps, r, and the number of remaining transmissions, b. In particular, the value function at index (b, r) is a function of the value function at indices $(b-1, r+1)$ and $(b, r+1)$. The functional dependencies are shown in figure 9.18. The arrows in this figure illustrate the functional dependencies implied by equation (9.28).

The initial conditions for this recursion are the value functions at the indices shaded in figure 9.18. The initial values for indices $(0, r)$ where $r \in [B, B+1, \ldots, M]$ are

$$V(\zeta, 0, r) = \begin{cases} \frac{Q(M+1-r)}{1-a^2} + \left(\overline{P}_r + \zeta^2 - \frac{Q}{1-a^2}\right) \frac{1-a^{2(M+1-r)}}{1-a^2} & \text{if } |a| \neq 1 \\ \frac{Q(M+r)(M+1-r)}{2} + \left(\overline{P}_r + \zeta^2 - rQ\right)(M+1-r) & \text{if } |a| = 1 \end{cases}. \quad (9.29)$$

These initial conditions are determined by recognizing that $V(\zeta, 0, r)$ is the cost assuming that no transmissions occur between steps r and M. The other set of initial conditions are marked by the indices in figure 9.18 that are located along the diagonal. In this case

$$V(\zeta, b, r) = \sum_{k=r}^{M} \overline{P}_k \quad (9.30)$$

for b ranging between 1 and B and $r = M+1-b$. This value function is the MSEE from time step r assuming there is a transmission in each of the remaining time steps.

The value function $V(\zeta, B, 0)$ can be computed by recursively determining the value function in sets $\mathscr{S}_0^B(k)$ for k starting at M and ranging backward to 0. The collection

$$\mathscr{S}_0^B(k) = \left\{S_k^{\max\{0, B-k\}}, \ldots, S_k^{\min\{B, M+1-k\}}\right\}$$

consists of the indices enclosed by the rectangles in figure (9.18). The value function in set $\mathscr{S}_0^B(M)$ is determined by the initial conditions described above. Using the order of computation implied by the arrows in figure 9.18, it should be apparent that the value function for all nodes in $\mathscr{S}_0^B(M-1)$ can be computed from the known values in $\mathscr{S}_0^B(M)$. In a similar way, one can see that the value function at indices in $\mathscr{S}_0^B(M-2)$ are computed from the values in $\mathscr{S}_0^B(M-1)$. One continues this computation recursively to obtain the value function in $\mathscr{S}_0^B(0)$.

Let's see what's involved in computing the value function and the trigger set S_k^b at index (b, k) which corresponds to time step k with b remaining transmissions. The trigger set, S_k^b, can be chosen to be the symmetric interval $[-\theta_k^b, \theta_k^b]$ where $\theta_k^b \in \mathbb{R}$ is a real positive number that must be computed. In particular, this leads to the MSEE event-trigger where a transmission occurs if $|e_k^-| > \theta_k^b$. If (b, k) are the initial indices shaded in figure 9.18, then the value function is given by the initial conditions in equations (9.29)-(9.30). For other indices, the value function, $V(\zeta,b,k)$, and associated threshold θ_k^b must be numerically computed using the recursion in equation (9.28).

$V(\zeta,b,k)$ is computed numerically at a number of discrete points in the real line. Recall that $V(\zeta,b,k)$ is determined as a choice between the functions $V_{NT}(\zeta,b,k)$ and $V_T(\zeta,b,k)$. These two functions satisfy

$$\begin{aligned} V_T(\zeta,b,k) &= \overline{P}_k + V_T(\zeta,b-1,k+1) \\ &\quad + \int_{-\theta_{k+1}^{b-1}}^{\theta_{k+1}^{b-1}} \left(V_T(x,b-1,k+1) - V_{NT}(x,b-1,k+1)\right) p(x|0)dx \\ V_{NT}(\zeta,b,k) &= \overline{P}_k + \zeta^2 + V_T(\zeta,b,k+1) \\ &\quad - \int_{-\theta_{k+1}^{b}}^{\theta_{k+1}^{b}} \left(V_T(x,b,k+1) - V_{NT}(x,b,k+1)\right) p(x|\zeta)dx \end{aligned}$$

where $p(x|y)$ is the probability density of e_{k+1}^- conditioned on $e_k = y$. The value of the optimal thresholds θ_{k+1}^b and θ_{k+1}^{b-1} are needed to evaluate these two functions. These thresholds can be computed in a recursive manner.

Given θ_{k+1}^b and θ_{k+1}^{b-1}, the next optimal threshold, θ_k^b, can be found using a bisection search. In particular, the optimal threshold occurs at ζ^* when $V_T(\zeta^*,b,k) = V_{NT}(\zeta^*,b,k)$. So the threshold must satisfy $\theta_k^b = |\zeta|^*$. Once this threshold is determined through the bisection search, then one can see that for $|\zeta| > \theta_k^b$ the value function $V(\zeta,b,k) = V_T(\zeta,b,k)$ and for $|\zeta| \leq \theta_k^b$, the value function must satisfy $V(\zeta,b,k) = V_{NT}(\zeta,b,k)$. This allows one to readily evaluate $V(\zeta,b,k)$ at a number of distinct points, ζ, along the real line.

An example of this computation is provided below for the system

$$\begin{aligned} x_k &= 1.2x_{k-1} + w_k \\ y_k &= x_k + v_k. \end{aligned} \tag{9.31}$$

The mean and covariance of the initial state are 1 and 2, respectively. The covariance of the noise terms, w and v, are both 1. Fix the horizon $M = 8$ and the total number of transmissions $B = 2$. Using the algorithm mentioned above, the value function is evaluated at various values of ζ. Figure 9.19 shows the resulting value function. The solid line in the figure is the value function for various values of time k. The right-hand plot shows $V(\zeta,1,k)$ and the left-hand plot shows $V(\zeta,2,k)$. The threshold θ_k^b is marked by the dots in the figure. Outside of the interval defined by the dots one finds that $V(\zeta,b,k) = V_T(\zeta,b,k)$ and this is a constant because $V_T(\zeta,b,k)$ is independent of ζ. Inside the region, the value function varies as a function of ζ.

To see how well the MSEE event-triggers perform, let's vary B from 0 to 9 and repeat the experiment $10,000$ times for both the optimal event-trigger and comparable periodic triggering of transmissions. The plot in figure (9.20) shows the MSEE for the optimal event-triggered and periodically triggered transmissions as a function of the total number of transmissions, B. The plot shows that the experimentally

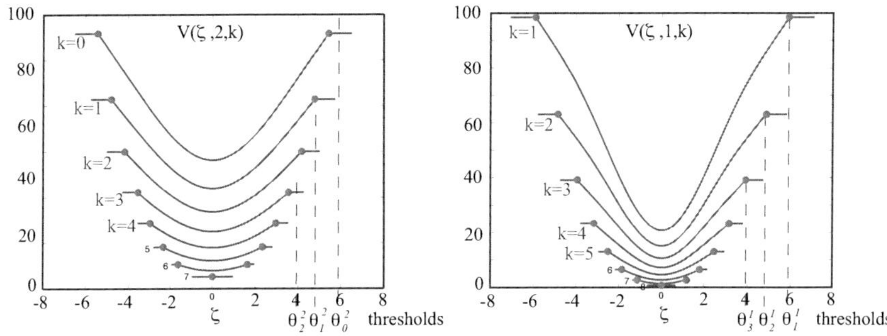

Fig. 9.19 Value functions $V(\zeta,1,k)$ and $V(\zeta,2,k)$ for sample system

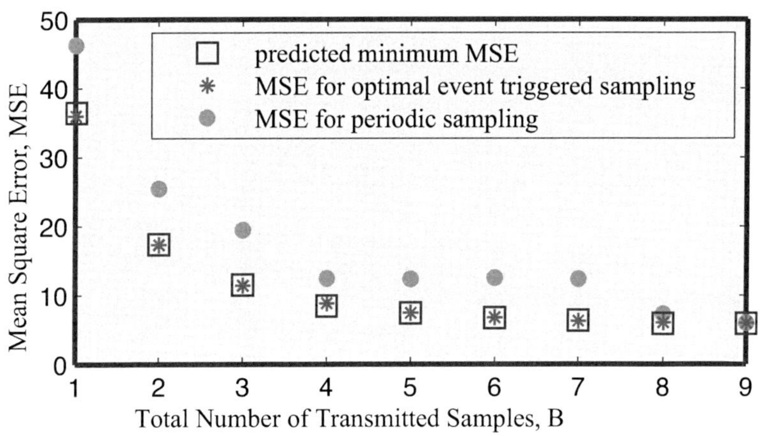

Fig. 9.20 MSE for periodic and event-triggered system

observed MSEE equals the MSEE predicted by the value function. The plot also shows that the optimal event-triggered transmission strategy always generates a smaller total MSEE than comparable periodically triggered systems.

This section has studied the design of MSEE event-triggers for a simple distributed estimation problem. This problem was solved in [30, 61, 58] under a variety of assumptions. The main contribution of this section was a direct derivation of the optimal event-triggers using dynamic programming concepts as well as an explicit method for computing the optimal event-triggering thresholds. The methods in this section recover the original results in [59]. The analysis may be general enough to suggest specific ways of extending the treatment of the scalar system to state estimation of more general linear vector processes.

This section has focused on the estimation problem, but the framework used here may also be extended to control problems. For control, one simply takes the output of the remote estimator and connects it back into the plant through a controller.

Real-life applications that fit into this model are found in wireless sensor-actuator networks. The associated control problem that seeks to maximize control performance subject to a communication usage constraint was solved in [87, 88] for the infinite horizon case. In general, the event-triggering thresholds solving the infinite horizon problem are constants. Finite horizon versions of this control problem were treated by [31] and [62]. In the finite horizon case, the event-triggering thresholds are time-varying functions of the initial system state. It has proven difficult, however, to apply this work to vector systems due to the computational complexity associated with solving the dynamic programming equations. Recent progress has been made in resolving the computational complexity issue for infinite horizon problems through the use of quadratic approximations for the value function [19, 18]. A related approach was used in [42] to address the complexity in the finite-horizon estimation problem.

9.6 Event-Triggered Approaches to Optimization

This section introduces an event-triggered distributed algorithm that solves network utility maximization (NUM) problems in large-scale networked systems [79, 78]. Existing distributed algorithms for the NUM problem are gradient-based schemes that converge to the optimal point provided the communication between subsystems is sufficiently frequent. Analytic bounds on the communication interval required to ensure convergence tend to be inversely proportional to certain measures of network complexity such as network diameter and connectivity. As a result, the total message passing complexity in such algorithms can be very great. The event-triggered algorithm presented in this section appears to reduce the message passing complexity by nearly two-orders of magnitude. Moreover, experimental results indicate that this complexity may be independent of network diameter and connectivity.

Related Work: Many problems in networked systems can be formulated as optimization problems. This includes estimation [56, 69, 33], source localization [56], data gathering [15, 14], routing [46], control [77], resource allocation [55, 89] in sensor networks, resource allocation in wireless communication networks [86, 16], congestion control in wired communication networks [36, 44], and optimal power dispatch [38] in electrical power grid. The consensus problem [52] can also be viewed as a distributed optimization problem where the objective function is the total state mismatch between neighboring agents. Many of these problems may be viewed as multi-agent optimization problems that can be solved by a distributed implementation of a subgradient algorithm [49]. In all of these problems, subsystems communicate with each other in order to collaboratively solve a network optimization problem.

Distributed algorithms that solve such network optimization problems include the center-free distributed algorithms [28], distributed asynchronous gradient-based algorithms [72] and distributed subgradient methods [49]. These early algorithms suggest that if the communication between adjacent subsystems is sufficiently frequent, then the state of the network will asymptotically converge to the optimal point. Later developments in such distributed algorithms may be found in the networking community. Most of these later algorithms focus on solving the *Network Utility Maximization* (NUM) problem. The NUM problem maximizes a global separable measure of network system performance subject to linear inequality constraints that are directly related to throughput constraints. This problem originates in congestion control for Internet traffic [36, 44]. The NUM problem, however, has a general form and many problems in other areas can be recast as a NUM problem with little or no variation. As a matter of fact, many of the aforementioned problems can be reformulated as NUM problems.

Among the existing algorithms [36, 44, 85, 53] solving the NUM problem, the dual-decomposition approach proposed by Low et al. [44] is the most widely used. Low et al. showed that their dual-decomposition algorithm was convergent for a step-size that was inversely proportional to two important measures of network size: the maximum path length $\overline{L}$ and the maximum number of neighbors $\overline{S}$. So as these two measures get large, the step size required to ensure convergence becomes extremely small. Step size, of course, determines the number of computations required for the algorithm's convergence. Under dual-decomposition, system agents exchange information at each iteration, so that step size, γ, also determines the message passing complexity of the algorithm. Therefore if one uses the *stabilizing* step size, dual-decomposition algorithms will have a message passing complexity that quickly scales to unreasonable levels as the network diameter, $\overline{L}$, or the neighborhoood size, $\overline{S}$, increases. In particular, it was shown in [44] that the dual-decomposition is convergent if the step size satisfies

$$0 < \gamma < \gamma^* = \frac{-2\max_{(i,x_i)} \nabla^2 U_i(x_i)}{\overline{L}\overline{S}} \tag{9.32}$$

where $\overline{L}$ is the maximum number of links any user uses, $\overline{S}$ is the maximum number of users any link has, and $U_i(x_i)$ is the utility user i receives for transmitting at rate x_i. For many networked systems this type of message passing complexity may be unacceptable. This is particularly true for systems communicating over a wireless network. In such networked systems, the energy required for communication can be significantly greater than the energy required to perform computation. As a result, it would be beneficial if one can somehow separate communication and computation in these distributed algorithms. This could reduce the message passing complexity of distributed algorithms such as dual-decomposition. This section shows how event-triggering can be used to realize the separation between communication and computation in a primal algorithm solving the NUM problem.

NUM Problem: The NUM problem consists of a network of N users and M links. Let $\mathscr{S} = \{1,\ldots,N\}$ denote the set of users and $\mathscr{L} = \{1,\ldots,M\}$ denote the set of links. Each user generates a flow with a specified data rate. Each flow may traverse several links (which together constitute a route) before reaching its destination. The set of links that are used by user $i \in \mathscr{S}$ will be denoted as $\mathscr{L}_i$ and the set of users that are using link $j \in \mathscr{L}$ will be denoted as $\mathscr{S}_j$. The NUM problem takes the form

$$\begin{array}{l} \text{maximize: } U(x) = \sum_{i \in \mathscr{S}} U_i(x_i) \\ \text{subject to: } Ax \leq c, \quad x \geq 0 \end{array} \tag{9.33}$$

where $x = \begin{bmatrix} x_1 & \cdots & x_N \end{bmatrix}^T$ and $x_i \in \mathbb{R}$ is user i's data rate. $A \in \mathbb{R}^{M \times N}$ is the routing matrix mapping users onto links and $c \in \mathbb{R}$ is a vector of link capacities. The ji^{th} component, A_{ji}, is 1 if user i's flow traverses link j and is zero otherwise. The jth row of Ax represents the total data rates going through link j. This rate cannot exceed the link's capacity c_j. The cost function $U : \mathbb{R}^N \to \mathbb{R}$ is the sum of the user *utility* functions, $U_i : \mathbb{R} \to \mathbb{R}$, for $i = 1, 2, \ldots, N$. These utility functions represent the reward, $U_i(x_i)$, (*i.e.* quality-of-service) that user i receives by transmitting at rate x_i.

A specific example of a NUM problem is shown in figure 9.21. This figure shows a linear network consisting of $M = 5$ links with $N = 3$ users. User 1's route includes links 1-4, user 2's route includes links 2-3, and user 3's route uses link 3-5. Assuming each link has a capacity limit of 1, the throughput constraint therefore becomes,

$$Ax = \begin{bmatrix} 1 & 0 & 0 \\ 1 & 1 & 0 \\ 1 & 1 & 1 \\ 1 & 0 & 1 \\ 0 & 0 & 1 \end{bmatrix} \begin{bmatrix} x_1 \\ x_2 \\ x_3 \end{bmatrix} \leq \begin{bmatrix} 1 \\ 1 \\ 1 \\ 1 \\ 1 \end{bmatrix} = c.$$

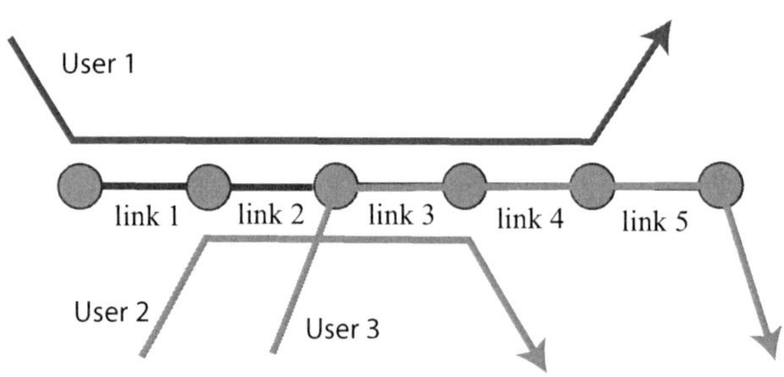

Fig. 9.21 Example of NUM Problem

The solution to the NUM problem maximizes the summed utility seen by all users in the network as a function of the users' transmission rates. These rates must clearly be non-negative. Moreover, if $U_i(x) = \alpha_i \log(x)$ where α_i is a positive constant, then it can be shown [36] that all the user rates in the optimal solution must be positive. In other words, the optimal solution does not result in certain users from being denied access to the network, thereby assuring that all users have *fair* access to network resources.

Augmented Lagrangian Algorithm: While early algorithms used methods based on the dual to the problem in equation (9.33), this section examines a primal *augmented* Lagrangian method. In particular, let's introduce a slack variable $s \in \mathbb{R}^M$ and replace the link constraints ($c_j - a_j^T x \geq 0$ for all $j \in \mathscr{L}$) by the following equality constraint

$$a_j^T x - c_j + s_j = 0, \quad s_j \geq 0, \quad \text{for all } j \in \mathscr{L}.$$

where s_j is called the *slack variable* for the jth constraint. The *augmented cost* then becomes

$$L(x,s;\lambda,w) = -\sum_{i\in\mathscr{S}} U_i(x_i) + \sum_{j\in\mathscr{L}} \lambda_j(a_j^T x - c_j + s_j) + \frac{1}{2}\sum_{j\in\mathscr{L}} \frac{1}{w_j}(a_j^T x - c_j + s_j)^2.$$

Here a penalty parameter w_j is associated with each link constraint and $w = [w_1,\ldots,w_M]$ is the vector of penalty parameters. The other variable λ_j is an estimate of the Lagrange multiplier, λ_j^*, associated with link j's constraint, $c_j - a_j^T x \geq 0$. A vector formed from these estimates is denoted as $\lambda = [\lambda_1,\ldots,\lambda_M]$. The vector $a_j^T = [A_{j1},\cdots,A_{jN}]$ is the jth row of the routing matrix A.

$L(x,s;\lambda,w)$ is a continuous function of x and s for fixed λ and w. The user rate, x^*, and the slack variable s^*, that minimizes the augmented cost satisfies the following equation

$$\min_{x\geq 0, s\geq 0} L(x,s;\lambda,w) = \min_{x\geq 0}\min_{s\geq 0} L(x,s;\lambda,w) = \min_{x\geq 0} L_p(x;\lambda,w) \tag{9.34}$$

where one defines the *augmented Lagragian function* associated with the NUM problem as

$$L_p(x;\lambda,w) = -\sum_{i\in\mathscr{S}} U_i(x_i) + \sum_{j\in\mathscr{L}} \psi_j(x;\lambda,w)$$

and where

$$\psi_j(x;\lambda,w) = \begin{cases} -\frac{1}{2}w_j\lambda_j^2 & \text{if } c_j - a_j^T x - w_j\lambda_j \geq 0 \\ \lambda_j(a_j^T x - c_j) + \frac{1}{2w_j}(a_j^T x - c_j)^2 & \text{otherwise} \end{cases}.$$

The optimization problem in equation (9.34) is then used as a starting point for developing a recursive procedure that asymptotically approaches the solution of the original NUM problem.

The original NUM problem's solution can be approached arbitrarily closely by solving an appropriately defined sequence of the optimization problems in equation (9.34). This sequence of problems involve minimizing $L_p(x;\lambda[k],w[k])$ for appropriately chosen sequences of penalty parameters, w, and multiplier estimates, λ. Let $x^*[k]$ denote the approximate minimizer for $L_p(x;\lambda[k],w[k])$. It has been shown [6] that for appropriately chosen sequences $\{w[k]\}_{k=0}^{\infty}$ and $\{\lambda[k]\}_{k=0}^{\infty}$, the sequence of approximate minimizers, $\{x^*[k]\}_{k=0}^{\infty}$ converges to the optimal point of the NUM problem. The appropriate choice for these sequences is that for all $j \in \mathscr{L}$

- the sequence of penalty parameters, $\{w_j[k]\}_{k=0}^{\infty}$, is monotone decreasing to zero
- and the sequence of Lagrange multiplier estimates, $\{\lambda_j[k]\}_{k=0}^{\infty}$, is a sequence where

$$\lambda_j[k+1] = \max\left\{0, \lambda_j[k] + \frac{1}{w_j[k]}\left(a_j^T x^*[k] - c_j\right)\right\}.$$

A detailed description of how the sequences $w[k]$ and $\lambda[k]$ are updated in a distributed manner will be found in [78].

A primal algorithm based on the augmented Lagrangian method was developed [78] that converges to the exact minimizer of the NUM problem. In many scenarios, however, it may suffice to obtain an approximate minimizer which can be obtained by considering the problem of minimizing $L_p(x;\lambda,w)$ for a fixed λ and w. In particular, if $\lambda_j = 0$ and w_j is sufficiently small, the minimizer of $L_p(x;\lambda,w)$ will be a good approximation to the solution of the original NUM problem. In this regard the *basic primal algorithm* can be stated as follows

1. **Initialization:** Select any initial user rate $x[0] > 0$. Set $\lambda_j = 0$ and select a sufficiently small $w_j > 0$ for all $j \in \mathscr{L}$.
2. **Recursive Loop:** Minimize $L_p(x;\lambda,w)$ by letting

$$x[k+1] = \max\left\{0, x[k] - \gamma\frac{\partial L_p}{\partial x}(x[k];\lambda,w)\right\} \tag{9.35}$$

for $k = 0, 1, \ldots, \infty$.

The smaller w is the more accurate our approximate solution is. The recursion shown in step 2 tries to minimize $L_p(x;\lambda,w)$ using a gradient following method in which γ is a sufficiently small step size. The computations shown above can be easily distributed among users and links.

Event-Triggered NUM Algorithm: Implementing the aforementioned primal algorithm in a distributed manner requires communication between users and links. An event-triggered implementation of the algorithm assumes that the transmission of messages between users and links is triggered by some local error signal crossing a state-dependent threshold. The main problem is to determine a threshold condition that results in message streams ensuring asymptotic convergence to the NUM problem's approximate solution.

The minimizer of the Lagrangian $L_p(x;\lambda,w)$ is searched for using the gradient following recursion in equation (9.35). Assuming that computation is *cheap*, one realizes this gradient recursion as a continuous-time system in which

$$\begin{aligned} x_i(t) &= -\int_0^t \left(\frac{\partial L_p}{\partial x_i}(x(\tau);\lambda,w)\right)^+_{x_i(\tau)} d\tau \\ &= \int_0^t \left(\frac{\partial U_i(x_i(\tau))}{\partial x_i} - \sum_{j\in\mathscr{L}_i}\mu_j(\tau)\right)^+_{x_i(\tau)} d\tau \end{aligned} \tag{9.36}$$

for each user $i \in \mathscr{S}$ where

$$\mu_j(t) = \max\left\{0, \lambda_j + \frac{1}{w_j}\left(\sum_{i \in \mathcal{S}_j} x_i(t) - c_j\right)\right\}. \tag{9.37}$$

The function $\mu_j : \mathbb{R} \to \mathbb{R}$ is a scalar characterizing how close the jth link constraint is to being active. The link constraint is active at time t when $\mu_j(t) = 0$. Given a function $f : \mathbb{R} \to \mathbb{R}$, its *positive projection* is defined as

$$(f(x))_x^+ = \begin{cases} 0 & \text{if } x = 0 \text{ and } f(x) < 0 \\ f(x) & \text{otherwise} \end{cases}.$$

The positive projection used above guarantees that the user rate, $x_i(t)$, is always non-negative along the trajectory.

Equation (9.36) is the continuous-time version of the discrete-time update shown in equation (9.35). Note that in equation (9.36), user i computes its rate based only on the information from itself and the information of μ_j from those links that are being used by user i. As noted above μ_j characterizes how close the j link constraint is to being active. One may think of μ_j as the jth link's local *state*. From equation (9.37), link j only needs to be able to measure the total flow that goes through itself. Since all of this information is locally available, the update of the user rate in equation (9.36) can be done in a distributed manner.

In equation (9.36), the link state information is available to the user in a continuous manner. Let's consider an *event-triggered* version of equation (9.36). This is done by allowing the user to only access a *sampled* version of the link state. In particular, let's associate a sequence of *sampling* instants, $\{T_j^L[\ell]\}_{\ell=0}^{\infty}$ with the jth link. The time $T_j^L[\ell]$ denotes the instant when the jth link samples its link state μ_j for the ℓth time and transmits that state to users $i \in \mathcal{S}_j$. One can see that at any time $t \in \mathbb{R}$, the sampled link state is a piecewise constant function of time in which

$$\hat{\mu}_j(t) = \mu_j(T_j^L[\ell])$$

for all $\ell = 0, 1, \ldots, \infty$ and any $t \in [T_j^L[\ell], T_j^L[\ell+1])$. In this regard, the event-triggered version of equation (9.36) takes the form

$$x_i(t) = \int_0^T \left(\frac{\partial U_i(x_i(\tau))}{\partial x_i} - \sum_{j \in \mathcal{L}_i} \hat{\mu}_j(\tau)\right)_{x_i(\tau)}^+ d\tau$$

for all ℓ and any $t \in [T_j^L[\ell], T_j^L[\ell+1])$.

Let's now try to establish conditions on the sampling times $\{T_j^L[\ell]\}$ that ensure the gradient update shown in equation (9.35) is convergent. For notational convenience let the time derivative of the user rate, $x_i(t)$, be denoted as $z_i(t)$. Referring to $z_i(t)$ as the *user state*, one sees that z_i satisfies the equation

$$z_i(t) = \dot{x}_i(t) = \left(\frac{\partial U_i(x_i(t))}{\partial x_i} - \sum_{j \in \mathcal{L}_i} \hat{\mu}_j(t)\right)_{x_i(t)}^+$$

for all $i \in \mathscr{S}$. Now we take $L_p(x;\lambda,w)$ as a candidate Lyapunov function. The directional derivative of L_p is

$$\begin{aligned}
\dot{L}_p(x;\lambda,w) &= \sum_{i=1}^{M} \frac{\partial L_p}{\partial x_i}\frac{dx_i}{dt} = -\sum_{i=1}^{N} z_i \left(\frac{\partial U_i(x_i(t))}{\partial x_i} - \sum_{j=1}^{M} \mu_j A_{ji} \right) \\
&\leq -\sum_{i=1}^{N} \left(\frac{1}{2} z_i^2 - \frac{1}{2} \left(\sum_{j=1}^{M} (\mu_j - \hat{\mu}_j) A_{ji} \right)^2 \right) \\
&\leq -\frac{1}{2}\sum_{i=1}^{N} z_i^2 + \frac{1}{2}\sum_{j=1}^{M} \overline{LS}(\mu_j - \hat{\mu}_j)^2.
\end{aligned}$$

To assure that $\dot{L}_p$ is negative definite, one needs to select the sampling times so that

$$\sum_{j=1}^{M} \overline{LS}(\mu_j - \hat{\mu}_j)^2 \leq \sum_{i=1}^{N} z_i^2.$$

This almost looks like one of the state-dependent event-triggers used earlier in section 9.4. Unfortunately, this trigger cannot be implemented in a distributed manner. While the left-hand side is separable over the links, the right-hand side is summed over the users. So the preceding analysis does not give rise to a distributed event triggered algorithm.

To develop local event-triggers that can be easily distributed across the network, let's consider another sequence of times $\{T_i^S[\ell]\}_{\ell=0}^{\infty}$ for each user $i \in \mathscr{S}$. The time $T_i^S[\ell]$ is the ℓth time when user i transmits its user state to all links $j \in \mathscr{L}_i$. One can therefore see that at any time $t \in \mathbb{R}$, the sampled user rate is a piecewise constant function of time satisfying

$$\hat{z}_i(t) = z_i(T_i^S[\ell])$$

for all $\ell = 0, 1, \ldots, \infty$ and any $t \in [T_i^S[\ell], T_i^S[\ell+1])$. One can now use this sampled user state in our earlier expression for $\dot{L}_p$ to show that

$$\dot{L}(x;\lambda,w) \leq -\frac{1}{2}\sum_{i=1}^{N}\left[z_i^2 - \rho \hat{z}_i^2\right] - \frac{1}{2}\sum_{j=1}^{M}\left[\rho \sum_{i \in \mathscr{S}_j} \frac{1}{L}\hat{z}_i^2 - \overline{LS}(\mu_j - \hat{\mu}_j)^2\right]$$

for some $\rho \in (0,1)$. The derivative, $\dot{L}_p$, is negative definite as long as

$$\begin{aligned}
0 &< \sum_{i=1}^{N}[z_i^2 - \rho \hat{z}_i^2] \\
0 &< \sum_{j=1}^{M}\left[\rho \sum_{i \in \mathscr{S}_j} \frac{1}{L}\hat{z}_i^2 - \overline{LS}(\mu_j - \hat{\mu}_j)^2\right].
\end{aligned}$$

In this case, both inequalities are separable. The first one is separable over the users and the second one is separable over the links. One can therefore ensure these conditions are satisfied if

$$z_i^2 - \rho \hat{z}_i^2 > 0 \tag{9.38}$$

for each $i \in \mathscr{S}$. This condition can be enforced by requiring that the user transmit z_i at those time instants when the inequality is about to be violated. The other condition is satisfied if

$$\overline{LS}(\mu_j - \hat{\mu}_j)^2 < \rho \sum_{i \in \mathscr{S}_j} \frac{1}{\overline{L}} \hat{z}_i^2 \tag{9.39}$$

for each $j \in \mathscr{L}$. This condition can be enforced by requiring that the link transmit μ_j at those time instants when the inequality is about to be violated. The informal discussion given above therefore establishes that if user/link transmissions are generated using the event-triggers in equation (9.38) and (9.39), then $L_p(x;\lambda,w)$ indeed becomes a Lyapunov function for this system and one can ensure that this system is convergent to a neighborhood of the optimal solution of the NUM problem.

Figure 9.22 shows the event-triggered optimization algorithm in graphical from. This figure uses the system network that was introduced in figure 9.21. In this case each link in the network has an associated router which monitors the total data flowing through the link ($\sum_{i \in \mathscr{S}_j} x_i(t) - c_j$). Attached to each router is a *price agent* that updates the link state μ_j and checks the event-trigger in equation (9.39) to determine whether or not it will transmit its local link state. In a dual manner, each user that is pumping data into the network has an associated *rate agent* that updates the user state $z_i(t)$ and checks the trigger in equation (9.38) to determine when to transmit to the links. One therefore see that the algorithm has both a feedback (link to user) and feedforward path (user to link) in which the information streams are both sporadic in nature.

Scaling of Event-Triggered Algorithm: Let's compare the number of message exchanges of the event-triggered algorithm against the dual-decomposition algorithm. Simulation results show that event-triggered algorithms reduce the number of message exchanges by up to two orders of magnitude when compared to dual-decomposition. Moreover, the event-triggered algorithm's message passing complexity scales in a way that appears to be independent of network diameter or connectivity.

Denote $s \in \mathscr{U}[a,b]$ if s is a random variable uniformly distributed on $[a,b]$. Given M, N, $\overline{L}$ and $\overline{S}$, the network used for simulation is generated in the following way. One randomly generates a network with M links and N users, where $|\mathscr{S}_j| \in \mathscr{U}[1,\overline{S}]$ for $j \in \mathscr{L}$ and $|\mathscr{L}_i| \in \mathscr{U}[1,\overline{L}]$ for $i \in \mathscr{S}$. One makes sure that at least one link has $\overline{S}$ users, and at least one user uses $\overline{L}$ links. After the network is generated, a utility function $U_i(x_i) = \alpha_i \log x_i$ is assigned to each user i, where $\alpha_i \in \mathscr{U}[0.8, 1.2]$. Link j is assigned capacity $c_j \in \mathscr{U}[0.8, 1.2]$. Once the network is generated, dual-decomposition and the event-triggered augmented Lagrangian

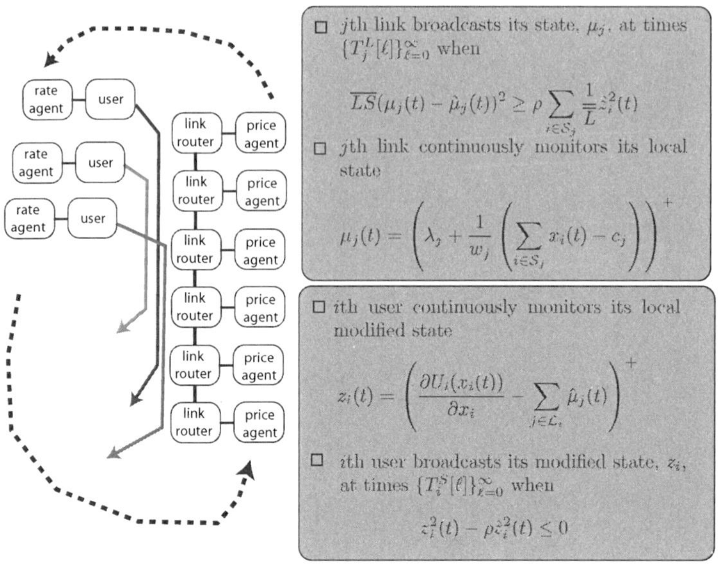

Fig. 9.22 Diagram of the event-triggered primal algorithm

algorithms are simulated. The optimal rate x^* and its corresponding utility U^* are calculated using a global optimization technique.

Define the error (for all algorithms) as

$$e(k) = \left| \frac{U(x(k)) - U^*}{U^*} \right|$$

where $x(k)$ is the rate at the kth iteration. $e(k)$ is the 'normalized deviation' from the optimal point at the kth iteration. In all algorithms, one counts the number of iterations K for $e(k)$ to decrease to and stay in the neighborhood $\{e(k)|e(k) \leq e_d\}$. In dual-decomposition, message passing from the links to the users occurs at each iteration synchronously. So K is a measure of the total number of message exchanges. In the event-triggered algorithms, events occur in a totally asynchronous way. So one adds the total number of triggered events, and divide this number by the link number M. This provides an equivalent iteration number K for the event-triggered algorithms, and is a measure of the total number of message exchanges. One should point out that since these simulations compare a primal algorithm and a dual algorithm, they run at different time scales. Iteration number is then a more appropriate measure of convergence than time [17, 34].

The default settings for the simulation are as follows: $M = 60$, $N = 150$, $\overline{L} = 8$, $\overline{S} = 15$, and $e_d = 3\%$. For all three algorithms, the initial condition is $x_i(0) \in \mathscr{U}[0.01, 0.05]$ for all $i \in \mathscr{S}$. In dual-decomposition, initial price $p_j = 0$ for $j \in \mathscr{L}$, and the step size γ is calculated using equation (9.32). In the event-triggered primal algorithm, the parameters are $\rho = 0.5$, $\lambda_j = 0$, and $w_j = 0.01$ for $j \in \mathscr{L}$.

We now consider a Monte Carlo simulation where M, N, and $\overline{L}$ are fixed and $\overline{S}$ is varied from 7 to 26. For each $\overline{S}$, all algorithms were run 1500 times, and each time a random network which satisfies the above specification is generated. The mean m_K and standard deviation σ_K of K are computed for each $\overline{S}$. m_K is used as the criterion for comparing the scalability of both algorithms. The left-hand plot in figure 9.23 plots the iteration number K on a logarithmic scale as a function of $\overline{S}$ for all algorithms. The circles represent m_K for dual-decomposition and the squares correspond to the primal algorithm.

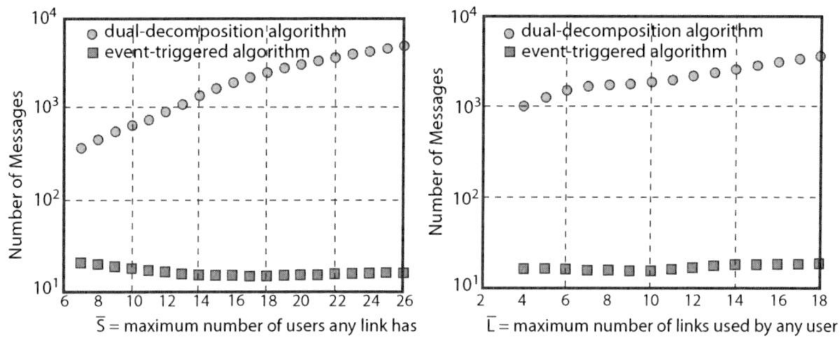

Fig. 9.23 Iteration number K as a function of $\overline{S}$ and $\overline{L}$ for all algorithms

For the primal algorithm, when $\overline{S}$ increases from 7 to 26, m_K does not show noticeable increase. For the primal algorithm, m_K varies between 15.1 and 21.1. For dual-decomposition, m_K increases from 0.3856×10^3 to 5.0692×10^3. Our event-triggered algorithm is up to two orders of magnitude faster than dual-decomposition. These results also show that the event triggered message passing complexity scales in a manner that is independent of $\overline{S}$. This is stands in stark contrast to the dual-decomposition algorithm which scales superlinearly with respect to $\overline{S}$.

These algorithms were also simulated as a function of $\overline{L}$. In particular, $\overline{L}$ was varied from 4 to 18. The right-hand plot in figure 9.23 plots K (on a logarithmic scale) as a function of $\overline{L}$ for all algorithms. For the primal algorithm, when $\overline{L}$ increases from 4 to 18, m_K increases slowly. In particular, m_K increases from 15.0 to 18.2. For dual-decomposition, m_K increases from 0.9866×10^3 to 3.5001×10^3. The event-triggered algorithm again is up to two orders of magnitude faster than the dual-decomposition.

This section presented a primal event-triggered distributed algorithm for solving network utility maximization problems based on augmented Lagrangian methods. Simulation results suggest that event-triggering greatly reduces the message passing complexity of such distributed optimization algorithms. Optimization of networked systems therefore represents another important application of event-triggering that can be applied to a wide range of applications ranging from traffic control to power dispatch in electrical power grids.

9.7 Research Issues

No chapter of this nature is complete without a discussion of future research issues. Event-triggering represents a new paradigm for real-time feedback control, but the topics covered in this chapter only touch upon what has recently been done. As is often the case, good preliminary work presents just as many questions as it answers and this is certainly the case for event-triggered research as of the writing of this chapter. To help motivate the research issues being raised in this section, let's consider a real-time implementation of a state-dependent event-triggered control system.

There are case studies examining the performance of event-driven control based on static thresholds [64, 63]. There is, however, very little experimental work examining the implementation of the state-dependent event-triggers introduced in sections 9.3 and 9.4. Early work in this direction will be found in [10] in which a self-triggered controller is implemented in a linear analog plant using a real-time kernel. Another early implementation will be found in [13] where the performance of different scheduling protocols for event-triggered controllers on a shared network is investigated. Finally, an experimental study directly comparing the *best* periodic controller to an event-triggered controller will be found in [26].

Figure 9.24 shows results from a recent experiment implementing state-dependent event-triggered feedback-linearizing controllers for the 3 degree-of-freedom (DOF) helicopter system. The plant is a Quanser© 3DOF helicopter controlled by a pentium III PC running the S.H.a.R.K. real-time kernel [22]. In this case, a feedback-linearizing controller was designed for the system with the objective of regulating the travel rate, $\dot{\tau}$, elevation, ε, and pitch ρ of the vehicle. Event-triggered and periodically triggered implementations of this system were implemented in the S.H.a.R.K. kernel and the results from one of these experiments is shown on the right-hand side of figure 9.24.

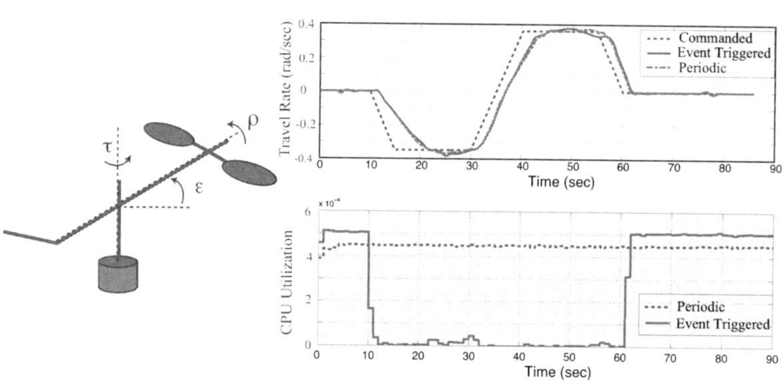

Fig. 9.24 Real-time Hardware Implementation of State-Dependent Event-Triggered System

The top plot in figure 9.24 shows the travel rate as a function of time. The commanded travel rate is shown by the solid dashed line and the other traces show the response of the event-triggered and periodically triggered controllers. What is important to note here is that the behavior is nearly identical in both cases. The bottom plot shows the normalized CPU utilization of the event-triggered and periodically triggered controllers. What one notices here is that when the vehicle is commanded to non-zero travel rates, the event-triggered task's utilization drops considerably. During those periods, however, when the commanded travel rate is near zero (*i.e.* the vehicle is hovering), the CPU utilization increases and actually exceeds the utilization of the periodically triggered controller.

These results actually confirm what the prior analysis in section 9.3 discovered. In particular, if one looks back at the results from [63] shown in figure (9.1), one sees that event-triggering indeed reduces the overall CPU utilization relative to comparable performing periodic controllers. For the case in [63, 24], however, a uniform event-triggering threshold is chosen so that the system demonstrates considerable chattering when the system is close to its equilibrium point. Under state-dependent event-triggering, however, this type of chattering in the system response does not appear. But because the experiment's input disturbance, w, is wideband sensor noise an excessive number of events are triggered, just as was shown earlier in figure 9.7 of section 9.3. What these results suggest is that state-dependent event-triggering can reduce the jerky behavior seen under the static thresholds used in [63]. The current theory, however, does not adequately balance that gain against the increased use of CPU resources.

With the findings from this experiment in hand, one can now identify a number of important issues that future research into event-triggered feedback must confront. Probably the most immediate is that we develop a better understanding of how to adequately *trade-off control system performance against the reduction in the use of computational or communication resources*. In particular, if one examines the ISS or $\mathscr{L}_2$ event-triggering concepts discussed in sections 9.3 and 9.4, one notes that while the analysis guarantees the preservation of some assumed stability concept, it says almost nothing about the message passing complexity. To be fair, these analyses do bound the minimum sampling period of state-dependent event-triggering. But these bounds are obtained as an afterthought, once the stability-preserving threshold has been determined. What is really needed is an analysis that treats both stability-preserving performance and communication (or computational) resource usage within the same analytical framework. To some extent, this approach was attempted in the event-triggered estimation scheme considered in section 9.5. In that case, the design of the event-trigger was posed as a minimization problem in which the transmission rate between sensor and remote observer was constrained. But that analysis is still far from being mature enough to be applied to real-life systems. The analysis constrains its attention to scalar linear systems and it is unclear how those results might be generalized to vector or nonlinear systems with real-life uncertainties.

Event-triggering samples the system state over time. The focus on constraining communication in section 9.5 can be seen as trying to identify fundamental limits on the rate at which information should be transmitted over the feedback channel. Sampling in *time*, however, is not the only way one can sample a signal. One may also sample the signal in *space*, *i.e.* quantization. This suggests there should be a close connection between results on minimum quantization feedback control [43] and event-triggered feedback. In particular, an important issue involves a unified approach to *quantization and sampling* in distributed control and estimation problems. Joint quantization and sampling issues were examined in [39], but a full understanding of this relationship has yet to be completed.

Another important issue concerns the development of *event-triggered output feedback controllers*. The experiment shown in figure 9.24 made use of state-dependent event-triggers that presume full access to the state. In the experimental system, however, the sensors only directly measure the travel angle, τ, elevation angle ε, and pitch angle ρ. For the experiment a periodic task was used to estimate the actual states of the system and then a separate event-triggering task was used to invoke separate controllers for the travel, elevation, and pitch dynamics of the vehicle. This implementation, however, is still far from what one would do in practice. Since most of the computational effort is actually done in the observer task, the true reduction in CPU utilization is very modest for this experiment. To truly realize the benefits of event-triggering, one would need an event-triggered output feedback controller, in which triggering is done solely on the basis of observed sensor measurements, rather than state estimates.

To some extent, the event-triggered estimation methods discussed in section 9.5 provide a first step at developing measurement-based event-triggers. But precisely how this might be integrated into an output feedback system is unclear. One might, for instance, implement an event-triggered observer, whose states are then used to trigger the control action. But this interconnection of an event-triggered estimator and event-triggered controller has not been studied at all. It is unclear whether one can invoke some event-triggered separation principle. As soon as issues regarding observer based control are raised, one must also confront traditional observability and controllability issues. We are aware of no recent work regarding these deeper *system theoretic properties of event-triggered systems*.

Finally, let's return to the implementation questions raised in the experiment. As noted above, the task set in this experiment consists of a hybrid combination of sporadic event-triggered tasks and periodically triggered tasks that work together to realize state-dependent event-triggered controllers. In realizing such hybrid task sets there are always *implementation issues regarding scheduling and fault-tolerance* that need to be addressed. In particular, it is still unclear how best to schedule this mixture of sporadic and periodically triggered tasks to ensure the determinism so often insisted upon in *safety-critical applications*. One reason for insisting on periodically driven task sets in control, is that they provide a highly predictable behavior. When faults do occur, the impact of those faults can be readily analyzed due to the highly deterministic nature of the resulting task environment. This type of deterministic modeling does not seem to be available for the task sets currently used to support event-triggered feedback and as a result it would be highly unlikely that anyone would choose event-triggering for safety-critical applications. This need not be the case, but to establish that event-triggering is suitable for safety-critical applications one must develop a modeling framework whose predictive abilities can provide broad assurances about the fault-tolerant properties of event-triggered systems.

Event-triggered feedback represents an exciting new approach to real-time networked control systems that has the potential of more efficiently using computational and communication resources while assuring high levels of application performance. These applications can be found in control, estimation, and optimization. While the promise of event-triggering is great, there is still significant work remaining to be done. A deeper understanding of the relationship between application performance and resource usage must be cultivated. In particular, a close examination must be made of the connection between quantized and event-triggered feedback. The current frameworks must be extended to event-triggered output controllers. This extension will require a deeper understanding of the fundamental system theoretic properties of event-triggered systems, especially as they pertain to the separation between control and estimation. Finally, we must more critically evaluate the scheduling and fault-tolerance of real-time implementations of event-triggered controllers, especially as they pertain to safety-critical applications. Much has already been done, but a great deal remains to be accomplished if event-triggering can indeed be used to build safety-critical real-time networked control systems.

Acknowledgements. The author gratefully acknowledges the partial financial support of the National Science Foundation (NSF-CNS-07-20457 and NSF-ECCS-0925229). This work grew out of discussions with P. Tabuada (UCLA) , M. Heemels (Eindhoven), A. Cervin (Lund), P. Marti (Catalunya), M. Johansson (KTH), K. Johansson (KTH), and X. Hu (Notre Dame) as well as the hard work of graduate students, X. Wang (UIUC), P. Wan, L. Li, and J. Viramontes-Perez. Finally, the author would like to thank A. Bemporad (Siena) and the European Union's WIDE project for supporting the presentation of this work at the Third WIDE Ph.D. School on Networked Control Systems.

References

1. Anta, A., Tabuada, P.: Self-triggered stabilization of homogeneous control systems. In: Proceedings of the American Control Conference, Seattle, Washington, USA, June 11-13, 2008, pp. 4129–4134 (2008)
2. Arzen, K.-E.: A simple event based PID controller. In: Procedings of the 14th World Congress of the International Federation of Automatic Control (IFAC), Beijing, P.R. China (1999)
3. Arzen, K.-E., Cervin, A., Eker, J., Sha, L.: An introduction to control and scheduling co-design. In: IEEE Conference on Decision and Control, Sydney, NSW, Australia, vol. 5, pp. 4865–4870 (December 2000)
4. Astrom, K.J., Bernhardsson, B.M.: Comparison of Riemann and Lebesgue sampling for first order stochastic systems. In: Proceedings of the 41st IEEE Conference on Decision and Control, Las Vegas, Nevada, USA, December 10-13, vol. 2, pp. 2011–2016 (2002)

5. Bao, L., Skoglund, M., Johansson, K.H.: Encoder-decoder design for event-triggered feedback control over bandlimited channels. In: American Control Conference, Minneapolis, Minnesota, USA (2006)
6. Bertsekas, D.P.: Nonlinear programming. Athena Scientific, Belmont (1999)
7. Bhattacharya, R., Balas, G.J.: Anytime control algorithm: model reduction approach. Journal of Guidance, Control and Dynamics 27(5), 767–776 (2004)
8. Buttazzo, G., Lipari, G., Abeni, L.: Elastic task model for adaptive rate control. In: IEEE Real-Time Systems Symposium (RTSS), pp. 286–295 (1998)
9. Caccamo, M., Buttazzo, G., Sha, L.: Elastic feedback control. In: IEEE Euromicro Conference on Real-Time Systems, ECRTS (2000)
10. Camacho, A., Marti, P., Velasco, M., Bini, E.: Demo abstract: Implementation of self-triggered controllers. In: Demo Session of 15th IEEE Real-time and Embedded Technology and Applications Symposium (RTAS 2009), San Francisco, California, USA (2009)
11. Carnevale, D., Teel, A.R., Nesic, D.: A Lyapunov proof of improved maximum allowable transfer interval for networked control systems. IEEE Transactions on Automatic Control 52, 892–897 (2007)
12. Cervin, A., Eker, J.: Control-scheduling codesign of real-time systems: the control server approach. Journal of Embedded Computing 1(2), 209–224 (2004)
13. Cervin, A., Henningsson, T.: Scheduling of event-triggered controllers on a shared network. In: Proceedings of the 47th IEEE Conference on Decision and Control, Cancun, Mexico (December 2008)
14. Chen, W.P., Hou, J.C., Sha, L., Caccamo, M.: A distributed, energy-aware, utility-based approach for data transport in wireless sensor networks. In: Proceedings of the IEEE Milcom (2005)
15. Chen, W.P., Sha, L.: An energy-aware data-centric generic utility based approach in wireless sensor networks. In: IPSN, pp. 215–224 (2004)
16. Chiang, M., Bell, J.: Balancing supply and demand of bandwidth in wireless cellular networks: utility maximization over powers and rates. In: Proc. IEEE INFOCOM, vol. 4, pp. 2800–2811 (2004)
17. Chiang, M., Low, S.H., Calderbank, A.R., Doyle, J.C.: Layering as optimization decomposition: A mathematical theory of network architectures. Proceedings of the IEEE 95(1), 255–312 (2007)
18. Cogill, R.: Event-based control using quadratic approximate value functions. In: IEEE Conference on Decision and Control, Shanghai, China (2009)
19. Cogill, R., Lall, S., Hespanha, J.P.: A constant factor approximation algorithm for event-based sampling. In: Proceedings of the American Control Conference, New York City, USA (July 2007)
20. Fleming, W.H., Rishel, R.W.: Deterministic and stochastic control. Springer, Heidelberg (1975)
21. Fontanelli, D., Greco, L., Bicchi, A.: Anytime control algorithms for embedded real-time systems. In: Hybrid Systems: computation and control (2008)
22. Gai, P., Abeni, L., Giorgi, M., Buttazzo, G.: A new kernel approach for modular real-time systems development. In: Proceedings of the 13th IEEE Euromicro Conference on Real-Time Systems (2001)
23. Heemels, W.P.M.H., Gorter, R.J.A., van Zijl, A., van den Bosch, P., Weiland, S.: Asynchronous measurement and control: a case study on motor synchronization. Control Engineering Practice 7, 1467–1482 (1999)
24. Heemels, W.P.M.H., Sandee, J.H., van den Bosch, P.P.J.: Analysis of event-driven controllers for linear systems. International Journal of Control 81(4), 571–590 (2008)

25. Heemels, W.P.M.H., Teel, A.R., van de Wouw, N., Nesic, D.: Networked control systems with communication constraints: tradeoffs between sampling intervals, delays and performance. Submitted to the 2009 European Control Conference, ECC (2009)
26. Henningsson, T., Cervin, A.: Comparison of LTI and event-based control for a moving cart with quantized position measurements. In: European Control Conference, Budapest, Hungary (August 2009)
27. Henningsson, T., Johannesson, E., Cervin, A.: Sporadic event-based control of first-order linear stochastic systems. Automatica 44(11), 2890–2895 (2008)
28. Ho, Y.C., Servi, L., Suri, R.: A class of center-free resource allocation algorithms. In: Large Scale Systems Theory and Applications: Proceedings of the IFAC Symposium, Toulouse, France, June 24-26, 1980, p. 475. Franklin Book Co. (1981)
29. Hristu-Varsakelis, D., Kumar, P.R.: Interrupt-based feedback control over shared communication medium. Technical Report TR 2003-34, University of Maryland, ISR (2003)
30. Imer, O.C., Basar, T.: Optimal estimation with limited measurements. In: Proceedings of the IEEE Conference on Decision and Control, Seville, Spain (2005)
31. Imer, O.C., Basar, T.: To measure or to control: optimal control of LTI systems with scheduled measurements and controls. In: American Control Conference (2006)
32. Isidori, A.: Nonlinear Control Systems II. Springer, Heidelberg (1999)
33. Johansson, B., Rabi, M., Johansson, M.: A simple peer-to-peer algorithm for distributed optimization in sensor networks. In: Proceedings of the 46th IEEE Conference on Decision and Control, pp. 4705–4710 (2007)
34. Johansson, B., Soldati, P., Johansson, M.: Mathematical Decomposition Techniques for Distributed Cross-Layer Optimization of Data Networks. IEEE Journal on Selected Areas in Communications 24(8), 1535–1547 (2006)
35. Karatzas, I., Wang, H.: Utility maximization with discretionary stopping. SIAM Journal on Control and Optimization 39(1), 306–329 (2000)
36. Kelly, F.P., Maulloo, A.K., Tan, D.K.H.: Rate control for communication networks: shadow prices, proportional fairness and stability. Journal of the Operational Research Society 49(3), 237–252 (1998)
37. Khalil, H.K.: Nonlinear Systems, 3rd edn. Prentice-Hall, Englewood Cliffs (2002)
38. Kim, B.H., Baldick, R.: A comparison of distributed optimal power flow algorithms. IEEE Transactions on Power Systems 15(2), 599–604 (2000)
39. Kofman, I., Braslavsky, J.H.: Level crossing sampling in feedback stabilization under data-rate constraints. In: IEEE Conference on Decision and Control, San Diego, CA, USA (2006)
40. Lehmann, D., Lunze, J.: Event-based control: a state-feedback approach. In: Proceedings of the European Control Conference, Budapest, Hungary, pp. 1716–1721 (2009)
41. Lemmon, M., Chantem, T., Hu, X.S., Zyskowski, M.: On self-triggered full-information h-infinity controllers. In: Hybrid Systems: computation and control, Pisa, Italy (July 2007)
42. Li, L., Lemmon, M.D.: Optimal event triggered transmission of information in distributed state estimation problems. In: American Control Conference, Baltimore, MD, USA (2010)
43. Liberzon, D.: On stabilization of linear systems with limited information. IEEE Transactions on Automatic Control 48, 304–307 (2003)
44. Low, S.H., Lapsley, D.E.: Optimization flow control, I: basic algorithm and convergence. IEEE/ACM Transactions on Networking (TON) 7(6), 861–874 (1999)
45. Lu, C., Stankovic, J.A., Son, S.H., Tao, G.: Feedback control real-time scheduling: Framework, modeling and algorithms. Real-time Systems 23(1-2), 85–126 (2002)

46. Madan, R., Lall, S.: Distributed algorithms for maximum lifetime routing in wireless sensor networks. In: IEEE GLOBECOM 2004, vol. 2 (2004)
47. Matveev, A., Savkin, A.: The problem of state estimation via asynchronous communication channels with irregular transmission times. IEEE Transactions on Automatic Control 48(4), 670–676 (2003)
48. Mazo, M., Tabuada, P.: On event-triggered and self-triggered control over sensor/actuator networks. In: Proceedings of the 47th IEEE Conference on Decision and Control, Cancun, Mexico (December 2008)
49. Nedic, A., Ozdaglar, A.: Distributed subgradient methods for multi-agent optimization. IEEE Transactions on Automatic Control 54(1), 48–61 (2009)
50. Nesic, D., Teel, A.R.: Input-output stability properties of networked control systems. IEEE Transactions on Automatic Control 49(10), 1650–1667 (2004)
51. Nesic, D., Teel, A.R.: Input-to-state stability of networked control systems. Automatica 40(12), 2121–2128 (2004)
52. Olfati-Saber, R., Fax, J.A., Murray, R.M.: Consensus and cooperation in networked multi-agent systems. Proceedings of the IEEE 95(1), 215–233 (2007)
53. Palomar, D.P., Chiang, M.: Alternative Distributed Algorithms for Network Utility Maximization: Framework and Applications. IEEE Transactions on Automatic Control 52(12), 2254–2269 (2007)
54. Polak, E.: Stability and graphical analysis of first order of pulse-width modulated sampled data regulator systems. IRE Trans. Automatic Control AC-6(3), 276–282 (1963)
55. Qiu, Y., Marbach, P.: Bandwidth allocation in ad hoc networks: A price-based approach. In: Proceedings of IEEE INFOCOM 2003, vol. 2, pp. 797–807 (2003)
56. Rabbat, M., Nowak, R.: Distributed optimization in sensor networks. In: Proceedings of the third international symposium on Information processing in sensor networks, pp. 20–27 (2004)
57. Rabi, M., Johansson, K.H., Johansson, M.: Optimal stopping for event-triggered sensing and actuation. In: Proceedings of the 47th IEEE Conference on Decision and Control, Cancun, Mexico (December 2008)
58. Rabi, M., Moustakides, G.V., Baras, J.S.: Efficient sampling for keeping track of an Ornstein-Uhlenbeck process. In: Proceedings of the Mediterranean conference on control and automation (2006)
59. Rabi, M., Moustakides, G.V., Baras, J.S.: Multiple sampling for estimation on a finite horizon. In: 45th IEEE Conference on Decision and Control, pp. 1351–1357 (2006)
60. Rabi, M., Moustakides, G.V., Baras, J.S.: Adaptive sampling for linear state estimation. Submitted to the SIAM journal on Control and Optimization (December 2008)
61. Rabi, M.: Packet based Inference and Control. PhD thesis, University of Maryland (2006)
62. Rabi, M., Baras, J.S.: Level-triggered control of a scalar linear system. In: Proceedings of the 16th Mediterranean Conference on Control and Automation, Athens, Greece (July 2007)
63. Sandee, J.H.: Event-driven Control in Theory and Practice: tradeoffs in software and control performance. PhD thesis, Technische Universiteit Eindhoven (2006)
64. Sandee, J.H., Heemels, W.P.M.H., van den Bosch, P.P.J.: Case studies in event-driven control. In: Hybrid Systems: computation and control, Pisa, Italy (April 2007)
65. Sandee, J.H., Visser, P.M., Heemels, W.P.M.H.: Analysis and experimental validation of processor load for event-driven controllers. In: IEEE Conference on Control and Applications (CCA), Munich, Germany, pp. 1879–1884 (2006)
66. Seto, D., Lehoczky, J.P., Sha, L., Shin, K.G.: On task schedulability in real-time control systems. In: IEEE Real-time Technology and Applications Symposium (RTAS), pp. 13–21 (1996)

67. Sijs, J., Lasar, M.: On event based state estimation. In: Majumdar, R., Tabuada, P. (eds.) HSCC 2009. LNCS, vol. 5469, pp. 336–350. Springer, Heidelberg (2009)
68. Sinopoli, B., Schenato, L., Franceschetti, M., Poolla, K., Jordan, M., Sastry, S.: Kalman filtering with intermittent observations. IEEE Transactions on Automatic Control 49(9), 1453–1464 (2004)
69. Speranzon, A., Fischione, C., Johansson, K.H.: Distributed and Collaborative Estimation over Wireless Sensor Networks. In: Proceedings of the IEEE Conference on Decision and Control, pp. 1025–1030 (2006)
70. Tabuada, P.: Event-triggered real-time scheduling of stabilizing control tasks. IEEE Transactions on Automatic Control 52(9), 1680–1685 (2007)
71. Tabuada, P., Wang, X.: Preliminary results on state-triggered scheduling of stabilizing control tasks. In: IEEE Conference on Decision and Control (2006)
72. Tsitsiklis, J., Bertsekas, D., Athans, M.: Distributed asynchronous deterministic and stochastic gradient optimization algorithms. IEEE Transactions on Automatic Control 31(9), 803–812 (1986)
73. Tsypkin, Y.Z.: Relay Control Systems. Cambridge University Press, Cambridge (1984)
74. van der Schaft, A.J.: L2-gain and passivity techniques in nonlinear control. Springer, Heidelberg (2000)
75. Velasco, M., Marti, P., Fuertes, J.M.: The self triggered task model for real-time control systems. In: Work-in-Progress Session of the 24th IEEE Real-time Systems Symposium (RTSS 2003), Cancun, Mexico (December 2003)
76. Voulgaris, P.: Control of asynchronous sampled data systems. IEEE Transactions on Automatic Control 39(7), 1451–1455 (1994)
77. Wan, P., Lemmon, M.: Distributed Flow Control using Embedded Sensor-Actuator Networks for the Reduction of Combined Sewer Overflow (CSO) Events. In: Proceedings of the 46th IEEE Conference on Decision and Control, pp. 1529–1534 (2007)
78. Wan, P., Lemmon, M.D.: Distributed network utility maximization using event-triggered augmented lagrangian methods. In: Proceedings of the American Control Conference, St. Louis, MO, USA (June 2009)
79. Wan, P., Lemmon, M.D.: Event-triggered distributed optimization in sensor networks. In: Information Processing in Sensor Networks (IPSN), San Francisco, California, USA (April 2009)
80. Wang, X., Lemmon, M.D.: Decentralized event-triggered broadcasts over networked control systems. In: Hybrid Systems: computation and control, St. Louis, Missouri (April 2008)
81. Wang, X., Lemmon, M.D.: Event-triggered broadcasting across distributed networked control systems. In: Proceedings of the American Control Conference, Seattle, Washington, USA (June 2008)
82. Wang, X., Lemmon, M.D.: Event-triggering in distributed networked control systems. Submitted to the IEEE Transactions on Automatic Control (February 2009)
83. Wang, X., Lemmon, M.D.: Self-triggered feedback control systems with finite-gain l2 stability. IEEE Transactions on Automatic Control 54(3), 452–467 (2009)
84. Wang, X., Lemmon, M.D.: Self-triggered feedback systems with state-independent disturbances. In: Proceedings of the American Control Conference, St. Louis Missouri, USA (June 2009)
85. Wen, J.T., Arcak, M.: A unifying passivity framework for network flow control. IEEE Transactions on Automatic Control 49(2), 162–174 (2004)
86. Xiao, L., Johansson, M., Boyd, S.P.: Simultaneous routing and resource allocation via dual decomposition. IEEE Transactions on Communications 52(7), 1136–1144 (2004)

87. Xu, Y., Hespanha, J.P.: Optimal communication logics in networked control systems. In: Proceedings of the IEEE Conference on Decision and Control, Nassau, Bahamas, vol. 4, pp. 3527–3532 (2004)
88. Xu, Y., Hespanha, J.P.: Communication logic design and analysis for networked control systems. In: Menini, L., Zaccarian, L., Abdallah, C.T. (eds.) Current Trends in Nonlinear Systems and Control, Systems and Control: Foundations and Applications, pp. 495–514. Birkhäuser, Boston (2006)
89. Xue, Y., Li, B., Nahrstedt, K.: Optimal resource allocation in wireless *ad hoc* networks: a price-based approach. IEEE Transactions on Mobile Computing 5(4), 347–364 (2006)
90. Zhang, W., Branicky, M.S., Phillips, S.M.: Stability of networked control systems. IEEE Control Systems Magazine 21(1), 84–99 (2001)
91. Zhu, B., Sinopoli, B., Poolla, K., Sastry, S.: Estimation over wireless sensor networks. In: American Control Conference, pp. 2732–2737 (2007)

Index

Lecture Notes in Control and Information Sciences

Edited by M. Thoma, F. Allgöwer, M. Morari

Further volumes of this series can be found on our homepage:
springer.com

Vol. 406: Bemporad, A., Heemels, M., Johansson, M.:
Networked Control Systems
363 p. 2010 [978-0-85729-032-8]

Vol. 405: Stefanovic, M., Safonov, M.G.:
Safe Adaptive Control
appro. 153 p. 2010 [978-1-84996-452-4]

Vol. 404: Giri, F.; Bai, E.-W. (Eds.):
Block-oriented Nonlinear System Identification
425 p. 2010 [978-1-84996-512-5]

Vol. 403: Tóth, R.;
Modeling and Identification of Linear Parameter-Varying Systems
319 p. 2010 [978-3-642-13811-9]

Vol. 402: del Re, L.; Allgöwer, F.; Glielmo, L.; Guardiola, C.; Kolmanovsky, I. (Eds.):
Automotive Model Predictive Control
284 p. 2010 [978-1-84996-070-0]

Vol. 401: Chesi, G.; Hashimoto, K. (Eds.):
Visual Servoing via Advanced Numerical Methods
393 p. 2010 [978-1-84996-088-5]

Vol. 400: Tomás-Rodríguez, M.; Banks, S.P.:
Linear, Time-varying Approximations to Nonlinear Dynamical Systems
298 p. 2010 [978-1-84996-100-4]

Vol. 399: Edwards, C.; Lombaerts, T.; Smaili, H. (Eds.):
Fault Tolerant Flight Control
appro. 350 p. 2010 [978-3-642-11689-6]

Vol. 398: Hara, S.; Ohta, Y.; Willems, J.C.; Hisaya, F. (Eds.):
Perspectives in Mathematical System Theory, Control, and Signal Processing
appro. 370 p. 2010 [978-3-540-93917-7]

Vol. 397: Yang, H.; Jiang, B.; Cocquempot, V.:
Fault Tolerant Control Design for Hybrid Systems
191 p. 2010 [978-3-642-10680-4]

Vol. 396: Kozlowski, K. (Ed.):
Robot Motion and Control 2009
475 p. 2009 [978-1-84882-984-8]

Vol. 395: Talebi, H.A.; Abdollahi, F.; Patel, R.V.; Khorasani, K.:
Neural Network-Based State Estimation of Nonlinear Systems
appro. 175 p. 2010 [978-1-4419-1437-8]

Vol. 394: Pipeleers, G.; Demeulenaere, B.; Swevers, J.:
Optimal Linear Controller Design for Periodic Inputs
177 p. 2009 [978-1-84882-974-9]

Vol. 393: Ghosh, B.K.; Martin, C.F.; Zhou, Y.:
Emergent Problems in Nonlinear Systems and Control
285 p. 2009 [978-3-642-03626-2]

Vol. 392: Bandyopadhyay, B.; Deepak, F.; Kim, K.-S.:
Sliding Mode Control Using Novel Sliding Surfaces
137 p. 2009 [978-3-642-03447-3]

Vol. 391: Khaki-Sedigh, A.; Moaveni, B.:
Control Configuration Selection for Multivariable Plants
232 p. 2009 [978-3-642-03192-2]

Vol. 390: Chesi, G.; Garulli, A.; Tesi, A.; Vicino, A.:
Homogeneous Polynomial Forms for Robustness Analysis of Uncertain Systems
197 p. 2009 [978-1-84882-780-6]

Vol. 389: Bru, R.; Romero-Vivó, S. (Eds.):
Positive Systems
398 p. 2009 [978-3-642-02893-9]

Vol. 388: Jacques Loiseau, J.; Michiels, W.; Niculescu, S-I.; Sipahi, R. (Eds.):
Topics in Time Delay Systems
418 p. 2009 [978-3-642-02896-0]

Vol. 387: Xia, Y.; Fu, M.; Shi, P.:
Analysis and Synthesis of Dynamical Systems with Time-Delays
283 p. 2009 [978-3-642-02695-9]

Vol. 386: Huang, D.;
Nguang, S.K.:
Robust Control for Uncertain
Networked Control Systems with
Random Delays
159 p. 2009 [978-1-84882-677-9]

Vol. 385: Jungers, R.:
The Joint Spectral Radius
144 p. 2009 [978-3-540-95979-3]

Vol. 384: Magni, L.; Raimondo, D.M.;
Allgöwer, F. (Eds.):
Nonlinear Model Predictive Control
572 p. 2009 [978-3-642-01093-4]

Vol. 383: Sobhani-Tehrani E.:
Khorasani K.;
Fault Diagnosis of Nonlinear Systems
Using a Hybrid Approach
360 p. 2009 [978-0-387-92906-4]

Vol. 382: Bartoszewicz A.;
Nowacka-Leverton A.:
Time-Varying Sliding Modes for Second
and Third Order Systems
192 p. 2009 [978-3-540-92216-2]

Vol. 381: Hirsch M.J.; Commander C.W.;
Pardalos P.M.; Murphey R. (Eds.):
Optimization and Cooperative
Control Strategies: Proceedings of the 8th
International Conference on Cooperative
Control and Optimization
459 p. 2009 [978-3-540-88062-2]

Vol. 380: Basin M.:
New Trends in Optimal Filtering and Control for
Polynomial and Time-Delay Systems
206 p. 2008 [978-3-540-70802-5]

Vol. 379: Mellodge P.; Kachroo P.:
Model Abstraction in Dynamical Systems:
Application to Mobile Robot Control
116 p. 2008 [978-3-540-70792-9]

Vol. 378: Femat R.; Solis-Perales G.:
Robust Synchronization of Chaotic Systems
Via Feedback
199 p. 2008 [978-3-540-69306-2]

Vol. 377: Patan K.:
Artificial Neural Networks for
the Modelling and Fault
Diagnosis of Technical Processes
206 p. 2008 [978-3-540-79871-2]

Vol. 376: Hasegawa Y.:
Approximate and Noisy Realization of
Discrete-Time Dynamical Systems
245 p. 2008 [978-3-540-79433-2]

Vol. 375: Bartolini G.; Fridman L.;
Pisano A.; Usai E. (Eds.):
Modern Sliding Mode Control Theory
465 p. 2008 [978-3-540-79015-0]

Vol. 374: Huang B.; Kadali R.:
Dynamic Modeling, Predictive Control
and Performance Monitoring
240 p. 2008 [978-1-84800-232-6]

Vol. 373: Wang Q.-G.; Ye Z.; Cai W.-J.;
Hang C.-C.:
PID Control for Multivariable Processes
264 p. 2008 [978-3-540-78481-4]

Vol. 372: Zhou J.; Wen C.:
Adaptive Backstepping Control of Uncertain
Systems
241 p. 2008 [978-3-540-77806-6]

Vol. 371: Blondel V.D.; Boyd S.P.;
Kimura H. (Eds.):
Recent Advances in Learning and Control
279 p. 2008 [978-1-84800-154-1]

Vol. 370: Lee S.; Suh I.H.;
Kim M.S. (Eds.):
Recent Progress in Robotics:
Viable Robotic Service to Human
410 p. 2008 [978-3-540-76728-2]

Vol. 369: Hirsch M.J.; Pardalos P.M.;
Murphey R.; Grundel D.:
Advances in Cooperative Control and
Optimization
423 p. 2007 [978-3-540-74354-5]

Vol. 368: Chee F.; Fernando T.
Closed-Loop Control of Blood Glucose
157 p. 2007 [978-3-540-74030-8]

Vol. 367: Turner M.C.; Bates D.G. (Eds.):
Mathematical Methods for Robust and
Nonlinear Control
444 p. 2007 [978-1-84800-024-7]

Vol. 366: Bullo F.; Fujimoto K. (Eds.):
Lagrangian and Hamiltonian Methods for
Nonlinear Control 2006
398 p. 2007 [978-3-540-73889-3]

Vol. 365: Bates D.; Hagström M. (Eds.):
Nonlinear Analysis and Synthesis
Techniques for Aircraft Control
360 p. 2007 [978-3-540-73718-6]